Localization and Perturbation of Zeros of Entire Functions

PURE AND APPLIED MATHEMATICS

A Program of Monographs, Textbooks, and Lecture Notes

LECTURE NOTES IN
PURE AND APPLIED MATHEMATICS

Recent Titles

Localization and Perturbation of Zeros of Entire Functions

Michael Gil'
Ben Gurion University of the Negev
Israel

CRC Press
Taylor & Francis Group
Boca Raton London New York

CRC Press is an imprint of the
Taylor & Francis Group an **informa** business

A CHAPMAN & HALL BOOK

Chapman & Hall/CRC
Taylor & Francis Group
6000 Broken Sound Parkway NW, Suite 300
Boca Raton, FL 33487-2742

© 2010 by Taylor and Francis Group, LLC
Chapman & Hall/CRC is an imprint of Taylor & Francis Group, an Informa business

No claim to original U.S. Government works

ISBN-13: 978-1-4398-0032-4 (hbk)

ISBN-13: 978-1-138-11678-8 (pbk)

Library of Congress Cataloging-in-Publication Data

Gil, M. I. (Mikhail Iosifovich)
   Localization and perturbation of zeros of entire functions / Michael Gil.
      p. cm. -- (Pure and applied mathematics recent titles ; 258)
   Includes bibliographical references and index.
   ISBN 978-1-4398-0032-4 (hardcover : alk. paper)
   1. Functions, Entire. 2. Localization theory. 3. Perturbation (Mathematics) I. Title.
II. Series.

QA353.E5G55 2009
515'.98--dc22

2009036640

Visit the Taylor & Francis Web site at
http://www.taylorandfrancis.com

and the CRC Press Web site at
http://www.crcpress.com

CONTENTS

Biography

Michael I. Gil' is a Professor at Ben Gurion University of the Negev, Beer-Sheva, Israel. The author of 7 books and about 200 scientific papers. He was born 10.03.1941 in Kharkov, Ukraine. Dr. Gil' is a member of the American Mathematical Society. He received the Ph.D. degree (1973) in mathematics from Voronejz State University, Russia, and the Doctor of Physical and Mathematical Sciences (fourth) degree (1990) from Moscow Institute of System Researches of the former Soviet Union's Academy of Sciences (now the Russian Academy of Sciences). His areas of research: operator functions, analytic functions, differential and functional differential equations, stability of systems.

Preface

1. One of the most important problems in the theory of entire functions is the problem of the distribution of the zeros of entire functions. Many other problems in fields close to the complex function theory lead to this problem. The connection between the growth of an entire function and the distribution of its zeros was investigated in the classical works of Borel, Hadamard, Jensen, Lindelöf, Nevanlinna, and other authors. Many excellent books are concerned with the zeros of entire functions but deal mainly with the asymptotic distributions of zeros. However, in various applications, for example, in numerical mathematics and system theory, the bounds for zeros are very important, but the bounds are investigated considerably less than the asymptotic distributions. The suggested book is devoted to the bounds for the zeros of entire functions and variations of zeros under perturbations.

The contents of the book are closely connected with the following well-known results: the Hadamard theorem on the series of powers of the zeros of finite order entire functions, the Jensen inequality for the counting function, the Ostrowski inequalities for the real and imaginary parts of the zeros of polynomials, and the Hurwitz theorem on the preservation of multiplicities of zeros. A considerable part of the book deals with the following problem: if the Taylor coefficients of two entire functions are close, how close are their zeros? In particular, we estimate the distance between the zeros of an entire function and its critical points, as well as the distance between the zeros of an entire function and the zeros of partial sums of its Taylor series.

The book suggests a new approach to the investigation of entire functions, which is particularly based on the recent estimates for the resolvents of compact operators.

2. The book consists of 12 chapters.

Chapters 1 and 2 are of a preliminary character. The material of these chapters is systematically used in the proofs of the results presented in the remaining chapters of the book. In Chapter 1, we accumulate results about finite matrices. In particular, we present various inequalities for the eigenvalues and singular values of matrices. In addition, we discuss estimates for the resolvents and determinants of finite matrices and variations of eigenvalues under perturbations. Chapter 1 also contains the well-known inequalities for convex functions. Chapter 2 discusses some concepts and results of the spectral theory of compact operators in a Hilbert space, which are used in this text. The major part of the chapter is devoted to the eigenvalues and singular values of the Schatten - von Neumann operators, as well as to estimates for the resolvents and spectrum perturbations of these operators. The regularized determinants are also considered.

Chapter 3 contains the background material. It is a collection of the basic conceptions and classical theorems of the theory of entire functions, such as the Weierstrass theorem on the representation of an arbitrary entire function by an infinite product and the Hadamard theorem on the representation of a finite-order entire function by a canonical product. The classical results of Borel, Hadamard, and Lindelöf on the connections between the growth of an entire function and the distribution of its zeros, and the Jensen theorem on the counting function of zeros are also included in that chapter.

In Chapter 4, various inequalities for the zeros of polynomials are presented. The material of this chapter is basic for the next chapters, where the corresponding results are derived for entire functions via limits of sequences of polynomials.

Chapter 5 is one of the main chapters of the present book. Let f be an entire function whose zeros $z_1(f), z_2(f), ...$ taken with their multiplicities are enumerated in the increasing order. If f has $l < \infty$ finite zeros, we put

$$\frac{1}{z_k(f)} = 0 \quad (k = l + 1, l + 2, ...).$$

In Chapter 5, estimates for the sums

$$\sum_{k=1}^{j} \frac{1}{|z_k(f)|} \quad (j = 1, 2, ...),$$

$$S_p(f) := \sum_{k=1}^{\infty} \frac{1}{|z_k(f)|^p} \quad \text{and} \quad J_p(f) := \sum_{k=1}^{\infty} |Im \ \frac{1}{z_k(f)}|^p \quad (p \geq \rho(f))$$

are derived. Here $\rho(f)$ is the order of f. These estimates enable us to obtain inequalities for the counting function of the zeros of f that supplement the Jensen inequality. We also establish relations between the series

$$\hat{s}_m(f) := \sum_{k=1}^{\infty} \frac{1}{z_k^m(f)} \quad (m \geq \rho(f))$$

and the traces of certain finite matrices.

In Chapter 6, we investigate the variations of the zeros of finite-order entire functions under perturbations. In particular, the distance between the zeros of an entire function and its critical points is estimated. Also the distance between the zeros of the Taylor series and the zeros of its tail is investigated.

In Chapter 7, in the case of entire functions whose order is less than two, we improve the perturbation results derived in Chapter 6. In addition, in Chapter 7, an identity between the sums

$$\sum_{k=1}^{\infty} \left(Im \ \frac{1}{z_k(f)} \right)^2 \quad \text{and} \quad \sum_{k=1}^{\infty} \left(Re \ \frac{1}{z_k(f)} \right)^2$$

is established.

An entire function f is said to be of the exponential type if it satisfies the inequality

$$|f(z)| \leq M e^{\alpha |z|} \quad (M, \alpha = const; \ z \in \mathbf{C}).$$

In Chapter 8, for the exponential type entire functions, some results from Chapters 5 and 7 are detailed. Besides, an essential role is played by the Borel transform.

A quasipolynomial is an entire function of the form

$$f(z) = \sum_{j=0}^{n} \sum_{k=1}^{m} c_{jk} z^j e^{\lambda_k z} \quad (c_{jk}, \lambda_k = const; \; z \in \mathbf{C}).$$

The Borel transform of a quasipolynomial is easily calculated. This fact enables us to obtain in Chapter 9 explicit bounds for the zeros of quasipolynomials. In addition, in that chapter two-sided estimates and positivity conditions for quasipolynomials on the positive half-line are derived. Applications to differential and functional differential equations are also discussed.

Chapter 10 contains additional results on the zeros of entire functions. In particular, the transforms that generalize the Borel transform are introduced. Also upper and lower bounds for the canonical products are established.

Chapter 11 is concerned with polynomials having matrix coefficients (polynomial matrix pencils). Bounds for the sums of characteristic values are established. In addition, variations of the characteristic values of polynomial matrix pencils under perturbations are investigated. Also coupled systems of polynomial equations and vector difference equations are considered.

Chapter 12 deals with entire matrix valued functions. In particular, bounds for characteristic values are derived in terms of the Taylor coefficients. Applications to vector differential equations are also discussed.

3. The present book is the first book that presents a systematic exposition of the bounds for the zeros of entire functions and variations of zeros under perturbations. It is also the first book that systematically deals with the operator approach to the theory of analytic functions.

The book is directed not only to specialists in the theory of analytic functions but to anyone interested in various applications who has had at least a first-year graduate-level course in analysis.

Acknowledgment. This work was supported by the Kamea Fund of Israel.

Chapter 1

Finite Matrices

This chapter is of a preliminary character. Here we accumulate results about finite matrices. In particular, we present various inequalities for the eigenvalues and singular values of matrices. In addition, we discuss estimates for the resolvents and determinants of finite matrices, as well as variations of eigenvalues under perturbations. The chapter also contains the well-known inequalities for convex functions. The material of this chapter is systematically used in the proofs of the results presented in the remaining chapters of the book.

1.1 Inequalities for eigenvalues and singular numbers

Let $\mathbf{C}^n$ be the complex n-dimensional Euclidean space with the unit matrix I, a scalar product $(.,.)$ and the Euclidean norm $\|h\| = \sqrt{(h,h)}$ $(h \in \mathbf{C}^n)$. For an $n \times n$-matrix A, $\lambda_k(A)$, $k = 1,...,n$ are the eigenvalues repeated according to their algebraic multiplicities and enumerated in the decreasing way: $|\lambda_k(A)| \geq |\lambda_{k+1}(A)|$; A^* is the adjoint matrix. The eigenvalues $s_k(A)$ $(k = 1,...,n)$ of matrix $(A^*A)^{1/2}$ taken their multiplicities and enumerated in the decreasing way are called the singular numbers (s-numbers) of A. $A_R = \frac{1}{2}(A + A^*)$ is the real component and $A_I = \frac{1}{2i}(A - A^*)$ is the imaginary component of A. In addition,

$$\sigma(A) = \{\lambda_k(A)\}_{k=1}^n$$

denotes the spectrum of A; $R_\lambda(A) = (A - I\lambda)^{-1}$ $(\lambda \notin \sigma(A))$ is the resolvent of A.

Lemma 1.1.1 *(Weyl's inequalities) For any $n \times n$-matrix A, the inequalities*

$$\sum_{j=1}^k |\lambda_j(A)| \le \sum_{j=1}^k s_j(A) \quad (k = 1, \dots, n)$$

are valid. They become equalities if and only if A is normal: $AA^* = A^*A$.

For the proof see [Gohberg and Krein, 1969, Section II.3.1], [Pietsch, 1987]. The following lemma is very useful.

Lemma 1.1.2 *Let A and B be $n \times n$-matrices. Then*

$$\sum_{k=1}^j s_k(A + B) \le \sum_{k=1}^j s_k(A) + s_k(B) \quad (j = 1, ..., n).$$

This result is proved in [Gohberg and Krein, 1969, Section II.4.2].
    Let $\{e_k\}$ be an orthogonal normal basis in $\mathbf{C}^n$. Then

$$Trace\ A = Tr\ A = \sum_{k=1}^n (Ae_k, e_k).$$

In addition,

$$N_p(A) := \sqrt[p]{Tr\ (AA^*)^{p/2}} = \left[\sum_{k=1}^n s_k^p(A) \right]^{1/p}$$

for a finite $p \ge 1$, is called the von Neumann - Schatten norm of A. In particular,

$$N_2(A) := \sqrt{Tr\ (AA^*)} = \left[\sum_{k=1}^n s_k^2(A) \right]^{1/2}$$

is *the Hilbert-Schmidt (Frobenius) norm*. This equality is equivalent to the following one:

$$N_2(A) = \left(\sum_{j,k=1}^n |a_{jk}|^2 \right)^{1/2},$$

where a_{jk} $(j, k = 1, \dots, n)$ are the entries of A in an orthogonal normal basis.
    Let

$$r_s(A) := \max_k |\lambda_k(A)|$$

be the *spectral radius* of A. The Gel'fand formula

$$r_s(A) = \lim_{k \to \infty} \sqrt[k]{\|A^k\|}$$

is valid. The limit always exists. Moreover,

$$r_s(A) \le \sqrt[k]{\|A^k\|}$$

for any integer $k \ge 1$. So

$$r_s(A) \le \|A\|.$$

We need also the following result.

Lemma 1.1.3 *The spectral radius $r_s(A)$ of a matrix $A = (a_{jk})_{j,k=1}^{n}$ satisfies the inequality*

$$r_s(A) \le \max_j \sum_{k=1}^{n} |a_{jk}|.$$

The proof of this result can be found in [Krasnosel'skij et al., 1989, Section 16]. The following result is proved in [Gohberg and Krein, 1969, Lemma II.6.1].

Lemma 1.1.4 *Let $\tilde{\lambda}_k(A)$, $k = 1, ..., m \le n$, be the nonreal eigenvalues of an $n \times n$-matrix A, repeated according to their algebraic multiplicities and enumerated in the decreasing way:*

$$|Im \, \tilde{\lambda}_k(A)| \ge |Im \, \tilde{\lambda}_{k+1}(A)| \quad (k = 1, ..., m - 1).$$

Then

$$\sum_{j=1}^{k} |Im \, \tilde{\lambda}_j(A)| \le \sum_{j=1}^{k} s_j(A_I) \quad (k = 1, ..., m).$$

1.2 Inequalities for convex functions

The following result is classical, cf. [Hardy, Litllwood, and Polya, 1934], [Gohberg and Krein, 1969, Lemma II.3.4], [Gohberg et al., 2000, p. 53].

Lemma 1.2.1 *Let $\phi(x)$ $(-\infty \le x \le \infty)$ be a convex continuous function, such that*

$$\phi(-\infty) = \lim_{x \to -\infty} \phi(x) = 0,$$

and

$$a_j, b_j \quad (j = 1, 2, ..., l \le \infty)$$

be two nonincreasing sequences of real numbers, such that

$$\sum_{k=1}^{j} a_k \leq \sum_{k=1}^{j} b_k \ \ (j = 1, 2, ..., l).$$

Then

$$\sum_{k=1}^{j} \phi(a_k) \leq \sum_{k=1}^{j} \phi(b_k) \ \ (j = 1, 2, ..., l).$$

The next result is also well known, cf. [Gohberg and Krein, 1969, Chapter II], [Gohberg et al., 2000, p. 53].

Lemma 1.2.2 *Let a scalar-valued function* $\Phi(t_1, t_2, ..., t_j)$ *with an integer* j *be defined on the domain*

$$-\infty < t_j \leq t_{j-1}... \leq t_2 \leq t_1 < \infty$$

and have continuous partial derivatives, satisfying the condition

$$\frac{\partial \Phi}{\partial t_1} > \frac{\partial \Phi}{\partial t_2} > ... > \frac{\partial \Phi}{\partial t_j} > 0 \ for \ t_1 > t_2 > ... > t_j,$$

and $a_k, b_k \ \ (k = 1, 2, ..., j)$ *be two nonincreasing sequences of real numbers satisfying the condition*

$$\sum_{k=1}^{m} a_k \leq \sum_{k=1}^{m} b_k \ \ (m = 1, 2, ..., j).$$

Then $\Phi(a_1, ..., a_j) \leq \Phi(b_1, ..., b_j)$.

1.3 Traces of powers of matrices

For a natural $n \geq 2$ introduce the $n \times n$-matrix

$$A_n = \begin{pmatrix} -a_1 & -a_2 & ... & -a_{n-1} & -a_n \\ 1 & 0 & ... & 0 & 0 \\ 0 & 1 & ... & 0 & 0 \\ . & . & ... & . & . \\ 0 & 0 & ... & 1 & 0 \end{pmatrix} \tag{3.1}$$

with given complex numbers $a_k, \ k = 1, ..., n$.

Lemma 1.3.1 *For any natural $m \leq n$, we have the equality*

$$Trace\ A_n^m = Trace\ A_m^m,$$

where A_m is defined by (3.1) with $n = m$.

Proof: Let $m < n$. Consider the $n \times n$-matrix

$$B_{n,m} = \begin{pmatrix} b_{11} & b_{12} & \cdots & b_{1,n-m} & \cdots & b_{1,n-1} & b_{1,n} \\ b_{21} & b_{22} & \cdots & b_{2,n-m} & \cdots & b_{2,n-1} & b_{2,n} \\ \cdot & \cdot & \cdots & \cdot & \cdots & \cdot & \cdot \\ b_{m1} & b_{m2} & \cdots & b_{m,n-m} & \cdots & b_{m,n-1} & b_{m,n} \\ 1 & 0 & \cdots & 0 & \cdots & 0 & \cdot \\ 0 & 1 & \cdots & 0 & \cdots & 0 & 0 \\ \cdot & \cdot & \cdots & \cdot & \cdots & 0 & 0 \\ 0 & 0 & \cdots & 1 & \cdots & 0 & 0 \end{pmatrix}$$

with some entries b_{jk}. The direct calculations show that the matrix $C_{n,m+1} := B_{n,m}A_n$ has the form

$$C_{n,m+1} = \begin{pmatrix} c_{11} & c_{12} & \cdots & c_{1,n-m-1} & \cdots & c_{1,n-1} & c_{1,n} \\ c_{21} & c_{22} & \cdots & c_{2,n-m-1} & \cdots & c_{2,n-1} & c_{2,n} \\ \cdot & \cdot & \cdots & \cdot & \cdots & \cdot & \cdot \\ c_{m1} & c_{m2} & \cdots & c_{m,n-m-1} & \cdots & c_{m,n-1} & c_{m,n} \\ c_{m+1,1} & c_{m+1,2} & \cdots & c_{m+1,n-m-1} & \cdots & c_{m+1,n-1} & c_{m+1,n} \\ 1 & 0 & \cdots & 0 & \cdots & 0 & 0 \\ 0 & 1 & \cdots & 0 & \cdots & 0 & 0 \\ \cdot & \cdot & \cdots & \cdot & \cdots & \cdot & \cdot \\ 0 & 0 & \cdots & 1 & \cdots & 0 & 0 \end{pmatrix}$$

where

$$c_{jk} = -b_{j1}a_k + b_{j,j+1} \ (1 \leq j \leq m, k < n),$$

$$c_{jn} = -b_{j1}a_n \ (1 \leq j \leq m), c_{m+1,k} = -a_k \ (1 \leq k \leq n). \qquad (3.2)$$

Hence

$$Trace\ C_{n,m+1} = \sum_{k=1}^{m+1} c_{kk} = -\sum_{k=1}^{m} b_{k1}a_k + b_{k,k+1} - a_{m+1}. \qquad (3.3)$$

Now put $B_{n,m} = A_n^m$, $m = 1, 2, ..., n-1$. Then $C_{n,m+1} = A_n^{m+1}$. Let $a_{jk}^{(m)}$ be the entries of A_n^m. Then according to (3.2) and (3.3),

$$a_{jk}^{(m+1)} = -a_{j1}^{(m)}a_k + a_{j,k+1}^{(m)} \ (k < n);$$

$$a_{jn}^{(m+1)} = -a_{j1}^{(m)}a_n \ (1 \leq j \leq m); \quad a_{m+1,k}^{(m+1)} = -a_k.$$

Thus,

$$Trace\ A_n^{m+1} = \sum_{k=1}^{m}(-a_{k1}^{(m)}a_k + a_{k,k+1}^{(m)}) - a_{m+1} \qquad (3.4)$$

Then taking $m = 2, 3, ...$, we can assert that $Trace\ A_n^m$ contains $a_1, ..., a_m$, only, and $Trace\ A_n^m = Trace\ A_m^m$. This proves the required result for $m < n$. The case $m = n$ is trivial. Q. E. D.

Remark 1.3.2 *From (3.4) we have*

$$Trace\ A_n^2 = a_1^2 - 2a_2\ (n \geq 2)$$

and

$$Trace\ A_n^3 = -a_1^3 + 3a_1a_2 - 3a_3\ (n \geq 3).$$

1.4   A relation between determinants and resolvents

In this section, for the brevity, we denote the Hilbert-Schmidt norm of a matrix A by $N(A)$. That is, $N(A) = N_2(A)$.

Theorem 1.4.1 *Let A be an $n \times n$ matrix and $I - A$ be invertible. Then*

$$\|(I - A)^{-1}\ det\ (I - A)\| \leq$$

$$[1 + \frac{1}{n-1}\ (N^2(A) - 2Re\ Trace\ (A) + 1)]^{(n-1)/2}. \qquad (4.1)$$

The proof of this theorem is presented in this section below. Let us mention the following consequence of this theorem.

Corollary 1.4.2 *Let A be an $n \times n$-matrix. Then*

$$\|(I\lambda - A)^{-1}\ det\ (\lambda I - A)\| \leq$$

$$[\frac{N^2(A) - 2Re\ (\bar{\lambda}\ Trace\ (A))\ + n|\lambda|^2}{n-1}]^{(n-1)/2}\ (\lambda \notin \sigma(A)). \qquad (4.2)$$

In particular, let V be a nilpotent matrix. Then

$$\|(I\lambda - V)^{-1}\| \leq$$

$$\frac{1}{|\lambda|}\ [1 + \frac{1}{n-1}\ (1 + \frac{N^2(V)}{|\lambda|^2})]^{(n-1)/2}\ (\lambda \neq 0). \qquad (4.3)$$

Indeed, inequality (4.2) is due to Theorem 1.4.1 with $\lambda^{-1}A$ instead of A, and the equality $|\lambda|^2\lambda^{-1} = \bar{\lambda}$ taken into account. If V is nilpotent, then

$$|det\ (\lambda I - V)|^2 = det\ (I\lambda - V)\ det\ (I\bar{\lambda} - V^*) = |\lambda|^{2n}.$$

Moreover, $Trace\ V = 0$. So (4.2) implies (4.3).

To prove Theorem 1.4.1, we need the following lemma.

Lemma 1.4.3 *Let A be a positive definite Hermitian $n \times n$ matrix: $A = A^* > 0$. Then*

$$\|A^{-1} \det A\| \leq \left[\frac{Trace\ A}{n-1}\right]^{n-1}.$$

Proof: Without loss of generality, assume that

$$\lambda_n(A) = \min_{k=1,\dots,n} \lambda_k(A).$$

Then $\|A^{-1}\| = \lambda_n^{-1}(A)$ and

$$\|A^{-1} \det A\| = \prod_{k=1}^{n-1} \lambda_k(A).$$

Hence, because of the inequality between the arithmetic and geometric mean values, we get

$$\|A^{-1} \det A\| \leq \left[\frac{1}{n-1} \sum_{k=1}^{n-1} \lambda_k\right]^{n-1} \leq \left[\frac{1}{n-1}\ Trace\ A\right]^{n-1},$$

since A is positive definite. As claimed. Q. E. D.

Proof of Theorem 1.4.1: For any $A \in B(\mathbf{C}^n)$, the operator

$$B := (I - A)(I - A^*)$$

is positive definite and

$$\det B = \det (I - A)(I - A^*) = \det (I - A)\det (I - A^*) =$$

$$|\det (I - A)|^2.$$

Moreover,

$$Trace\ [(I - A)(I - A^*)] = Trace\ I - Trace\ (A + A^*) + Trace\ (AA^*) =$$

$$n - 2Re\ Trace\ A + N^2(A).$$

But

$$\|B^{-1}\| = \|(I - A)^{-1}(I - A^*)^{-1}\| = \|(I - A)^{-1}\|^2.$$

Now Lemma 1.4.3 yields

$$\|B^{-1}\det B\| = \|(I - A)^{-1}\ \det (I - A)\|^2 \leq$$

$$\left[\frac{1}{n-1}\ (n + N^2(A) - 2Re\ Trace\ (A))\right]^{n-1} = .$$

$$\left[1 + \frac{1}{n-1}\ (N^2(A) - 2Re\ Trace\ (A) + 1)\right]^{n-1},$$

as claimed. Q. E. D.

1.5 Estimates for norms of resolvents in terms of the distance to spectrum

Let $A = (a_{jk})$ be an $n \times n$-matrix $(n \geq 2)$. The following quantity plays an essential role in this section:

$$g(A) = \left(N_2^2(A) - \sum_{k=1}^{n} |\lambda_k(A)|^2\right)^{1/2}. \qquad (5.1)$$

Since

$$\sum_{k=1}^{n} |\lambda_k(A)|^2 \geq |Trace\ A^2|,$$

we get

$$g^2(A) \leq N_2^2(A) - |Trace\ A^2|. \qquad (5.2)$$

Theorem 1.5.1 *Let A be an $n \times n$-matrix. Then for any regular λ of A,*

$$\|(I\lambda - A)^{-1})\| \leq \frac{1}{\rho(A,\lambda)}\left[1 + \frac{1}{n-1}\left(1 + \frac{g^2(A)}{\rho^2(A,\lambda)}\right)\right]^{(n-1)/2}$$

where

$$\rho(A,\lambda) := \min_{k=1,\dots,n} |\lambda - \lambda_k(A)|$$

is the distance between $\sigma(A)$ and a complex point λ.

To prove this theorem, consider the Schur triangular representation

$$A = D + V, \qquad (5.3)$$

where D is the diagonal (normal) matrix and V is the nilpotent (upper triangular) matrix. Besides the spectra of A and D coincide:

$$\sigma(A) = \sigma(D). \qquad (5.4)$$

About the triangular representation see for instance [Gantmacher, 1967]. We will call D and V the diagonal part and nilpotent one of A, respectively.

First, let us prove the following:

Lemma 1.5.2 *For any linear operator A in $\mathbf{C}^n$, we have*

$$N_2^2(V) = g^2(A) := N_2^2(A) - \sum_{k=1}^{n} |\lambda_k(A)|^2,$$

where V is the nilpotent part of A.

Proof: Let D be the diagonal part of A. Then, as it is simple to check, both matrices V^*D and D^*V are nilpotent. Therefore,

$$Trace \ (D^*V) = 0 \ \text{and} \ Trace \ (V^*D) = 0.$$

It is easy to see that

$$Trace \ (D^*D) = \sum_{k=1}^{n} |\lambda_k(A)|^2.$$

Hence due to (5.3)

$$N^2(A) = Trace \ (D+V)^*(V+D) = Trace \ (V^*V + D^*D) =$$

$$N^2(V) + \sum_{k=1}^{n} |\lambda_k(A)|^2,$$

and the required equality is proved. Q. E. D.

Proof of Theorem 1.5.1: From (5.3) it follows

$$(I\lambda - A)^{-1} = (I\lambda - D - V)^{-1} = (I\lambda - D)^{-1}(I + B_\lambda)^{-1}, \qquad (5.5)$$

where $B_\lambda := -V(I\lambda - D)^{-1}$. But operator B_λ is a nilpotent one. So Theorem 1.4.1 implies

$$\|(I + B_\lambda)^{-1}\| \le [1 + \frac{1 + N^2(B_\lambda)}{n-1}]^{(n-1)/2}. \qquad (5.6)$$

Take into account that

$$N_2(B_\lambda) = N_2(V(I\lambda - D)^{-1}) \le \|(I\lambda - D)^{-1}\| N_2(V) = \frac{1}{\rho(D, \lambda)} N_2(V).$$

Moreover, because to Lemma 1.5.2, $N_2(V) = g(A)$. Thus,

$$N_2(B_\lambda) \le \frac{g(A)}{\rho(D, \lambda)} = \frac{g(A)}{\rho(A, \lambda)}.$$

Now (5.5) and (5.6) imply the required result. Q. E. D.

Note that from Lemma 1.5.2 it follows

$$g(e^{i\tau} A + zI) = g(A)$$

for all $\tau \in \mathbf{R}$ and $z \in \mathbf{C}$, since the nilpotent parts of $e^{i\tau} A + zI$ and A coincide. Note also that from Theorem 1.9.1 proved below it follows the useful inequality

$$g^2(A) \le \frac{1}{2} N^2(A^* - A). \qquad (5.7)$$

For example, consider the $n \times n$-matrix

$$A_n(\gamma) = \begin{pmatrix} -a_1 & \cdots & -a_{n-1} & -a_n \\ 1/2^\gamma & \cdots & 0 & 0 \\ \cdot & \cdots & \cdot & \cdot \\ 0 & \cdots & 1/n^\gamma & 0 \end{pmatrix}$$

with complex numbers a_k and a constant $\gamma \geq 0$. Take into account that

$$A_n^2(\gamma) = \begin{pmatrix} a_1^2 - a_2/2^\gamma & \cdots & a_1 a_{n-1} & a_1 a_n \\ -a_1/2^\gamma & \cdots & -a_{n-1}/2^\gamma & -a_n/2^\gamma \\ 1/6^\gamma & \cdots & 0 & 0 \\ \cdot & \cdots & \cdot & \cdot \\ 0 & \cdots & 0 & 0 \end{pmatrix}.$$

Thus, we obtain

$$Trace\, A_n^2(\gamma) = a_1^2 - 2^{1-\gamma} a_2.$$

Inequality (5.2) implies

$$g^2(A_n(\gamma)) \leq N^2(A_n(\gamma)) - |Trace\, A_n^2(\gamma)|.$$

Hence we get

Lemma 1.5.3 *The inequality*

$$g^2(A_n(\gamma)) \leq \sum_{k=1}^{n} \left(|a_k|^2 + \frac{1}{k^{2\gamma}}\right) - 1 - |a_1^2 - 2^{1-\gamma} a_2|$$

is valid.

1.6   Bounds for roots of some scalar equations

Let us consider the algebraic equation

$$z^n = P(z) \quad (n > 1), \text{ where } P(z) = \sum_{j=0}^{n-1} c_j z^{n-j-1} \tag{6.1}$$

with non-negative coefficients c_j $(j = 0, ..., n-1)$.

Lemma 1.6.1 *The extreme right-hand root z_0 of equation (6.1) is non-negative and the following estimates are valid:*

$$z_0 \leq [P(1)]^{1/n} \quad \text{if } P(1) \leq 1, \tag{6.2}$$

and

$$1 \leq z_0 \leq P(1) \quad \text{if } P(1) \geq 1. \tag{6.3}$$

Proof: Since all the coefficients of $P(z)$ are non-negative, it does not decrease as $z > 0$ increases. From this it follows that if $P(1) \leq 1$, then $z_0 \leq 1$. So $z_0^n \leq P(1)$, as claimed.

Now let $P(1) \geq 1$, then due to (6.1) $z_0 \geq 1$ because $P(z)$ does not decrease. It is clear that

$$P(z_0) \leq z_0^{n-1} P(1)$$

in this case. Substituting this inequality in (6.1), we get (6.3). Q. E. D.

Setting $z = ax$ with a positive constant a in (6.1), we obtain

$$x^n = \sum_{j=0}^{n-1} \frac{c_j}{a^{j+1}} x^{n-j-1}. \tag{6.4}$$

Let

$$a = 2 \max_{j=0,\ldots,n-1} {}^{j+1}\!\sqrt{c_j}.$$

Then

$$\sum_{j=0}^{n-1} \frac{c_j}{a^{j+1}} \leq \sum_{j=0}^{n-1} 2^{-j-1} = 1 - 2^{-n} < 1.$$

Let x_0 be the extreme right-hand root of equation (6.4), then by (6.2) we have $x_0 \leq 1$. Since $z_0 = ax_0$, we have derived the following result.

Corollary 1.6.2 *The extreme right-hand root z_0 of equation (6.1) is non-negative. Moreover,*

$$z_0 \leq 2 \max_{j=0,\ldots,n-1} {}^{j+1}\!\sqrt{c_j}.$$

Furthermore, consider the scalar equation

$$\sum_{k=1}^{\infty} a_k z^k = 1 \tag{6.5}$$

where the complex, in general, coefficients $a_k, k = 1, 2, \ldots$ have the property

$$\gamma_0 := 2 \sup_{k=1,2,\ldots} \sqrt[k]{|a_k|} < \infty.$$

Lemma 1.6.3 *Any root z_0 of equation (6.5) satisfies the inequality $|z_0| \geq 1/\gamma_0$.*

Proof: Substitute in (6.5) $z_0 = x\gamma_0^{-1}$. We have

$$1 = \sum_{k=1}^{\infty} \frac{a_k}{\gamma_0^k} x^k \leq \sum_{k=1}^{\infty} \frac{|a_k|}{\gamma_0^k} |x|^k.$$

But

$$\sum_{k=1}^{\infty} \frac{|a_k|}{\gamma_0^k} \leq \sum_{k=1}^{\infty} 2^{-k} = 1$$

and therefore, $|x| \geq 1$. Hence,

$$|z_0| = \frac{1}{\gamma_0}|x| \geq \frac{1}{\gamma_0}.$$

As claimed. Q. E. D.

The following lemma will be used to investigate perturbations of finite-order entire functions.

Lemma 1.6.4 *For any integer $p \geq 1$, the extreme right (unique positive) root z_a of the equation*

$$\sum_{j=0}^{p-1} \frac{1}{y^{j+1}} \, exp \, \left[\frac{1}{2}\left(1 + \frac{1}{y^{2p}}\right)\right] = a \quad (a = const > 0) \tag{6.6}$$

satisfies the inequality $z_a \leq \delta_p(a)$, where

$$\delta_p(a) := \begin{cases} pe/a & \text{if } a \leq pe, \\ [ln \, (a/p)]^{-1/2p} & \text{if } a > pe \end{cases}.$$

Proof: Assume that

$$pe \geq a. \tag{6.7}$$

Since the function

$$f(y) = \sum_{j=0}^{p-1} \frac{1}{y^{j+1}} \, exp \, \left[\frac{1}{2}\left(1 + \frac{1}{y^{2p}}\right)\right]$$

is nonincreasing and $f(1) = pe$, we have $z_a \geq 1$. But because of (6.6),

$$z_a = \frac{1}{a}\sum_{j=0}^{p-1} \frac{1}{z_a^j} \, exp \, \left[\frac{1}{2}\left(1 + z_a^{-2p}\right)\right] \leq \frac{pe}{a}.$$

So in the case (6.7), the lemma is proved. Now let

$$pe < a. \tag{6.8}$$

Then $z_a \leq 1$. But

$$\sum_{j=0}^{p-1} x^{j+1} \leq px^p \leq p \, exp \, [x^p - 1] \leq p \, exp \, \left[\frac{1}{2}\left(x^{2p} + 1\right) - 1\right]$$

$$= p \, exp \, \left[\frac{1}{2}(x^{2p} - 1)\right] \quad (x \geq 1).$$

So

$$f(y) = \sum_{j=0}^{p-1} \frac{1}{y^{j+1}} \, exp \, \left[\frac{1}{2}\left(1 + \frac{1}{y^{2p}}\right)\right] \leq p \, exp \, \left[\frac{1}{y^{2p}}\right] \quad (y \leq 1).$$

But $z_a \leq 1$ under (6.8). We thus have

$$a = f(z_a) \leq p \, exp \, \Big[\frac{1}{z_a^{2p}}\Big].$$

Or

$$z_a^{2p} \leq \frac{1}{ln \, (a/p)}.$$

This finishes the proof. Q. E. D.

We need also the following simple result.

Lemma 1.6.5 *The unique positive root z_0 of the equation*

$$ze^z = a \quad (a = const > 0) \tag{6.9}$$

satisfies the estimate

$$z_0 \geq ln \, \Big[\frac{1}{2} + \sqrt{\frac{1}{4} + a}\Big]. \tag{6.10}$$

If, in addition, the condition $a \geq e$ holds, then

$$z_0 \geq ln \, a - ln \, ln \, a. \tag{6.11}$$

Proof: Since $z \leq e^z - 1$ $(z \geq 0)$, we arrive at the relation $a \leq e^{2z_0} - e^{z_0}$. Hence, $e^{z_0} \geq r_{1,2}$, where $r_{1,2}$ are the roots of the polynomial $y^2 - y - a$. This proves inequality (6.10).

Furthermore, if the condition $a \geq e$ holds, then $z_0 e^{z_0} \geq e$ and $z_0 \geq 1$. Now (6.9) yields $e^{z_0} \leq a$ and $z_0 \leq ln \, a$. So

$$a = z_0 e^{z_0} \leq e^{z_0} ln \, a.$$

Hence, inequality (6.11) follows. Q. E. D.

1.7 Perturbations of matrices

Let A and B be $n \times n$-matrices. The following quantity

$$sv_A(B) := \max_{k=1,\dots,n} \; \min_{j=1,\dots,n} |\lambda_j(A) - \lambda_k(B)|$$

is called the spectral variation of B with respect to A, see [Stewart and Sun, 1990].

Denote

$$q = \|A - B\|,$$

where $\|.\|$ is the Euclidean norm, again.

Recall also that $\sigma(A)$ denotes the spectrum of A.

Lemma 1.7.1 *For any $\mu \in \sigma(B)$, we have either $\mu \in \sigma(A)$, or*

$$q\|R_\mu(A)\| \geq 1.$$

Proof: Suppose that the inequality

$$q\|R_\mu(A)\| < 1. \tag{7.1}$$

holds. We can write $R_\mu(A) - R_\mu(B) = R_\mu(B)(B - A)R_\mu(A)$. This yields

$$\|R_\mu(A) - R_\mu(B)\| \leq \|R_\mu(B)\|q\|R_\mu(A)\|.$$

Thus, (7.1) implies

$$\|R_\mu(B)\| \leq \|R_\mu(A)\|(1 - q\|R_\mu(A)\|)^{-1}.$$

That is, μ is a regular point of B. This contradiction proves the result.
   Q. E. D.

Lemma 1.7.2 *Assume that*

$$\|R_\lambda(A)\| \leq \phi\left(\frac{1}{\rho(A,\lambda)}\right) \text{ for all regular } \lambda \text{ of } A, \tag{7.2}$$

where $\phi(x)$ is a monotonically increasing non-negative function of a non-negative variable x, such that $\phi(0) = 0$ and $\phi(\infty) = \infty$. Then the inequality

$$sv_A(B) \leq z(\phi, q)$$

is true, where $z(\phi, q)$ is the unique (positive) root of the equation

$$q\phi\left(\frac{1}{z}\right) = 1. \tag{7.3}$$

Proof: This lemma is a particular case of Lemma 2.12.2 proved in the next chapter. Q. E. D.

Lemma 1.7.3 *Let A and B be $n \times n$-matrices. Then*

$$sv_A(B) \leq y(q, A),$$

where $y(q, A)$ is the unique nonnegative root of the algebraic equation

$$1 = \frac{q}{y}\left[1 + \frac{1}{n-1}\left(1 + \frac{g^2(A)}{y^2}\right)\right]^{(n-1)/2}. \tag{7.4}$$

Proof: This result immediately follows from Lemma 1.7.2 and Theorem 1.5.1. Q. E. D.

Put in (7.4),

$$x = \frac{y}{g(A)}.$$

Then (7.4) takes the form

$$\frac{q}{xg(A)}\left[1 + \frac{1}{n-1}\left(1 + \frac{1}{x^2}\right)\right]^{(n-1)/2} = 1.$$

Or

$$x^n = \frac{qx}{g(A)}\left[x^2\left(1 + \frac{1}{n-1}\right) + \frac{1}{n-1}\right]^{(n-1)/2}.$$

Denote

$$v_n = \left[1 + \frac{2}{n-1}\right]^{(n-1)/2} = \left[\frac{n+1}{n-1}\right]^{(n-1)/2}.$$

Lemma 1.6.1 implies $y(q, A) \le \delta(A, q)$, where

$$\delta(A, q) := \begin{cases} g^{1-1/n}(A)\sqrt[n]{qv_n} & \text{if } v_n q \le g(A) \\ qv_n & \text{if } v_n q \ge g(A) \end{cases}.$$

Now from Lemma 1.7.3 it follows

Corollary 1.7.4 *Let A and B be $n \times n$-matrices. Then the inequality $sv_A(B) \le \delta(A, q)$ is valid.*

Now let us consider perturbations of determinants.

Theorem 1.7.5 *Let A and B be linear operators in $\mathbf{C}^n$. Then for any integer $p \ge 1$, one has*

$$|\det(A) - \det(B)| \le \frac{N_p(A - B)}{n^{n/p}}\left(1 + \frac{1}{2}(N_p(A + B) + N_p(A - B))\right)^n.$$

Proof: The function $\det(A + \lambda B)$ is a polynomial in λ. Besides the operators A and B can be considered as elements of the space $C^{n \times n}$ with norm N_p. Moreover, making use of the inequality for the arithmetic and geometric mean values, we have

$$|\det(A)|^p = \prod_{k=1}^{n} |\lambda_k(A)|^p \le \left(\frac{1}{n}\sum_{k=1}^{n} |\lambda_k(A)|^p\right)^n \le \frac{1}{n^n}N_p^{pn}(A).$$

Now the required result is due to Lemma 2.14.1 proved below. Q. E. D.

1.8   Preservation of multiplicities of eigenvalues

For a $b \in \mathbf{C}$ and an $r > 0$, put

$$\Omega(b, r) := \{z \in \mathbf{C} : |z - b| \leq r\}.$$

Lemma 1.8.1 *Let A and B be $n \times n$-matrices. Assume that*

$$A \text{ has an eigenvalue } \lambda(A) \text{ of the algebraic multiplicity } \nu. \qquad (8.1)$$

Let condition (7.2) hold. In addition, with the notations $q = \|A - B\|$ and

$$\beta(A) := \min_{t \in \sigma(A), t \neq \lambda(A)} |t - \lambda(A)|/2,$$

for a positive $r < \beta(A)$, let

$$q\phi\left(\frac{1}{r}\right) < 1. \qquad (8.2)$$

Then the set $\sigma(B) \cap \Omega(\lambda(A), r)$ consists of eigenvalues, whose total algebraic multiplicity is equal to ν.

Proof: This lemma is a particular case of Lemma 2.13.1 proved in the next chapter. Q. E. D.

Since ϕ monotonically increases, from the latter lemma we obtain the following result.

Corollary 1.8.2 *Under conditions (8.1) and (7.2), let $z(\phi, q)$ be the unique positive root of equation (7.3). Then the set $\sigma(B) \cap \Omega(\lambda(A), r(\phi, q))$ consists of eigenvalues of B, whose total algebraic multiplicity is equal to ν, provided $z(\phi, q) \leq \beta(A)$.*

From Lemma 1.8.1 and Theorem 1.5.1, we at once obtain the following result.

Corollary 1.8.3 *Let A and B be $n \times n$-matrices, and the conditions (8.1) and*

$$\frac{q}{r}\left[1 + \frac{1}{n-1}\left(1 + \frac{g^2(A)}{r^2(A, \lambda)}\right)\right]^{(n-1)/2} < 1$$

hold for a positive $r < \beta(A)$. Then the set $\sigma(B) \cap \Omega(\lambda(A), r)$ consists of eigenvalues, whose total algebraic multiplicity is equal to ν.

1.9 An identity for imaginary parts of eigenvalues

Theorem 1.9.1 *For any $n \times n$-matrix A, we have*

$$N_2^2(A) - \sum_{k=1}^{n} |\lambda_k(A)|^2 = 2N_2^2(A_I) - 2\sum_{k=1}^{n} |Im\ \lambda_k(A)|^2.$$

To prove this result, first let us prove the following lemma.

Lemma 1.9.2 *For any linear operator A in $\mathbf{C}^n$, the inequality*

$$N_2^2(V) = 2N_2^2(A_I) - 2\sum_{k=1}^{n} |Im\ \lambda_k(A)|^2$$

holds, where V is the nilpotent part of A.

Proof: Clearly,

$$-4(A_I)^2 = (A - A^*)^2 = AA - AA^* - A^*A + A^*A^*.$$

But because of the triangular representation,

$$Trace\ (A - A^*)^2 = Trace\ (V + D - V^* - D^*)^2$$

$$= Trace\ [(V - V^*)^2 + (V - V^*)(D - D^*)$$

$$+(D - D^*)(V - V^*) + (D - D^*)^2]$$

$$= Trace\ (V - V^*)^2 + Trace\ (D - D^*)^2$$

since D^*V, VD^* are nilpotent operators. Hence,

$$N^2(A_I) = N^2(V_I) + N^2(D_I),$$

where $N(.) = N_2(.)$,

$$V_I = (V - V^*)/2i \text{ and } D_I = (D - D^*)/2i.$$

It is not hard to see that

$$N^2(V_I) = \frac{1}{2}\sum_{m=1}^{n}\sum_{k=1}^{m-1} |a_{km}|^2 = \frac{1}{2}N^2(V),$$

where a_{jk} are the entries of V in the Schur basis (the basis of the triangular representation). Thus,

$$N^2(V) = 2N^2(A_I) - 2N^2(D_I).$$

But

$$N^2(D_I) = \sum_{k=1}^{n} |Im\ \lambda_k(A)|^2.$$

Thus, we arrive at the required equality. Q. E. D.

The assertion of Theorem 1.9.1 follows from the previous lemma and Lemma 1.5.2. Q. E. D.

1.10   Additional estimates for resolvents

Recall that $\rho(A, \lambda) = \min_{k=1,...,n} |\lambda - \lambda_k(A)|$ and $g(A)$ is defined in Section 1.5. To formulate the result, for a natural $n > 1$ introduce the numbers

$$\gamma_{n,k} = \sqrt{\frac{C_{n-1}^k}{(n-1)^k}}\ \ (k = 1, ..., n-1)\ \text{and}\ \gamma_{n,0} = 1.$$

Here

$$C_{n-1}^k = \frac{(n-1)!}{(n-k-1)!k!}$$

are the binomial coefficients. Evidently, for all $n > 2$,

$$\gamma_{n,k}^2 = \frac{(n-1)(n-2)\ldots(n-k)}{(n-1)^k k!} \leq \frac{1}{k!}\ \ (k = 1, 2, ..., n-1). \tag{10.1}$$

Theorem 1.10.1 *Let A be a linear operator in $\mathbf{C}^n$. Then its resolvent $R_\lambda(A) = (A - \lambda I)^{-1}$ satisfies the inequality*

$$\|R_\lambda(A)\| \leq \sum_{k=0}^{n-1} \frac{g^k(A)\gamma_{n,k}}{\rho^{k+1}(A,\lambda)}\ \text{for any regular point}\ \lambda\ \text{of}\ A.$$

We will prove this theorem in the present section below. *The theorem is sharp:* if A is a normal matrix, then $g(A) = 0$ and

$$\|R_\lambda(A)\| = \frac{1}{\rho(A,\lambda)}\ \ (\lambda \notin \sigma(A)).$$

Moreover, the latter theorem and inequalities (10.1) imply

Corollary 1.10.2 *Let A be a linear operator in $\mathbf{C}^n$. Then*

$$\|R_\lambda(A)\| \leq \sum_{k=0}^{n-1} \frac{g^k(A)}{\sqrt{k!}\rho^{k+1}(A,\lambda)}\ \ (\lambda \notin \sigma(A)).$$

To prove Theorem 1.10.1 we need the following lemma.

Lemma 1.10.3 *For arbitrary positive numbers $a_1, ..., a_n$ and $m = 1, ..., n$, we have*

$$\sum_{1 \leq k_1 < k_2 < ... < k_m \leq n} a_{k_1} ... a_{k_m} \leq n^{-m} C_n^m \left[\sum_{k=1}^{n} a_k \right]^m.$$

Proof: Consider the following function of n positive variables $y_1, ..., y_n$:

$$R_m(y_1, ..., y_n) := \sum_{1 \leq k_1 < k_2 < ... < k_m \leq n} y_{k_1} y_{k_2} \cdots y_{k_m}.$$

Let us prove that under the condition

$$\sum_{k=1}^{n} y_k = n\, b, \tag{10.2}$$

where b is a given positive number, function R_m has a unique conditional maximum. To this end, denote

$$F_j(y_1, ..., y_n) := \frac{\partial R_m(y_1, ..., y_n)}{\partial y_j} = \frac{R_m(y_1, ..., y_n)}{y_j}.$$

Obviously, $F_j(y_1, ..., y_n)$ does not depend on y_j, symmetrically depends on other variables, and monotonically increases with respect to each of its variables. The conditional extremum of R_m under (10.2) are the roots of the equations

$$F_j(y_1, ..., y_n) - \lambda \frac{\partial}{\partial y_j} \sum_{k=1}^{n} y_k = 0 \ (j = 1, ..., n),$$

where λ is the Lagrange factor. Therefore,

$$F_j(y_1, ..., y_n) = \lambda \ (j = 1, ..., n).$$

Since $F_j(y_1, ..., y_n)$ does not depend on y_j, and $F_k(y_1, ..., y_n)$ does not depend on y_k, the equalities

$$F_j(y_1, ..., y_n) = F_k(y_1, ..., y_n) = \lambda,$$

or equivalently, the equalities

$$\frac{R_m(y_1, ..., y_n)}{y_j} = \frac{R_m(y_1, ..., y_n)}{y_k} = \lambda,$$

for all $k \neq j$ are possible if and only if $y_j = y_k$.

Thus R_m has under (10.2) a unique extremum when

$$y_1 = y_2 = ... = y_n = b. \tag{10.3}$$

But

$$R_m(b, ..., b) = b^m \sum_{1 \le k_1 < k_2 < ... < k_m \le n} 1 = b^m C_n^m. \tag{10.4}$$

Let us check that (10.3) gives us the maximum. Letting

$$y_1 \to nb \text{ and } y_k \to 0 \ (k = 2, ..., n),$$

we get

$$R_m(y_1, ..., y_n) \to 0.$$

Since the extremum (10.3) is unique, it is the maximum. Thus, under (10.2)

$$R_m(y_1, ..., y_n) \le b^m C_n^m \ (y_k \ge 0, \ k = 1, ..., n).$$

Take $y_j = a_j$ and

$$b = \frac{a_1 + ... + a_n}{n}.$$

Then

$$R_m(a_1, ..., a_n) \le C_n^m n^{-m} \Big[\sum_{k=1}^{n} a_k \Big]^m.$$

We thus get the required result. Q. E. D.

Again let $N(A) = N_2(A)$ be the Hilbert-Schmidt norm of A.

Lemma 1.10.4 *For any nilpotent operator V in $\mathbf{C}^n$, the inequalities*

$$\|V^p\| \le \gamma_{n,p} N_2^p(V) \ (p = 1, ..., n-1)$$

are valid.

Proof: Since V is nilpotent, because of the Schur theorem, we can represent it by an upper-triangular matrix with the zero diagonal:

$$V = (a_{jk})_{j,k=1}^{n} \text{ with } a_{jk} = 0 \ (j \ge k).$$

Denote

$$\|x\|_m = \Big(\sum_{k=m}^{n} |x_k|^2 \Big)^{1/2} \text{ for } m < n,$$

where x_k are coordinates of a vector x. We can write

$$\|Vx\|_m^2 = \sum_{j=m}^{n-1} \Big| \sum_{k=j+1}^{n} a_{jk} x_k \Big|^2 \text{ for all } m \le n-1.$$

Now we have (by Schwarz's inequality) the relation

$$\|Vx\|_m^2 \le \sum_{j=m}^{n-1} h_j \|x\|_{j+1}^2, \tag{10.5}$$

where

$$h_j = \sum_{k=j+1}^{n} |a_{jk}|^2 \quad (j < n).$$

Further, by Schwarz's inequality

$$\|V^2 x\|_m^2 = \sum_{j=m}^{n-1} \Big| \sum_{k=j+1}^{n} a_{jk}(Vx)_k \Big|^2 \le \sum_{j=m}^{n-1} h_j \|Vx\|_{j+1}^2.$$

Here $(Vx)_k$ are coordinates of Vx. Taking into account (10.5), we obtain

$$\|V^2 x\|_m^2 \le \sum_{j=m}^{n-1} h_j \sum_{k=j+1}^{n-1} h_k \|x\|_{k+1}^2 = \sum_{m \le j < k \le n-1} h_j h_k \|x\|_{k+1}^2.$$

Hence,

$$\|V^2\|^2 \le \sum_{1 \le j < k \le n-1} h_j h_k.$$

Repeating these arguments, we arrive at the inequality

$$\|V^p\|^2 \le \sum_{1 \le k_1 < k_2 < ... < k_p \le n-1} h_{k_1} ... h_{k_p}.$$

Therefore, because of the previous lemma,

$$\|V^p\|^2 \le \gamma_{n,p}^2 \Big(\sum_{j=1}^{n-1} h_j \Big)^p.$$

But

$$\sum_{j=1}^{n-1} h_j \le N_2^2(V).$$

This proves required result. Q. E. D.

Proof of Theorem 1.10.1: Let D and V be the diagonal and nilpotent parts of A, respectively. Since $R_\lambda(D)V$ is a nilpotent operator, by virtue of the previous lemma,

$$\|(R_\lambda(D)V))^k\| \le N_2^k(R_\lambda(D)V)\gamma_{n,k} \quad (k = 1, ..., n-1). \tag{10.6}$$

Since D is a normal operator, we can write down $\|R_\lambda(D)\| = \rho^{-1}(D, \lambda)$. It is clear that

$$N(R_\lambda(D)V) \le N(V)\|R_\lambda(D)\| = N(V)\rho^{-1}(D, \lambda).$$

But

$$A - \lambda I = D + V - \lambda I = (D - \lambda I)(I + R_\lambda(D)V).$$

We thus have

$$R_\lambda(A) = (I + R_\lambda(D)V)^{-1}R_\lambda(D) = \sum_{k=0}^{n-1}(R_\lambda(D)V)^k R_\lambda(D).$$

Now (10.6) yields the inequality

$$\|R_\lambda(A)\| \le \sum_{k=0}^{n-1} N^k(V)\frac{\gamma_{n,k}}{\rho^{k+1}(D,\lambda)}.$$

This relation proves the stated result, since A and D have the same eigenvalues and $N(V) = g(A)$, because of to Lemma 1.5.2. Q. E. D.

The just proven theorem, Lemmas 1.7.2 and 1.8.1, enable us to state additional perturbation results for matrices.

1.11   Gerschgorin's circle theorem

Let A be a complex $n \times n$-matrix, with entries a_{jk}. For $j = 1, ..., n$ write

$$R_j = \sum_{k=1,k\neq j}^{n} |a_{jk}|.$$

Let $\Omega(b,r)$ be the closed disc centered at $b \in \mathbf{C}$ with a radius r.

Theorem 1.11.1 *(Gerschgorin). Every eigenvalue of A lies within at least one of the discs $\Omega(a_{jj}, R_j)$.*

Proof: Let λ be an eigenvalue of A and let $x = (x_j)$ be the corresponding eigenvector. Let i be chosen so that $|x_i| = max_j|x_j|$. Then $|x_i| > 0$, otherwise $x = 0$. Since x is an eigenvector, $Ax = \lambda x$ or equivalent

$$\sum_{k=1}^{n} a_{ik} = \lambda x_i$$

so, splitting the sum, we get

$$\sum_{k=1,k\neq i}^{n} a_{ik}x_k = \lambda x_i - a_{ii}x_i.$$

We may then divide both sides by x_i (choosing i as we explained we can be sure that $x_i \neq 0$) and take the absolute value to obtain

$$|\lambda - a_{ii}| \le \sum_{k=1,k\neq i}^{n} |a_{jk}|\frac{|x_k|}{|x_i|} \le R_i,$$

where the last inequality is valid because

$$\frac{|x_k|}{|x_l|} \le 1.$$

As claimed. Q. E. D.

Note that for a diagonal matrix the Gerschgorin discs $\Omega(a_{jj}, R_j)$ coincide with the spectrum. Conversely, if the Gerschgorin discs coincide with the spectrum, the matrix is diagonal.

1.12   Cassini ovals and related results

Brauer and Ostrovski independently observed that *each eigenvalue of an $n \times n$ matrix $A = (a_{jk})$, $n \ge 2$, is contained in the Cassini ovals*

$$\Omega(m, j) := \left\{ \lambda \in \mathbf{C} : |\lambda - a_{mm}||\lambda - a_{jj}| \le \left(\sum_{k=1, k \ne m}^{n} |a_{mk}| \right)\left(\sum_{k=1, k \ne j}^{n} |a_{jk}| \right) \right\}$$

$$(j, m = 1, ..., n; \ j \ne m) \tag{12.1}$$

(see [Marcus and Minc, 1966, Section 3.2]).

That result leads to a better localization of the spectrum of a matrix than Gerschgorin's theorem; compare [Varga, 2004] and references therein.

To formulate our next result, for an integer $m \le n$, put

$$G_m(j) := \left\{ z \in \mathbf{C} : \ |(z - a_{mm})(z - a_{jj}) - a_{jm}a_{mj}| \right.$$

$$\le \sum_{k=1, k \ne m, k \ne j}^{n} |a_{jm}a_{mk} - (a_{mm} - z)a_{jk}| \right\}$$

$$(j = 1, ..., n; \ j \ne m).$$

Theorem 1.12.1 *For any integer $m \in \{1, ..., n\}$, each eigenvalue $\lambda(A)$ of a matrix $A = (a_{jk})_{j,k=1}^{n}$ either lies in the set*

$$\cup_{j=1, j \ne m}^{n} G_m(j),$$

or $\lambda(A) = a_{mm}$.

Proof: The proof is based on the Gaussian elimination. Note that using the Gaussian elimination in the considered area is new.

First, let us prove that, if $a_{mm} \neq 0$ and

$$|a_{mm}a_{jj} - a_{jm}a_{mj}| > \sum_{k=1, k \neq m, k \neq j}^{n} |a_{jm}a_{mk} - a_{mm}a_{jk}| \quad (j = 1, ..., n; \ j \neq m)$$

$$(12.2)$$

for an $m \leq n$, then matrix $A = (a_{jk})_{j,k=1}^{n}$ is invertible. Indeed, the invertibility of A is equivalent to the existence of a nontrivial solution $x = (x_k)$ to the system

$$\sum_{k=1}^{n} a_{jk}x_k = f_j \quad (f_j \in \mathbf{C}; \ j = 1, ..., n) \tag{12.3}$$

for all right-hand parts $f = (f_k)_{k=1}^{n} \neq 0$. Since $a_{mm} \neq 0$, we can write out

$$x_m = \left(- \sum_{k=1, k \neq m}^{n} a_{mk}x_k + f_m \right) \frac{1}{a_{mm}}. \tag{12.4}$$

Substitute this equality into (12.3). Then we get the system

$$-\frac{a_{jm}}{a_{mm}} \sum_{k=1, k \neq m}^{n} a_{mk}x_k + \sum_{k=1, k \neq m}^{n} a_{jk}x_k = F_j \quad (j \neq m), \tag{12.5}$$

where

$$F_j = f_j - \frac{a_{jm}f_m}{a_{mm}}.$$

If (12.5) has a nontrivial solution, then (12.3) also has a nontrivial solution. Rewrite (12.5) as

$$\sum_{k=1, k \neq m}^{n} b_{jk}x_k = F_j \quad (j \neq m), \tag{12.6}$$

where

$$b_{jk} = -\frac{a_{jm}a_{mk}}{a_{mm}} + a_{jk} \quad (j \neq m, k \neq m).$$

By the Hadamard criterion, system (12.6) is solvable for all right parts, provided

$$|b_{jj}| > \sum_{k=1, k \neq m, k \neq j}^{n} |b_{jk}|.$$

Or

$$\left| -\frac{a_{jm}a_{mj}}{a_{mm}} + a_{jj} \right| > \sum_{k=1, k \neq m, k \neq j}^{n} \left| -\frac{a_{jm}}{a_{mm}}a_{mk} + a_{jk} \right| \quad (j = 1, ..., n; \ j \neq m).$$

But this is equivalent to (12.2). So under (12.2) A is really invertible and the solution of (12.5) is nontrivial, provided

$$F_j = f_j - \frac{a_{jm}f_m}{a_{mm}} \neq 0$$

for at least one index $j \neq m$. If $F_j = 0$ for all $j \neq m$, then $x_j = 0, j \neq m$ and by (12.4),

$$x_m = \frac{f_m}{a_{mm}}.$$

So in this case (12.4) also has a nontrivial solution if $f_m \neq 0$. Now replacing A by $A - \lambda I$ we can assert that $A - \lambda I$ is invertible if $\lambda \neq a_{mm}$ and

$$|(\lambda - a_{mm})(\lambda - a_{jj}) - a_{jm}a_{mj}| > \sum_{k=1, k \neq m, k \neq j}^{n} |a_{jm}a_{mk} - (a_{mm} - \lambda)a_{jk}|$$

$$(j = 1, ..., n; \ j \neq m).$$

This proves the theorem. Q. E. D.

Below we also prove that $G_m(j) \subset \Omega(j, m)$. In addition, we can choose a fixed m, and we are not forced to check the corresponding inequalities for all integers $m \leq n$ as in (12.1). If $n = 2$, then Theorem 1.12.1 gives us the sharp invertibility condition

$$|a_{11}a_{22} - a_{21}a_{12}| > 0.$$

One can take $a_{11} = 0$ or $a_{22} = 0$. So Theorem 1.12.1 does not require that A should be diagonally dominant. Note that the Hadamard criterion does not give a result for matrices with the zero diagonal entries. To compare Theorem 1.12.1 with the Cassini ovals, note that by the Gerschgorin criterion, any given eigenvalue $\lambda = \lambda(A)$ of A satisfies the inequality

$$|\lambda - a_{mm}| \leq \sum_{k=2}^{n} |a_{mk}| \tag{12.7}$$

for at least one $m \leq n$. Without loss of generality assume that $m = 1$. Thanks to Theorem 1.12.1 either $\lambda = a_{11}$ or λ satisfies the inequalities

$$|(\lambda - a_{11})(z - a_{jj}) - a_{j1}a_{1j}| \leq \sum_{k=2, k \neq j}^{n} |a_{j1}a_{1k} - (a_{11} - \lambda)a_{jk}| \quad (j = 2, ...n).$$

Hence it follows that

$$|(a_{11} - \lambda)(a_{jj} - \lambda)| \leq |a_{j1}a_{1j}| + \sum_{k=2, k \neq j}^{n} |a_{j1}a_{1k}| + |(a_{11} - \lambda)a_{jk}|$$

$$= |a_{j1}| \sum_{k=2}^{n} |a_{1k}| + \sum_{i=2, i \neq j}^{n} |(a_{11} - \lambda)a_{ji}|.$$

Taking into account (12.7) with $m = 1$, we thus get

$$|(a_{11} - \lambda)(a_{jj} - \lambda)| \leq |a_{j1}| \sum_{k=2}^{n} |a_{1k}| + \sum_{i=2, i \neq j}^{n} |a_{ji}| \sum_{k=2}^{n} |a_{1k}|$$

$$= \Big(\sum_{k=2}^{n} |a_{1k}| \Big) \Big(\sum_{i=1, i \neq j}^{n} |a_{ji}| \Big) \ (j = 2, ...n).$$

So the sets $G_1(j)$ are contained in the Cassini ovals.

Example 1.12.2 *Let us consider the matrix*

$$T = \begin{pmatrix} -10 & -35 & -50 & -24 \\ 1 & 0 & 0 & 0 \\ 0 & 1 & 0 & 0 \\ 0 & 0 & 1 & 0 \end{pmatrix}.$$

This matrix has the eigenvalues

$$-1, -2, -3, \text{ and } -4.$$

We have

$$G_1(2) = \{z \in \mathbf{C} : |(10 + z)z + 35| \leq 74\}$$

and

$$G_1(3) = G_1(4) = \{z \in \mathbf{C} : |(z + 10)z| \leq |10 + z|\}.$$

Thanks to Theorem 1.12.1, hence it follows that any eigenvalue λ of T satisfies one of the following relations:

$$\lambda = -10, |\lambda| \leq 1 \text{ or } |\lambda^2 + 10\lambda + 35| \leq 74. \tag{12.8}$$

Clearly $\lambda^2 + 10\lambda + 35 = (\lambda - r_1)(\lambda - r_2)$, where

$$r_{1,2} = -5 \pm i\sqrt{10}.$$

Hence

$$74 \geq |\lambda - r_1||\lambda - r_2| \geq (|\lambda + 5| - \sqrt{10})^2.$$

Thus

$$|\lambda + 5| \leq \sqrt{10} + \sqrt{74}. \tag{12.9}$$

At the same time, the Gerschgorin sets for matrix T are defined by the inequalities

$$|\lambda + 10| \leq 109$$

(if we take the rows), or $|\lambda| \leq 50$ (if we take the columns). Clearly, the sets defined by (12.9) and therefore by (12.8) are included in the Gerschgorin sets.

1.13 The Brauer and Perron theorems

Again $A = (a_{jk})$ is a complex $n \times n$-matrix and for $j = 1, ..., n$ we write

$$R_j = \sum_{k=1, k \neq j}^{n} |a_{jk}|.$$

Recall that $\Omega(b, r)$ is the closed disc centered at $b \in \mathbf{C}$ with radius r.
We have the following result.

Theorem 1.13.1 *([Brauer, 1947]). For a given* $m \in \{1, ..., n\}$, *let*

$$|a_{jj} - a_{mm}| > R_j + R_m \tag{13.1}$$

for all $j \neq m$. *Then one and only one eigenvalue lies in the disk* $\Omega(a_{mm}, R_m)$.

Proof: We introduce the $n \times n$-matrix $B(t) = (b_{jk})$, where $b_{mm} = a_{mm}; b_{ij} = a_{ij}$ for $i \neq m$; and $b_{mj} = t a_{mj}$, the parameter t being real with $0 \leq t \leq 1$. All the eigenvalue of $B(t)$ lie in the union K of the disks $\Omega(a_{jj}, R_j)$, and the disc

$$\gamma_m(t) = |z - a_{mm}| < t \sum_{k=1, k \neq m}^{n} |a_{mk}| = t R_m,$$

which clearly is contained in $\gamma_m(1) = \Omega(a_{mm}, R_m)$. By (13.1), $\gamma_m(1) \cap K = 0$ as t varies continuously from 1 to 0, no eigenvalues of $B(t)$ can enter or leave $\gamma_m(1)$. But the eigenvalues for $B(0)$ are a_{mm} and the eigenvalues of an $(n-1) \times (n-1)$ matrix whose eigenvalues lie in K. Since $B(0)$ has only one eigenvalue in $\gamma_m(1)$, we infer by continuity that $B(1) = A$ has exactly one eigenvalue in $\gamma_m(1)$. Q. E. D.

The previous theorem remains valid if the R_j are replaced by

$$Q_i = \sum_{j=1, j \neq i}^{n} |a_{ji}|;$$

that is, if rows and columns are interchanged. Theorem 1.11.1 remains valid also when the R_i are replaced by $R_i^s Q_i^{1-s}$ for any $s \in [0, 1]$; cf. [Marcus and Minc, 1966, Section 3.2].
Another valuable result is the following Perron theorem.

Theorem 1.13.2 *If* $A = (a_{jk})$ *is a non-negative matrix, then the spectral radius* $r_s(A) = \max_k |\lambda_k(A)|$ *of* A *is its eigenvalue.*

For the proof of this theorem see [Gantmacher, 1967, Chapter 13, Section 2].
We also state the following comparison theorem.

Theorem 1.13.3 *If a matrix $A = (a_{jk})_{j,k=1}^n$ is non-negative and if in matrix*
$C = (c_{jk})_{j,k=1}^n$,

$$|c_{ji}| \le a_{ji}, i, j = 1, 2, \ldots n,$$

then any eigenvalue $\lambda(C)$ of C satisfies the inequality $|\lambda(C)| \le r_s(A)$.

Proof: By the Gel'fand formula

$$r_s(A) = \lim_{n \to \infty} \sqrt[n]{\|A^n\|},$$

where the norm is assumed to be Euclidean. But clearly, $\|C^n\| \le \|A^n\|$. So
$r_s(C) \le r_s(A)$. This proves the result. Q. E. D.

1.14   Comments to Chapter 1

As it was above mentioned, the material of Sections 1.1 and 1.2 is classical.
For more information about the inequalities for the eigenvalues of matrices,
see for instance [Marcus and Minc, 1964].

Theorem 1.3.1 has been proved in [Gil', 2007a]. The material of Sections
1.4 - 1.10 is adopted from [Gil', 2003]. For more information about perturbations of matrices, we refer the reader to the book [Stewart and Sun, 1990].
Many interesting results connected with Gerschgorin's circle theorem (Theorem 1.12.1) can be found in [Gantmaher, 1967] and [Varga, 2004]. Theorem
1.12.3 is probably new.

Chapter 2

Eigenvalues of Compact Operators

Chapter 2 is also auxiliary. It discusses some concepts and results of the spectral theory of compact operators in a Hilbert space, which will be used in this text. The major part of the chapter is devoted to the eigenvalues and singular values of Schatten - von Neumann operators, as well as to estimates for the resolvents and spectrum perturbations of these operators. The regularized characteristic determinants are also considered. By these determinants below, we investigate perturbations of the associate infinite products.

2.1 Banach and Hilbert spaces

In this section, we recall very briefly some basic notions of the theory of Banach and Hilbert spaces. More details can be found in any textbook on Banach and Hilbert spaces (e.g., [Ahiezer and Glazman, 1981], [Dunford and Schwartz, 1966]).

A linear space X over the set of complex numbers is called a *(complex) linear normed space,* if for any $x \in X$ a non-negative number $\|x\|_X = \|x\|$ is defined, called the norm of x, having the following properties:

1. $\|x\| = 0$ iff $x = 0$,
2. $\|\alpha x\| = |\alpha| \|x\|$,

3. $\|x + y\| \leq \|x\| + \|y\|$ for every $x, y \in X$, $\alpha \in \mathbf{C}$.

A sequence $\{h_n\}_{n=1}^{\infty}$ of elements of X converges *strongly* (in the norm) to $h \in X$ if

$$\lim_{n \to \infty} \|h_n - h\| = 0.$$

A sequence $\{h_n\}$ of elements of X is called the fundamental (Cauchy) one if

$$\|h_n - h_m\| \to 0 \text{ as } m, n \to \infty.$$

If any fundamental sequence converges to an element of X, then X is called *a (complex) Banach space.*

Let in a linear space H over $\mathbf{C}$ for all $x, y \in H$ a number (x, y) be defined, such that

1. $(x, x) > 0$, if $x \neq 0$, and $(x, x) = 0$, if $x = 0$,
2. $(x, y) = \overline{(y, x)}$,
3. $(x_1 + x_2, y) = (x_1, y) + (x_2, y)$ $(x_1, x_2 \in H)$,
4. $(\lambda x, y) = \lambda(x, y)$ $(\lambda \in \mathbf{C})$.

Then $(.,.)$ is called the scalar product. Define in H the norm by

$$\|x\| = \sqrt{(x, x)}.$$

If H is a Banach space with respect to this norm, then it is called *a Hilbert space.* The Schwarz inequality

$$|(x, y)| \leq \|x\| \, \|y\|$$

is valid.

If, in an infinite dimensional Hilbert space, there is a countable set whose closure coincides with the entire space, then that space is said to be *separable.* Any separable Hilbert space H possesses an orthonormal basis. This means that there is a sequence $\{e_k \in H\}_{k=1}^{\infty}$ such that

$$(e_k, e_j) = 0 \text{ if } j \neq k \text{ and } (e_k, e_k) = 1 \quad (j, k = 1, 2, ...)$$

and any $h \in H$ can be represented as

$$h = \sum_{k=1}^{\infty} c_k e_k$$

with $c_k = (h, e_k)$, $k = 1, 2, \ldots$. Besides the series strongly converges.

The following spaces are examples of normed spaces.

The space l^p is defined for $1 \leq p < \infty$ as the linear space of all sequences $x = \{x_k\}_{k=1}^{\infty}$ of scalars for which the norm

$$\|x\| = \left(\sum_{k=1}^{\infty} |x_k|^p \right)^{1/p}$$

is finite.

The space l^∞ is the linear space of all bounded sequences $x = \{x_k\}_{k=1}^\infty$ of scalars. The norm is given by the equation

$$\|x\| = \sup_k |x_k|.$$

The space $L^p(S)$ $(1 \le p < \infty)$ is a complex Banach space of scalar-valued functions defined on a compact set S of $\mathbf{R}^n$ with the norm

$$\|f\| = \left[\int_S |f(s)|^p ds \right]^{1/p}.$$

Note that the Hilbert space has been defined by a set of abstract axioms. It is noteworthy that some of the concrete spaces defined above satisfy these axioms, and hence are special cases of abstract Hilbert space. Thus, for instance, complex l^2 space is a Hilbert space if the scalar product (x, y) of the vectors $x = \{x_k\}$ and $y = \{y_k\}$ is defined by the formula

$$(x, y) = \sum_{k=1}^\infty x_k \bar{y}_k.$$

Also the complex space $L^2(S)$ is a Hilbert space with the scalar product

$$(f, g) = \int_S f(s)\bar{g}(s)ds.$$

2.2 Linear operators

An operator A, acting from a Banach space X into a Banach space Y, is called a linear one if

$$A(\alpha x_1 + \beta x_2) = \alpha A x_1 + \beta A x_2$$

for any $x_1, x_2 \in X$ and $\alpha, \beta \in \mathbf{C}$. If there is a constant a, such that the inequality

$$\|Ah\|_Y \le a\|h\|_X \text{ for all } h \in X$$

holds, then the operator is said to be bounded. The quantity

$$\|A\|_{X \to Y} := \sup_{h \in X} \frac{\|Ah\|_Y}{\|h\|_X}$$

is called the norm of A. If $X = Y$, we will write $\|A\|_{X \to X} = \|A\|_X$ or simply $\|A\|$.

A linear operator *is said to be completely continuous (compact)* if it is bounded and maps each bounded set in X into a compact one in Y.

Under the natural definitions of addition and multiplication by a scalar, and the norm, the set $B(X, Y)$ of all bounded linear operators acting from X into Y becomes a Banach space. If $Y = X$, we will write $B(X, X) = B(X)$. A sequence $\{A_n\}$ of bounded linear operators from $B(X, Y)$ converges *in the uniform operator topology* (in the operator norm) to an operator A if

$$\lim_{n \to \infty} \|A_n - A\|_{X \to Y} = 0.$$

A sequence $\{A_n\}$ of bounded linear operators *converges strongly* to an operator A if the sequence of elements $\{A_n h\}$ strongly converges to Ah for every $h \in X$.

If ϕ is a linear operator, acting from X into $\mathbf{C}$, then it is called a linear functional. It is bounded (continuous) if $\phi(x)$ is defined for any $x \in X$, and there is a constant a such that the inequality

$$|\phi(h)| \leq a \|h\|_X \text{ for all } h \in X$$

holds. The quantity

$$\|\phi\|_X := \sup_{h \in X} \frac{|\phi(h)|}{\|h\|_X}$$

is called *the norm of the functional* ϕ. All linear bounded functionals on X form a Banach space with that norm. This space is called the space *dual* to X and is denoted by X^*.

In the sequel $I_X = I$ is the identity operator in $X : Ih = h$ for any $h \in X$.

The operator A^{-1} is the inverse one to $A \in B(X, Y)$ if $AA^{-1} = I_Y$ and $A^{-1}A = I_X$.

Let $A \in B(X, Y)$. Consider a linear bounded functional f defined on Y. Then on X the linear bounded functional $g(x) = f(Ax)$ is defined. The operator realizing the relation $f \to g$ is called the operator A^* *dual (adjoint)* to A. By the definition

$$(A^* f)(x) = f(Ax) \ (x \in X).$$

The operator A^* is a bounded linear operator acting from Y^* to X^*.

Theorem 2.2.1 *(Banach - Steinhaus) Let $\{A_k\}$ be a sequence of linear operators acting from a Banach space X to a Banach space Y. Let for each $h \in X$,*

$$\sup_k \|A_k h\|_X < \infty.$$

Then the operator norms of $\{A_k\}$ are uniformly bounded. Moreover, if $\{A_n\}$ strongly converges to a (linear) operator A, then

$$\|A\|_{X \to Y} = \underline{\lim}_{n \to \infty} \|A_n\|_{X \to Y}.$$

For details see, for example, [Krein, 1972, p. 58].

A point λ of the complex plane is said to be a regular point of an operator A, if the operator $R_\lambda(A) := (A - I\lambda)^{-1}$ (the resolvent) exists and is bounded. The complement of all regular points of A in the complex plane is the *spectrum* of A. The spectrum of A is denoted by $\sigma(A)$.

In the infinite dimensional case, the *spectral radius* of A is defined by

$$r_s(A) = sup_{s \in \sigma(A)} |s|.$$

As in the finite dimensional space, the Gel'fand formula

$$r_s(A) = \lim_{k \to \infty} \sqrt[k]{\|A^k\|}$$

is valid. The limit always exists and

$$r_s(A) \leq \sqrt[k]{\|A^k\|} \quad (k = 1, 2, ...).$$

2.3 Classification of spectra

The material of this section can be found, for example, in the book [Dunford and Schwartz, 1966, Chapter VII].

Let X be a Banach space with the unit operator I, and A be a bounded linear operator in X. Recall that the resolvent $R_\lambda(A)$ and the spectrum $\sigma(A)$ of A are defined in the previous section.

It is convenient for many purposes to introduce a rough classification of the points of the spectrum.

(a) The set of $\lambda \in \sigma(A)$ such that $A - \lambda I$ is not one-to-one is called the point spectrum of A and is denoted by $\sigma_p(A)$. Thus $\lambda \in \sigma_p(A)$ if and only if $Ax = \lambda x$ for some non-zero $x \in X$.

(b) The set of $\lambda \in \sigma(A)$ such that $A - \lambda I$ is one-to-one and $(A - \lambda I)X$ is dense in X, but such that $(A - \lambda I)X \neq X$ is called the continuous spectrum of A and is denoted by $\sigma_c(A)$.

(c) The set of $\lambda \in \sigma(A)$ such that $A - \lambda I$ is one-to-one but such that $(A - \lambda I)X$ is not dense in X is called the residue spectrum of A and is denoted by $\sigma_r(A)$.

$\sigma_r(A), \sigma_c(A)$ and $\sigma_p(A)$ are disjoint and

$$\sigma(A) = \sigma_r(A) \cup \sigma_p(A) \cup \sigma_c(A).$$

For all regular λ, μ of A, the Hilbert identity

$$R_\lambda(A) - R_\mu(A) = (\lambda - \mu)R_\mu(A)R_\lambda(A)$$

is valid. Moreover, for all bounded linear operators A and B, we have

$$R_\lambda(A) - R_\lambda(B) = R_\lambda(A)(B - A)R_\lambda(B)$$

for any λ regular for A and B. Note also that

$$R_\lambda(A) = -\sum_{k=0}^{\infty} \frac{1}{\lambda^{k+1}} A^k \quad (A^{k+1} = A^k A, k = 1, 2, ...)$$

provided $|\lambda| > r_s(A)$, where $r_s(A)$ is the spectral radius. Besides the series converges in the operator norm.

Furthermore, an operator P is called *a projection* if $P^2 = P$.

If there is a nontrivial solution e of the equation

$$Ae = \lambda(A)e,$$

where $\lambda(A)$ is a number, then this number is called an eigenvalue of operator A, and $e \in X$ is an eigenvector corresponding to the eigenvalue $\lambda(A)$. Any eigenvalue is a point of the spectrum. An eigenvalue $\lambda(A)$ has the (algebraic) multiplicity $r \leq \infty$ if

$$dim(\cup_{k=1}^{\infty} ker(A - \lambda(A)I)^k) = r.$$

In the sequel $\lambda_k(A)$, $k = 1, 2, ...$ are the eigenvalues of A repeated according to their multiplicities. A vector v satisfying $(A - \lambda(A)I)^n v = 0$ for a natural n is a root vector of operator A corresponding to $\lambda(A)$. Any eigenvalue having the finite multiplicity belongs to the point spectrum.

Recall that an operator is compact if it maps bounded sets into compact ones. The spectrum of a linear compact operator is either finite, or the sequence of the eigenvalues of A converges to zero, any nonzero eigenvalue has the finite multiplicity.

An operator V is called *a quasinilpotent* one, if its spectrum consists of zero, only. So $r_s(V) = 0$ and

$$\lim_{k \to \infty} \sqrt[k]{\|V^k\|} = 0.$$

Now let $X = H$ be a Hilbert space with a scalar product $(.,.)$ and A be a bounded linear operator in H. Then a (bounded) linear operator A^* acting in H is adjoint to A if

$$(Af, g) = (f, A^*g) \text{ for every } h, g \in H.$$

The relation $\|A\| = \|A^*\|$ is true. A bounded operator A *is a self-adjoint* one if $A = A^*$. A is *a unitary operator* if $AA^* = A^*A = I$. A bounded linear operator satisfying the relation $AA^* = A^*A$ is called *a normal operator*. It is clear that unitary and self-adjoint operators are examples of normal ones.

A self-adjoint operator is positive definite if $(Ah, h) \geq 0$ $(h \in H)$. It is strongly positive definite if $(Ah, h) \geq c(h, h)$ for some positive c.

The spectrum of a self-adjoint operator is real; the spectrum of a positive definite self-adjoint operator is nonnegative; the spectrum of a strongly positive definite self-adjoint operator is positive; the spectrum of a unitary operator lies on $\{z \in \mathbf{C} : |z| = 1\}$.

If A is a normal operator, then

$$r_s(A) = \|A\|.$$

If P is a projection and $P^* = P$, then P is called *an orthoprojection* (an orthogonal projection).

2.4 Compact operators in a Hilbert space

In this section, H is a separable Hilbert space and A is a linear completely continuous (compact) operator in H. All the results, presented in this section can be found, for instance, in [Gohberg and Krein, 1969, Chapters 2 and 3], [Pietsch, 1987].

Any normal compact operator A can be represented in the form

$$A = \sum_{k=1}^{\infty} \lambda_k(A) E_k, \tag{4.1}$$

where E_k are eigenprojections of A, i.e., the projections defined by $E_k h = (h, d_k) d_k$ for all $h \in H$. Here d_k are the eigenvectors of A corresponding to $\lambda_k(A)$ with $\|d_k\| = 1$.

Vectors $v, h \in H$ are orthogonal if $(h, v) = 0$. The eigenvectors of normal operators are mutually orthogonal. Let a compact operator A be positive definite and represented by (4.1). Then we write

$$A^\beta := \sum_{k=1}^{\infty} \lambda_k^\beta(A) E_k \quad (\beta > 0).$$

A compact quasinilpotent operator is called *a Volterra operator*.

Let $\{e_k\}$ be an orthogonal normal basis in H and the series

$$\sum_{k=1}^{\infty} (A e_k, e_k)$$

converge. Then the sum of this series is called *the trace of A:*

$$Trace\ A = Tr\ A = \sum_{k=1}^{\infty} (A e_k, e_k).$$

An operator A satisfying the condition

$$Tr\ (A^*A)^{1/2} < \infty$$

is called *a nuclear operator*. Recall that A^* is adjoint operator. An operator A, satisfying the relation

$$Tr\ (A^*A) < \infty$$

is said to be *a Hilbert-Schmidt operator*.

As in the finite dimensional case, the eigenvalues $\lambda_k((A^*A)^{1/2})$ $(k = 1, 2, ...)$ of the operator $(A^*A)^{1/2}$ are called *the singular numbers (s-numbers)* of A and are denoted by $s_k(A)$. That is,

$$s_k(A) := \lambda_k((A^*A)^{1/2})\ (k = 1, 2, ...).$$

Enumerate singular numbers of A taking into account their multiplicity and in decreasing order. As in the finite dimensional case, the Weyl inequalities

$$\sum_{j=1}^{n} |\lambda_j(A)| \le \sum_{j=1}^{n} s_j(A)$$

and

$$\sum_{j=1}^{n} |Im\ \lambda_j(A)| \le \sum_{j=1}^{n} |\lambda_j(A_I)|\ \ (A_I = (A - A^*)/2i;\ n = 1, 2, ...)$$

are valid.

The set of completely continuous operators acting in a Hilbert space and satisfying the condition

$$N_p(A) := \Big[\sum_{k=1}^{\infty} s_k^p(A) \Big]^{1/p} < \infty,$$

for some $p \ge 1$, is called *the Schatten - von Neumann ideal and is denoted by SN_p. $N_p(.)$ is called the norm of ideal SN_p*. It is not hard to show that

$$N_p(A) = \sqrt[p]{Tr\ (AA^*)^{p/2}}.$$

Thus, SN_1 is the ideal of nuclear operators (*the Trace class*) and SN_2 is the ideal of Hilbert-Schmidt operators. $N_2(A)$ is called the *Hilbert-Schmidt norm*. Sometimes we will omit index 2 of the Hilbert-Schmidt norm, i.e.,

$$N(A) = N_2(A) = \sqrt{Tr\ (A^*A)}.$$

For all $p \ge 1$, the following propositions are true (the proofs can be found in the books [Gohberg and Krein, 1969, Section III.7], and [Pietsch, 1987].

If $A \in SN_p$, then also $A^* \in SN_p$. If $A \in SN_p$ and B is a bounded linear operator, then both AB and BA belong to SN_p. Moreover,

$$N_p(AB) \le N_p(A)\|B\|\ \text{and}\ N_p(BA) \le N_p(A)\|B\|.$$

Lemma 2.4.1 *If $A \in SN_p$ and $B \in SN_q$ $(1 < p, q < \infty)$, then $AB \in SN_s$ with*

$$\frac{1}{s} = \frac{1}{p} + \frac{1}{q}.$$

Moreover,

$$N_s(AB) \leq N_p(A)N_q(B).$$

For the proof of this lemma, see [Gohberg and Krein, 1969, Section III.7]. Let us point also the following result (Lidskij's theorem).

Theorem 2.4.2 *Let $A \in SN_1$. Then*

$$Tr \, A = \sum_{k=1}^{\infty} \lambda_k(A).$$

The proof of this theorem can be found in [Gohberg and Krein, 1969, Section III.8].

2.5 Compact matrices

We begin with the following important result.

Theorem 2.5.1 *Let T be a linear operator in a separable Hilbert space H, and $\{e_k\}_{k=1}^{\infty}$ be an arbitrary orthogonal normal basis for H.*
a) If $1 \leq p \leq 2$ and

$$\sum_{k=1}^{\infty} \|Te_k\|^p < \infty,$$

then T belongs to ideal SN_p and

$$N_p(T) \leq [\sum_{k=1}^{\infty} \|Te_k\|^p]^{1/p}.$$

b) If $2 \leq p < \infty$ and T belongs to ideal SN_p, then

$$[\sum_{k=1}^{\infty} \|Te_k\|^p]^{1/p} \leq N_p(T).$$

For the proof, see [Diestel et al., 1995, p. 82, Theorem 4.7].
The case $p = 2$ of the preceding theorem is of fundamental importance:

Corollary 2.5.2 *A linear operator T acting in H is a Hilbert-Schmidt operator if and only if there is an orthogonal normal basis $\{e_k\}_{k=1}^{\infty}$ in H such that*

$$\sum_{k=1}^{\infty} \|Te_k\|^2 < \infty.$$

In this case, the quantity

$$\sum_{k=1}^{\infty} \|Te_k\|^2$$

is independent of the choice of orthogonal normal basis $\{e_k\}_{k=1}^{\infty}$; in fact, for any orthogonal normal basis $\{e_k\}_{k=1}^{\infty}$ in H,

$$\left[\sum_{k=1}^{\infty} \|Te_k\|^2\right]^{1/2} = N_2(T),$$

cf. [Diestel et al., 1995, p. 83, Corollary 4.8].

Now, let $A = (a_{jk})_{j,k=1}^{\infty}$ be an infinite matrix. We consider here an operator in $l^2 = l^2(\mathbf{C})$ defined by $h \to Ah$ ($h \in l^2$). This operator is denoted again by A.

Theorem 2.5.1 implies

Corollary 2.5.3 *Let*

$$\sum_{j,k=1}^{\infty} |a_{jk}| < \infty.$$

Then the operator A generated by matrix (a_{jk}) is a nuclear operator in l^2. Moreover,

$$N_1(A) \leq \sum_{j,k=1}^{\infty} |a_{jk}|.$$

Indeed, let $\{e_k\}_{k=1}^{\infty}$ be a standard orthonormal basis in l^2. That is,

$$e_k = column \ (\delta_{jk})_{j=1}^{\infty},$$

where δ_{jk} is the Kronecker symbol: $\delta_{jk} = 0 \ k \neq j$, $\delta_{jj} = 1$. Then

$$Ae_k = \sum_{j=1}^{\infty} a_{jk}e_j. \tag{5.1}$$

So

$$\|Ae_k\| \leq \sum_{j=1}^{\infty} |a_{jk}|,$$

and we get the required result by Theorem 2.5.1.

Moreover, we can assert the following result.

Corollary 2.5.4 *Let*

$$\sum_{j,k=1}^{\infty} |a_{jk}|^2 < \infty.$$

Then the operator A generated by matrix (a_{jk}) is a Hilbert-Schmidt one in l^2. Moreover,

$$N_2(A) = \left[\sum_{j,k=1}^{\infty} |a_{jk}|^2 \right]^{1/2}.$$

Indeed, let $\{e_k\}_{k=1}^{\infty}$ be the above-defined standard orthonormal basis in l^2. Then (5.1) holds. So

$$\|Ae_k\|^2 = (Ae_k, Ae_k) = \left(\sum_{j=1}^{\infty} a_{jk}e_j, \sum_{m=1}^{\infty} a_{mk}e_m \right)$$

$$= \sum_{j=1}^{\infty} a_{jk} \sum_{m=1}^{\infty} \bar{a}_{mk}(e_j, e_m) = \sum_{j,k=1}^{\infty} |a_{jk}|^2,$$

and we get the required result, thanks to Corollary 2.5.2.

Now let us assume that, for some $1 < p, q < \infty$, the inequality

$$\left(\sum_{j=1}^{\infty} \left[\sum_{k=1}^{\infty} |a_{jk}|^q \right]^{p/q} \right)^{1/p} < \infty$$

is fulfilled. Then A is said to be of the *Hille-Tamarkin type* $[l_p, l_q]$ matrix, cf. [Pietsch, 1987, p. 230].

The case $q = \infty$, that is the *Hille-Tamarkin type* $[l_p, l_\infty]$ means that

$$\sum_{j=1}^{\infty} [\sup_k |a_{jk}|^p]^{1/p} < \infty.$$

The following fundamental result is valid.

Theorem 2.5.5 *Let*

$$\frac{1}{p} + \frac{1}{q} = 1. \tag{5.2}$$

In addition, let A be of the Hille-Tamarkin type $[l_p, l_q]$. Then the eigenvalues $\lambda_k(A)$ of A satisfy the inequality

$$\sum_{k=1}^{\infty} |\lambda_k(A)|^\nu < \infty$$

with $\nu = max\,\{2, p\}$. This result is best possible.

For the proof see [Pietsch, 1987, p. 232, Theorem 5.3.6].

Let us point an important consequence of this theorem.

Corollary 2.5.6 *Let the matrix A^*A be of the Hille-Tamarkin type $[l_p, l_q]$.*
Let condition (5.2) hold. Then $A \in SN_\mu$ with $\mu = max\{2, p\}/2$.

Indeed, thanks to the preceding theorem

$$\sum_{k=1}^{\infty} |\lambda_k(A^*A)|^{\mu/2} < \infty.$$

This proves the required result.

The following remarkable result can be found in the book [Pietsch, 1988, p. 325].

Theorem 2.5.7 *A matrix $A = (a_{jk})_{j,k=1}^{\infty}$ satisfies the inequality*

$$\sum_{k=1}^{\infty} |\lambda_k(A)|^p \le \sum_{k=1}^{\infty} \Big[\sum_{j=1}^{\infty} |a_{jk}|^q\Big]^{p/q}$$

$$(2 \le p < \infty)$$

provided (5.2) holds and the right-hand side of the inequality is finite.

2.6 Resolvents of Hilbert-Schmidt operators

Let A be a Hilbert-Schmidt operator. The following quantity plays a key role in this section:

$$g(A) = [N_2^2(A) - \sum_{k=1}^{\infty} |\lambda_k(A)|^2]^{1/2}, \qquad (6.1)$$

where $N_2(A)$ is the Hilbert-Schmidt norm of A. Since

$$\sum_{k=1}^{\infty} |\lambda_k(A)|^2 \ge |\sum_{k=1}^{\infty} \lambda_k^2(A)| = |Trace\ A^2|,$$

one can write

$$g^2(A) \le N_2^2(A) - |Trace\ A^2| \le N_2^2(A). \qquad (6.2)$$

If A is a normal Hilbert-Schmidt operator, then $g(A) = 0$, since

$$N_2^2(A) = \sum_{k=1}^{\infty} |\lambda_k(A)|^2$$

in this case. Let $A_I = (A - A^*)/2i$. As in the finite dimensional case, the inequality

$$g^2(A) \leq \frac{N_2^2(A - A^*)}{2} = 2N_2^2(A_I) \tag{6.3}$$

is valid (see Lemma 2.10.2 below). Set

$$\rho(A, \lambda) := \inf_{t \in \sigma(A)} |\lambda - t|.$$

Theorem 2.6.1 *Let A be a Hilbert-Schmidt operator. Then the inequalities*

$$\|R_\lambda(A)\| \leq \sum_{k=0}^{\infty} \frac{g^k(A)}{\rho^{k+1}(A, \lambda)\sqrt{k!}} \tag{6.4}$$

and

$$\|R_\lambda(A)\| \leq \frac{1}{\rho(A, \lambda)} exp\left[\frac{1}{2} + \frac{g^2(A)}{2\rho^2(A, \lambda)}\right] \quad (\lambda \notin \sigma(A)) \tag{6.5}$$

are true.

This theorem is proved in Section 2.11.

Theorem 2.6.1 is sharp. Inequality (6.4) becomes the equality

$$\|R_\lambda(A)\| = \rho^{-1}(A, \lambda),$$

if A is a normal operator, since $g(A) = 0$ in this case.

2.7 Operators with Hilbert-Schmidt powers

Assume that for some positive integer $p > 1$,

$$A^p \text{ is a Hilbert-Schmidt operator.} \tag{7.1}$$

Note that under (7.1) A can, in general, be a noncompact operator. Below in this section we will present a relevant example.

Corollary 2.7.1 *Let condition (7.1) hold for some integer $p > 1$. Then*

$$\|R_\lambda(A)\| \leq \|T_{\lambda,p}\| \sum_{k=0}^{\infty} \frac{g^k(A^p)}{\rho^{k+1}(A^p, \lambda^p)\sqrt{k!}} \quad (\lambda^p \notin \sigma(A^p)), \tag{7.2}$$

where

$$T_{\lambda,p} = \sum_{k=0}^{p-1} A^k \lambda^{p-k-1},$$

and

$$\rho(A^p, \lambda^p) = \inf_{t \in \sigma(A)} |t^p - \lambda^p|$$

is the distance between $\sigma(A^p)$ and the point λ^p.

Indeed, the identity

$$A^p - I\lambda^p = (A - I\lambda)\sum_{k=0}^{p-1} A^k \lambda^{p-k-1} = (A - I\lambda)T_{\lambda,p}$$

implies

$$(A - I\lambda)^{-1} = T_{\lambda,p}(A^p - I\lambda^p)^{-1}.$$

Thus,

$$\|(A - I\lambda)^{-1}\| \leq \|T_{\lambda,p}\| \, \|(A^p - I\lambda^p)^{-1}\|. \tag{7.3}$$

Applying Theorem 2.6.1 to the resolvent

$$(A^p - I\lambda^p)^{-1} = R_{\lambda^p}(A^p),$$

we obtain:

$$\|R_{\lambda^p}(A^p)\| \leq \sum_{k=0}^{\infty} \frac{g^k(A^p)}{\rho^{k+1}(A^p,\lambda^p)\sqrt{k!}} \quad (\lambda^p \notin \sigma(A^p)).$$

This yields the required result.

Theorem 2.6.1 and (7.3) give us the inequality

$$\|R_\lambda(A)\| \leq \frac{\|T_{\lambda,p}\|}{\rho(A^p,\lambda^p)} \, exp \left[\frac{1}{2} + \frac{g^2(A^p)}{2\rho^2(A^p,\lambda^p)}\right] \quad (\lambda^p \notin \sigma(A^p)),$$

provided condition (7.1) hold for some integer $p > 1$.

Example 2.7.2 *Consider a noncompact operator satisfying condition (7.1).*

Let H be an orthogonal sum of Hilbert spaces H_1 and H_2: $H = H_1 \oplus H_2$, and let A be a linear operator defined in H by the operator matrix

$$A = \begin{pmatrix} B_1 & W \\ 0 & B_2 \end{pmatrix},$$

where B_1 and B_2 are bounded linear operators acting in H_1 and H_2, respectively, and a bounded linear operator W maps H_2 into H_1. Evidently, A^2 is defined by the matrix

$$A^2 = \begin{pmatrix} B_1^2 & B_1W + WB_2 \\ 0 & B_2^2 \end{pmatrix}.$$

If B_1, B_2 are compact operators and W is a noncompact one, then A^2 is compact, while A is a noncompact operator.

2.8   Resolvents of Schatten - von Neumann operators

Let
$$A \in SN_{2p} \text{ for some integer } p > 1.$$

That is,
$$N_{2p}(A) = [\, Trace(A^*A)^p \,]^{1/2p} < \infty.$$

In this case, one can directly apply the results of the previous section, but in appropriate situations, the following result is more convenient.

Theorem 2.8.1 *Let* $A \in SN_{2p}$ *(*$p = 1, 2, ...$*). Then the inequalities*

$$\|R_\lambda(A)\| \le \sum_{m=0}^{p-1} \sum_{k=0}^{\infty} \frac{(2N_{2p}(A))^{pk+m}}{\rho^{pk+m+1}(A, \lambda)\sqrt{k!}} \qquad (8.1)$$

and

$$\|R_\lambda(A)\| \le \sum_{m=0}^{p-1} \frac{(2N_{2p}(A))^m}{\rho^{m+1}(A, \lambda)} \, exp \, [\frac{1}{2} + \frac{(2N_{2p}(A))^{2p}}{2\rho^{2p}(A, \lambda)}] \quad (\lambda \notin \sigma(A)) \qquad (8.2)$$

hold.

The proof of this theorem is presented in Section 2.11. Put

$$\theta_j^{(p)} = \frac{1}{\sqrt{[j/p]!}},$$

where $[x]$ means the integer part of a number $x > 0$. Now inequality (8.1) implies

Corollary 2.8.2 *Let* $A \in SN_{2p}$ *(*$p = 2, 3, ...$*). Then*

$$\|R_\lambda(A)\| \le \sum_{j=0}^{\infty} \frac{\theta_j^{(p)}(2N_{2p}(A))^j}{\rho^{j+1}(A, \lambda)} \quad (\lambda \notin \sigma(A)).$$

2.9   Auxiliary results

Let R_0 be a set in the complex plane and let $\epsilon > 0$. By $S(R_0, \epsilon)$, we denote the ϵ-neighborhood of R_0. That is,

$$dist\{R_0, S(R_0, \epsilon)\} \le \epsilon.$$

Lemma 2.9.1 *Let A be a bounded operator and let $\epsilon > 0$. Then there is a $\delta > 0$, such that, if a bounded operator B satisfies the condition $\|A - B\| \leq \delta$, then $\sigma(B)$ lies in $S(\sigma(A), \epsilon)$ and*

$$\|R_\lambda(A) - R_\lambda(B)\| \leq \epsilon$$

for any λ, which does not belong to $S(\sigma(A), \epsilon)$.

For the proof of this lemma, we refer the reader to the book [Dunford and Schwartz, 1966, p. 585].

A subspace $H_1 \subset H$ is an invariant subspace of an operator A, if $AH_1 \subseteq H_1$.

Recall that a linear operator is called a *Volterra one* if it is quasinilpotent and compact.

Lemma 2.9.2 *Let $V \in SN_p$, $p \geq 1$ be a Volterra operator. Then there is a sequence of nilpotent operators, having finite dimensional ranges and converging to V in the norm $N_p(.)$.*

Proof: Let

$$V = \sum_{k=1}^{\infty} s_k(V)(., e_k)v_k$$

be the polar representation of V, cf. [Pietsch, 1987, p. 79]. Here e_k and $v_k, k = 1, 2, ...$ are orthonormal sequences. Put

$$B_n = \sum_{k=1}^{n} s_k(V)(., e_k)v_k.$$

Denote by D_n and V_n the diagonal and nilpotent parts of B_n. Since $\sigma(V) = \{0\}$ and $\sigma(D_n) = \sigma(B_n)$, by the previous lemma, we have $\|D_n\| \to 0, n \to \infty$. But

$$N_p^p(D_n) = \sum_{k=1}^{n} |\lambda_k(D_n)|^p,$$

where the eigenvalues are enumerated in the decreasing way. Use the inequality

$$|\lambda_{k+m-1}(D_n)| = s_{k+m-1}(B_n - V_n) = s_{k+m-1}(V - V_n) \leq s_k(V) + s_m(V_n)$$

$$\leq s_k(V) + s_m(V) \ (k + m - 1 < n)$$

cf. [Pietsch, 1987, p. 79]. Hence,

$$|\lambda_{2j}(D_n)| \leq s_k(V) + s_{2j+1-k}(V)$$

for a $k < 2j \leq n$. Take $k = j$. Then

$$|\lambda_{2j+1}(D_n)| \leq |\lambda_{2j}(D_n)| \leq s_j(V) + s_{j+1}(V) \leq 2s_j(V).$$

So by the Lebesgues theorem

$$\lim_{n \to \infty} N_p^p(D_n) = \lim_{n \to \infty} \sum_{k=1}^{n} |\lambda_k(D_n)|^p = 0,$$

since $\sup_k |\lambda_k(D_n)| \to 0$. Hence,

$$N_p(V_n - V) = N_p(B_n - V - D_n) \le N_p(B_n - V) + N_p(D_n) \to 0$$

as $n \to \infty$. As claimed. Q. E. D.

We recall the following well-known result, cf. [Gohberg and Krein, 1969, Lemma I.4.2].

Lemma 2.9.3 *Let $M \ne H$ be the closed linear span of all the root vectors of a linear compact operator A, and let Q_A be the orthogonal projection of H onto $M^\perp$, where $M^\perp$ is the orthogonal complement of M in H. Then $Q_A A Q_A$ is a Volterra operator.*

The previous lemma means that A can be represented by the matrix

$$A = \begin{pmatrix} B_A & A_{12} \\ 0 & V_1 \end{pmatrix} \qquad (9.1)$$

acting in $M \oplus M^\perp$. Here $B_A = A(I - Q_A)$, $V_1 = Q_A A Q_A$ is a Volterra operator in $Q_A H$, and $A_{12} = (I - Q_A)A Q_A$.

Lemma 2.9.4 *Let A be a compact linear operator in H. Then there are a normal operator D and a Volterra operator V, such that*

$$A = D + V \text{ and } \sigma(D) = \sigma(A). \qquad (9.2)$$

Moreover, A, D, and V have the same invariant subspaces.

Proof: Let M be the linear closed span of all the root vectors of A, and P_A is the projection of H onto M. So the system of the root vectors of the operator $B_A = A P_A$ is complete in M. Thanks to the well-known Lemma I.4.1 from [Gohberg and Krein, 1969], there is an orthonormal basis (Schur's basis) $\{e_k\}$ in M, such that

$$B_A e_j = A e_j = \lambda_j(B_A) e_j + \sum_{k=1}^{j-1} a_{jk} e_k \quad (j = 1, 2, ...). \qquad (9.3)$$

We have $B_A = D_B + V_B$, where $D_B e_k = \lambda_k(B_A) e_k$, $k = 1, 2, ...$ and $V_B = B_A - D_B$ is a quasinilpotent operator. But according to (9.1) $\lambda_k(B_A) = \lambda_k(A)$, since V_1 is a quasinilpotent operator. Moreover, D_B and V_B have the same invariant subspaces. Take the following operator matrices acting in $M \oplus M^\perp$:

$$D = \begin{pmatrix} D_B & 0 \\ 0 & 0 \end{pmatrix} \text{ and } V = \begin{pmatrix} V_B & A_{12} \\ 0 & V_1 \end{pmatrix}.$$

Since the diagonal of V contains V_B and V_1 only, $\sigma(V) = \sigma(V_B) \cup \sigma(V_1) = \{0\}$. So V is quasinilpotent and (9.2) is proved. From (9.1) and (9.3), it follows that A, D, and V have the same invariant subspace, as claimed. Q. E. D.

Equality (9.2) is said to be *the triangular representation of* A. Besides, D and V will be called the diagonal part and nilpotent part of A, respectively.

Lemma 2.9.5 *Let $A \in SN_p$, $p \geq 1$. Let V be the nilpotent part of A. Then there exists a sequence $\{A_n\}$ of operators, having n-dimensional ranges, such that*

$$\sigma(A_n) \subseteq \sigma(A), \tag{9.4}$$

and

$$\sum_{k=1}^{n} |\lambda(A_n)|^p \to \sum_{k=1}^{\infty} |\lambda(A)|^p \ as \ n \to \infty. \tag{9.5}$$

Moreover,

$$N_p(A_n - A) \to 0 \ and \ N_p(V_n - V) \to 0 \ as \ n \to \infty, \tag{9.6}$$

where V_n are the nilpotent parts of A_n $(n = 1, 2, ...)$.

Proof: Let M be the linear closed span of all the root vectors of A, and P_A the projection of H onto M. So the system of root vectors of the operator $B_A = AP_A$ is complete in M. Let D_B and V_B be the nilpotent parts of B_A, respectively. According to (9.3), put

$$P_n = \sum_{k=1}^{n} (., e_k) e_k.$$

Then

$$\sigma(B_A P_n) = \sigma(D_B P_n) = \{\lambda_1(A), ..., \lambda_n(A)\}. \tag{9.7}$$

In addition, $D_B P_n$ and $V_B P_n$ are the diagonal and nilpotent parts of $B_A P_n$, respectively. Because of Lemma 2.9.2, there exists a sequence $\{W_n\}$ of nilpotent operators having n-dimensional ranges and converging in N_p to the operator V_1. Put

$$A_n = \begin{pmatrix} B_A P_n & P_n A_{12} \\ 0 & W_n \end{pmatrix}.$$

Then the diagonal part of A_n is

$$D_n = \begin{pmatrix} D_B P_n & 0 \\ 0 & 0 \end{pmatrix},$$

and the nilpotent part is

$$V_n = \begin{pmatrix} V_B P_n & P_n A_{12} \\ 0 & W_n \end{pmatrix}.$$

So relations (9.6) are valid. According to (9.7), relation (9.4) holds. Moreover $N_p(D_n - D_B) \to 0$. So relation (9.5) is also proved. This finishes the proof. Q. E. D.

2.10 Equalities for eigenvalues

Lemma 2.10.1 *Let V be a Volterra operator and $V_I := (V - V^*)/2i \in SN_2$. Then $V \in SN_2$. Moreover, $N_2^2(V) = 2N_2^2(V_I)$.*

Proof: We have the equality

$$Trace \ V^2 = Trace \ (V^*)^2 = 0,$$

because V is a Volterra operator. Hence,

$$N_2^2(V - V^*) = Trace \ (V - V^*)^2 = Trace \ (V^2 + VV^* + V^*V + (V^*)^2)$$
$$= Trace \ (VV^* + V^*V) = 2Trace \ (VV^*).$$

We arrive at the result. Q. E. D.

Lemma 2.10.2 *Let $A \in SN_2$. Then*

$$N_2^2(A) - \sum_{k=1}^{\infty} |\lambda_k(A)|^2 = 2N_2^2(A_I) - 2\sum_{k=1}^{\infty} |Im \ \lambda_k(A)|^2 = N_2^2(V),$$

where V is the nilpotent part of A.

Proof: Let D be the diagonal part of A. Thanks to Lemma 7.3.3 from the book [Gil', 2003a] VD^* is a Volterra operator:

$$Trace \ VD^* = Trace \ V^*D = 0. \tag{10.1}$$

Hence, thanks to the triangular representation (9.2) it follows that

$$Tr \ AA^* = Tr \ (D + V)(D^* + V^*) = Tr \ (DD^*) + Tr \ (VV^*).$$

Besides, because of (9.2) $\sigma(A) = \sigma(D)$. Thus,

$$N_2^2(D) = \sum_{k=1}^{\infty} |\lambda_k(A)|^2.$$

So the relation

$$N_2^2(V) = N_2^2(A) - \sum_{k=1}^{\infty} |\lambda_k(A)|^2$$

is proved. Furthermore, from the triangular representation (9.2) it follows that

$$-4Tr\ A_I^2 = Tr\ (A - A^*)^2 = Tr\ (D + V - D^* - V^*)^2.$$

Hence, thanks to (10.1), we obtain

$$-4Tr\ A_I^2 = Tr\ (D - D^*)^2 + Tr\ (V - V^*)^2.$$

That is,

$$N_2^2(A_I) = N_2^2(V_I) + N_2^2(D_I),$$

where

$$V_I = (V - V^*)/2i, D_I = (D - D^*)/2i.$$

Taking into account Lemma 2.10.1, we arrive at the equality

$$2N_2^2(A_I) - 2N_2^2(D_I) = N_2^2(V).$$

Moreover,

$$N_2^2(D_I) = \sum_{k=1}^{\infty} |Im\ \lambda_k(A)|^2.$$

This proves the required result. Q. E. D.

2.11   Proofs of Theorems 2.6.1 and 2.8.1

Proof of Theorem 2.6.1:   Because of Lemma 2.9.5 there exists a sequence $\{A_n\}$ of operators, having n-dimensional ranges, such that the relations (9.4) - (9.6) are valid. But because of Corollary 1.10.2,

$$\|R_\lambda(A_n)\| \leq \sum_{k=0}^{n-1} \frac{g^k(A_n)}{\rho^{k+1}(A_n, \lambda)\sqrt{k!}} \quad (\lambda \notin \sigma(A_n)).$$

Clearly, $\rho(A_n, \lambda) \geq \rho(A, \lambda)$. Now, letting $n \to \infty$ in the latter relation, we arrive at inequality (6.4). Inequality (6.5) follows from Theorem 1.5.1 when $n \to \infty$. This proves the stated result. Q. E. D.

Lemma 2.11.1 *Let* $V \in SN_2$ *be a quasinilpotent operator. Then*

$$\|V^k\| \leq \frac{N_2^k(V)}{\sqrt{k!}} \quad (k = 1, 2, ...). \tag{11.1}$$

Proof: Let V_n be an n-dimensional nilpotent operator. Then by Lemma 1.10.4

$$\|V_n^k\| \leq \gamma_{n,k} N_2^k(V_n) \quad (k = 1, \ldots, n-1)$$

But

$$\gamma_{n,k} \leq \frac{1}{\sqrt{k!}}.$$

So

$$\|V_n^k\| \leq \frac{N_2^k(V_n)}{\sqrt{k!}} \quad (k = 1, 2, \ldots).$$

Hence, letting $n \to \infty$, according to (9.4), (9.6) we arrive at the required result. Q. E. D.

Lemma 2.11.2 *For some integer $p \geq 1$, let $V \in SN_{2p}$ be a Volterra operator. Then*

$$\|V^{kp}\| \leq \frac{N_{2p}^{pk}(V)}{\sqrt{k!}} \quad (k = 1, 2, \ldots).$$

Proof: Since $V \in SN_{2p}$, we have $N_2(V^p) \leq N_{2p}^p(V)$. So $V^p \in SN_2$ and thus by (11.1),

$$\|V^{pk}\| \leq \frac{N_2^k(V^p)}{\sqrt{k!}} \leq \frac{N_{2p}^{kp}(V)}{\sqrt{k!}}.$$

As claimed. Q. E. D.

The previous lemma yields

Corollary 2.11.3 *For some integer $p \geq 1$, let $V \in SN_{2p}$ be a Volterra operator. Then*

$$\|V^j\| \leq \theta_j^{(p)} N_{2p}^j(V) \quad (j = 1, 2, \ldots).$$

Recall that numbers $\theta_j^{(p)}$ are introduced in Section 2.8.

Proof of Theorem 2.8.1: Thanks to the triangular representation, for an n-dimensional operator A_n, the relations

$$A_n - \lambda I = D_n + V_n - \lambda I = (D_n - \lambda I)(I + R_\lambda(D_n)V_n)$$

hold, where V_n and D_n are the nilpotent and diagonal parts of A_n, respectively. We thus have

$$R_\lambda(A_n) = R_\lambda(D_n)(I + V_n R_\lambda(D_n))^{-1} = \sum_{k=0}^{n-1}(V_n R_\lambda(D_n))^k.$$

So for $n = p(l+1)$ with an integer l, we get

$$R_\lambda(A_n) = R_\lambda(D_n)\sum_{k=0}^{n-1}(V_n R_\lambda(D_n))^k$$

$$= R_\lambda(D_n) \sum_{m=0}^{p-1} \sum_{k=0}^{l} (V_n R_\lambda(D_n))^{pk+m}. \tag{11.2}$$

Hence,

$$\|R_\lambda(A_n)\| \le \frac{1}{\rho(A_n, \lambda)} \sum_{m=0}^{p-1} \sum_{k=0}^{l} \|(V_n R_\lambda(D_n))^{pk+m}\|.$$

The operator $V_n R_\lambda(D_n)$ is nilpotent. Hence, the previous lemma implies

$$\|R_\lambda(A_n)\| \le \sum_{m=0}^{p} \sum_{k=0}^{l} \frac{N_{2p}^{pk+m}(V_n R_\lambda(D_n))}{\sqrt{k!}}$$

$$\le \sum_{j=0}^{p-1} \sum_{k=0}^{\infty} \frac{N_{2p}^{pk+m}(V_n)}{\rho^{pk+m+1}(A_n, \lambda)\sqrt{k!}} \quad (\lambda \notin \sigma(A)). \tag{11.3}$$

But the triangular representation yields the inequality

$$N_{2p}(V_n) \le N_{2p}(A_n) + N_{2p}(D_n) \le 2N_{2p}(A_n). \tag{11.4}$$

So

$$\|R_\lambda(A_n)\| \le \sum_{m=0}^{p-1} \sum_{k=0}^{\infty} \frac{(2N_{2p}(A_n))^{pk+m}}{\rho^{pk+m+1}(A_n, \lambda)\sqrt{k!}} \quad (\lambda \notin \sigma(A)).$$

Hence, letting $n \to \infty$, according to (9.4), (9.6), we arrive at (8.1). Furthermore, thanks to (6.5),

$$\|(I - W^p)^{-1}\| \le exp \left[\frac{1}{2} + \frac{N_2^2(W^p)}{2}\right]$$

for any quasinilpotent operator $W \in SN_{2p}$. Now (11.2) implies

$$R_\lambda(A_n) = R_\lambda(D_n) \sum_{m=0}^{p-1} (V_n R_\lambda(D_n))^m (I - (V_n R_\lambda(D_n))^p)^{-1}.$$

Consequently,

$$\|R_\lambda(A_n)\| \le \frac{1}{\rho(A_n, \lambda)} \sum_{k=0}^{p-1} \|(V_n R_\lambda(D_n))^k\| exp \left[\frac{1}{2} + \frac{N_2^2((V_n R_\lambda(D_n))^p)}{2}\right].$$

Take into account that

$$N_2((V_n R_\lambda(D_n))^p) \le N_{2p}^p(V_n R_\lambda(D_n))$$

and

$$N_{2p}(V_n R_\lambda(D_n)) \le \frac{N_{2p}(V_n)}{\rho(A_n, \lambda)}.$$

We thus get

$$\|R_\lambda(A_n)\| \leq \sum_{k=0}^{p-1} \frac{N_{2p}^k(V_n)}{\rho^{k+1}(A_n,\lambda)} exp\left[\frac{1}{2} + \frac{N_{2p}^{2p}(V_n)}{2\rho^{2p}(A_n,\lambda)}\right].$$

So according to (11.4)

$$\|R_\lambda(A_n)\| \leq \sum_{k=0}^{p-1} \frac{(2N_{2p}(A_n))^k}{\rho^{k+1}(A_n,\lambda)} exp\left[\frac{1}{2} + \frac{(2N_{2p}(A_n))^{2p}}{2\rho^{2p}(A_n,\lambda)}\right].$$

Hence, letting $n \to \infty$, according to (9.4), (9.6) we arrive at (8.2). This proves the stated result. Q. E. D.

2.12   Spectral variations

Definition 2.12.1 *Let A and B be linear operators in a Banach space. Then the quantity*

$$sv_A(B) := \sup_{\mu \in \sigma(B)} \inf_{\lambda \in \sigma(A)} |\mu - \lambda|$$

is called the spectral variation of B with respect to A. In addition,

$$hd(A,B) := \max\{sv_A(B), sv_B(A)\}$$

is the Hausdorff distance between the spectra of A and B.

First, we will prove the following technical lemma

Lemma 2.12.2 *Let A and B be bounded linear operators in a Banach space and*

$$\|R_\lambda(A)\| \leq F\left(\frac{1}{\rho(A,\lambda)}\right) \quad (\lambda \notin \sigma(A)) \tag{12.1}$$

where $F(x)$ is a monotonically increasing non-negative function of a non-negative variable x, such that $F(0) = 0$ and $F(\infty) = \infty$. Then with the notation

$$q := \|A - B\|,$$

we have

$$sv_A(B) \leq z(F,q),$$

where $z(F,q)$ is the unique positive and simple root of the equation

$$1 = qF(1/z). \tag{12.2}$$

Proof: Because of (12.1), we easily have that

$$1 \leq qF\left(\frac{1}{\rho(A,\mu)}\right) \text{ for all } \mu \in \sigma(B). \qquad (12.3)$$

Compare this inequality with (12.2). Since $F(x)$ monotonically increases, $z(F,q)$ is a unique positive root of (11.2), and $\rho(A,\mu) \leq z(F,q)$. This proves the required result. Q. E. D.

Again let SN_p denote the Schatten - von Neumann ideal in a separable Hilbert space H. First assume that

$$A \in SN_2. \qquad (12.4)$$

By virtue of the previous lemma and Theorem 2.8.1, we arrive at the following result.

Theorem 2.12.3 *Let a compact operator A acting in a separable Hilbert space H satisfy condition (12.4) and B be a bounded operator in H. Then*

$$sv_A(B) \leq \tilde{z}_1(A,q),$$

where $\tilde{z}_1(A,q)$ is the unique positive and simple root of the equation

$$1 = \frac{q}{z} \exp\left[\frac{1}{2} + \frac{g^2(A)}{2z^2}\right]. \qquad (12.5)$$

Furthermore, substitute in (12.5) the equality $z = xg(A)$. Then we arrive at the equation

$$\frac{1}{z} \exp\left[\frac{1}{2} + \frac{g^2(A)}{2z^2}\right] = \frac{g(A)}{q}.$$

Applying Lemma 1.6.4 to this equation, we get $\tilde{z}_1(A,q) \leq \tilde{\Delta}_1(A,q)$, where

$$\tilde{\Delta}_1(A,q) := \begin{cases} qe & \text{if } g(A) \leq eq \\ g(A)\,[\ln\,(g(A)/q)]^{-1/2} & \text{if } g(A) > eq \end{cases}.$$

Now Theorem 2.12.3 yields the inequality

$$sv_A(B) \leq \tilde{\Delta}_1(A,q). \qquad (12.6)$$

Now let

$$A \in SN_{2p} \quad (p = 2, 3, ...). \qquad (12.7)$$

Then by virtue of Lemma 2.12.2 and Theorem 2.8.1 we arrive at the following result.

Theorem 2.12.4 *Let condition (12.6) hold and B be a bounded operator in H. Then $sv_A(B) \leq \tilde{y}_p(A,q)$, where $\tilde{y}_p(A,B)$ is the unique positive root of the equation*

$$1 = q \sum_{m=0}^{p-1} \frac{(2N_{2p}(A))^m}{z^{m+1}} \exp\left[\frac{1}{2} + \frac{(2N_{2p}(A))^{2p}}{2z^{2p}}\right]. \qquad (12.8)$$

Furthermore, substitute in (12.8) the equality $z = x2N_{2p}(A)$ and apply Lemma 1.6.4. Then we get $\tilde{y}_p(A, q) \leq \Delta_p(A, q)$, where

$$\Delta_p(A,q) := \begin{cases} qpe & \text{if } 2N_{2p}(A) \leq epq \\ 2N_{2p}(A) \left[ln \; (2N_{2p}(A)/qp)\right]^{-1/2p} & \text{if } 2N_{2p}(A) > epq \end{cases}.$$

Now the previous theorem yields the inequality $sv_A(B) \leq \Delta_p(A, q)$.

2.13 Preservation of multiplicities of eigenvalues

For a $b \in \mathbf{C}$ and an $r > 0$, again put

$$\Omega(b,r) := \{z \in \mathbf{C} : |z - b| \leq r\}.$$

Lemma 2.13.1 *Let A and B be bounded linear operators acting in a Banach space, and*

A has an isolated eigenvalue $\lambda(A)$ of the algebraic multiplicity ν. (13.1)

In addition, with the notations $q = \|A - B\|$ and

$$\beta(A) := \inf_{t \in \sigma(A), t \neq \lambda(A)} |t - \lambda(A)|/2,$$

for a positive $r < \beta(A)$, let

$$q \sup_{|\lambda(A)-z|=r} \|R_z(A)\| < 1. \tag{13.2}$$

Then the set $\sigma(B) \cap \Omega(\lambda(A), r)$ consists of eigenvalues whose total algebraic multiplicity is equal to ν.

This result is a particular case of the well-known Theorem 4.3.18 [Kato, 1966, p. 214]. Again let $\rho(A, \lambda)$ be the distance between $\sigma(A)$ and a complex point λ. Again assume that there is a monotonically increasing function $F(y)$ of $y > 0$, such that the conditions $F(0) = 0$, $F(\infty) = \infty$ hold.

Lemma 2.13.2 *Under conditions (13.1) and (12.1), let $z(F, q)$ be the unique positive root of equation (12.2). Then the set $\sigma(B) \cap \Omega(\lambda(A), z(F, q))$ consists of eigenvalues of B, whose total algebraic multiplicity is equal to ν, provided $z(F, q) \leq \beta(A)$.*

Proof: Since F monotonically increases, comparing (13.2) and (12.2), we have the required result due to the previous lemma. Q. E. D.

From Lemma 2.13.2 and Theorem 2.8.1, it follows

Corollary 2.13.3 *Let $A \in S_{2p}$ and B be a compact operator in H, and the conditions (13.1) and*

$$q \sum_{m=0}^{p-1} \frac{(2N_{2p}(A))^m}{r^{m+1}} \, exp \, [\frac{1}{2} + \frac{(2N_{2p}(A))^{2p}}{2r^{2p}}] < 1$$

hold for a positive $r < \beta(A)$. Then the set $\sigma(B) \cap \Omega(\lambda(A), r)$ consists of eigenvalues, whose total algebraic multiplicity is equal to ν.

2.14   Entire Banach-valued functions and regularized determinants

Let X and Y be complex normed spaces with norms $\|.\|_X$ and $\|.\|_Y$, respectively, and F a Y-valued function defined on X. Assume that $F(C + \lambda \tilde{C})$ $(\lambda \in \mathbf{C})$ is an entire function for all $C, \tilde{C} \in X$. That is, for any $\phi \in Y^*$, the functional $< \phi, F(C + \lambda \tilde{C}) >$ defined on Y is an entire scalar-valued function. Let us prove the following technical lemma.

Lemma 2.14.1 *Let $F(C + \lambda \tilde{C})$ $(\lambda \in \mathbf{C})$ be an entire function for all $C, \tilde{C} \in X$ and there be a monotone non-decreasing function $W : [0, \infty) \to [0, \infty)$, such that*

$$\|F(C)\|_Y \le W(\|C\|_X) \; (C \in X).$$

Then

$$\|F(C) - F(\tilde{C})\|_Y \le \|C - \tilde{C}\|_X \, W \big(1 + \frac{1}{2}\|C + \tilde{C}\|_X + \frac{1}{2}\|C - \tilde{C}\|_X \big).$$

Proof: Put

$$g(\lambda) = F(\frac{1}{2}(C + \tilde{C}) + \lambda(C - \tilde{C})).$$

Then $g(\lambda)$ is an entire function and

$$F(C) - F(\tilde{C}) = g\big(\frac{1}{2}\big) - g\big(-\frac{1}{2}\big).$$

Thanks to the Cauchy integral,

$$g\big(\frac{1}{2}\big) - g\big(-\frac{1}{2}\big) = \frac{1}{2\pi i} \oint_{|z|=1/2+r} \frac{g(z)dz}{(z - \frac{1}{2})(z + \frac{1}{2})} \quad (r > 0).$$

Hence,

$$\|g\big(\frac{1}{2}\big) - g\big(-\frac{1}{2}\big)\|_Y \le \big(\frac{1}{2} + r\big) \sup_{|z|=1/2+r} \frac{\|g(z)\|_Y}{|z^2 - 1/4|}.$$

But

$$|z^2 - 1/4| \geq |z|^2 - 1/4 = (r + 1/2)^2 - 1/4 = r^2 + r$$

$$(z = (r + \frac{1}{2})e^{it}, \ 0 \leq t < 2\pi).$$

In addition,

$$\|g(z)\|_Y = \|F(\frac{1}{2}(C + \tilde{C}) + z(C - \tilde{C}))\|_Y = \|F(\frac{1}{2}(C + \tilde{C})$$

$$+ (r + \frac{1}{2})e^{it}(C - \tilde{C}))\|_Y \leq W(\frac{1}{2}\|C + \tilde{C}\|_X$$

$$+ (\frac{1}{2} + r)\|C - \tilde{C}\|_X) \ (|z| = 1/2 + r).$$

Therefore,

$$\|F(C) - F(\tilde{C})\|_Y = \|g(1/2) - g(-1/2)\|_Y$$

$$\leq (\frac{1}{2} + r) \frac{1/2 + r}{r^2 + r} W(\frac{1}{2}\|C + \tilde{C}\|_X + (\frac{1}{2} + r)\|C - \tilde{C}\|_X).$$

Hence,

$$\|F(C) - F(\tilde{C})\|_Y \leq \frac{1}{r} W(\frac{1}{2}\|C + \tilde{C}\|_X + (\frac{1}{2} + r)\|C - \tilde{C}\|_X).$$

Taking

$$r = \frac{1}{\|C - \tilde{C}\|_X},$$

we get the required result. Q. E. D.

For an $A \in SN_1$, the determinant is defined by

$$det_1 \ (I - A) := \prod_{k=1}^{\infty} (1 - \lambda_k(A)).$$

If $A \in SN_p$ with an integer $p \geq 2$, then the *regularized determinant* is defined as

$$det_p \ (I - A) := \prod_{k=1}^{\infty} (1 - \lambda_k(A)) exp \ [\sum_{j=1}^{p-1} \frac{\lambda_k^j(A)}{j}].$$

Put

$$G(u, p) := (1 - u)e^{u + \frac{u^2}{2} + \ldots + \frac{u^p}{p}} \ (p = 1, 2, \ldots), G(u, 0) = 1 - u.$$

Then

$$det_p \ (I - A) = \prod_{k=1}^{\infty} G(\lambda_k(A), p - 1).$$

The functions $G(u, p)$ are called the *primary factors*.

Lemma 2.14.2 *For any integer $p \geq 2$ and all complex numbers u, there is a constant*

$$\beta_p \leq 3e(2 + \ln(p-1)),$$

such that

$$\ln |G(u, p-1)| \leq \beta_p \frac{|u|^p}{1 + |u|}.$$

Proof: If

$$|u| \leq \frac{p}{p+1},$$

then

$$\ln |G(u,p)| \leq Re\left\{\ln(1-u) + u + \frac{u^2}{2} + ... + \frac{u^p}{p}\right\}$$

$$= -Re \sum_{k=p+1}^{\infty} \frac{u^k}{k} \leq \sum_{k=p+1}^{\infty} \frac{|u|^k}{k} < \frac{|u|^{p+1}}{(p+1)(1-|u|)} \leq |u|^{p+1}.$$

For

$$|u| > \frac{p}{p+1},$$

the inequality

$$\ln(1 + |u|) < |u|$$

leads to the following:

$$\ln |G(u,p)| \leq 2|u| + \frac{|u|^2}{2} + ... + \frac{|u|^p}{p}$$

$$= |u|^p \left(\frac{1}{p} + \frac{1}{p-1}\frac{1}{|u|} + ... + \frac{1}{2|u|^{p-2}} + 2\frac{1}{|u|^{p-1}}\right)$$

$$< |u|^p \left[2\left(1+\frac{1}{p}\right)^{p-1} + \sum_{k=2}^{p} \frac{1}{k}\left(1+\frac{1}{p}\right)^{p-k}\right]$$

$$< |u|^p\left(1+\frac{1}{p}\right)^p\left(2 + \sum_{k=2}^{p}\frac{1}{k}\right) < |u|^p\left(1+\frac{1}{p}\right)^p\left(2 + \int_1^p \frac{dx}{x}\right)$$

$$< |u|^p e(2 + \ln p) < \frac{|u|^{p+1}}{1+|u|} 3e(2 + \ln p).$$

As claimed. Q. E. D.

Lemma 2.14.3 *Let $A \in SN_p, p = 2, 3, ... $ Then*

$$|det_p(I - A)| \leq e^{\beta_p N_p^p(A)}.$$

If $A \in SN_1$. Then

$$|det_1(I - A)| \leq e^{N_1(A)}.$$

Proof: By the previous lemma,

$$|det_p\ (I - A)| \leq \prod_{k=1}^{\infty} e^{\beta_p |\lambda_k(A)|^p}.$$

Hence,

$$|det_p\ (I - A)| \leq exp\ [\beta_p \sum_{k=1}^{\infty} |\lambda_k(A)|^p].$$

Now the required result for $p \geq 2$ is due to Weyl's inequalities. The case $p = 1$ is obvious. Q. E. D.

We put $\beta_1 = 1$. Moreover, since

$$|1 - z|^2 = 1 - 2Re\ z + |z|^2 \leq e^{-2Re\ z + |z|^2}$$

we have

$$|(1 - z)e^z|^2 \leq e^{|z|^2}.$$

So one can take $\beta_2 = 1/2$.

The latter lemma and Lemma 2.14.1 imply

Corollary 2.14.4 *Let $A, B \in SN_p$ for an integer $p \geq 1$. Then*

$$|det_p\ (I - A) - det_p\ (I - B)| \leq$$

$$N_p(A - B)exp\ [\beta_p(1 + \frac{1}{2}N_p(A + B) + \frac{1}{2}N_p(A - B))^p].$$

2.15   Comments to Chapter 2

Sections 2.1 - 2.5 contain the well-known results that can be found in many books on the theory of linear operators, such as [Dunford and Schwartz, 1966], [Istratescu,1981], [Krein, S. G., 1972], etc. For more information about linear compact operators, we refer the reader to the excellent books [Diestel, Jarchow, and Tonge, 1995] and [Pietsch, 1987].

The material of Sections 2.6 - 2.13 is adopted from [Gil', 2003a]. About the classical results on spectrum perturbations, see for instance [Kato, 1966] and [Baumgartel, 1985].

Theorem 2.14.1 has been proved in [Gil', 2008b]. Lemma 2.14.2 is taken from the book [Levin, 1980]. The perturbation results for determinants, similar to Corollary 2.14.4 can be found in [Gohberg, Golberg, and Krupnik, 2000].

Chapter 3

Some Basic Results of the Theory of Analytic Functions

Chapter 3 contains the background material. It is a collection of basic conceptions and classical theorems of the theory of entire functions, part of which is given without proofs. In particular, we present the Weierstrass theorem on the representation of an arbitrary entire function by an infinite product and the Hadamard theorem on the representation of a finite-order entire function by a canonical product. The classical results of Borel, Hadamard, and Lindelöf on the connections between the growth of an entire function and the distribution of its zeros and the Jensen theorem on the counting function of zeros are also included in this chapter.

3.1 The Rouché and Hurwitz theorems

Theorem 3.1.1 *(Rouché's theorem). If $f_1(z)$ and $f_2(z)$ are analytic interior to a simple closed Jordan curve C and if they are continuous on C and for all $z \in C$,*

$$|f_2(z)| < |f_1(z)|,$$

then the function $f_1(z) + f_2(z)$ has the same number of zeros interior to C as does $f_1(z)$.

For the proof, see for instance the books [Marden, 1985, p. 2] and [Saks and Zigmund, 1965]. By the Rouché theorem, we derive the following result, which is called *the Hurwitz theorem.*

Theorem 3.1.2 *Let $f_n(z)$ $(n = 1, 2, ...)$ be a sequence of functions that are analytic in a region D and that converge uniformly to a function $f(z)$ in every closed subregion of D. Let z_0 be an interior point of D. If z_0 is a limit point of the zeros of the $f_n(z)$, then z_0 is a zero of $f(z)$. Conversely, if z_0 is an m-fold zero of $f(z)$, every sufficiently small neighborhood of z_0 contains exactly m zeros (counted with their multiplicities) of each f_n with $n > N$ for a sufficiently large integer N.*

Proof: Let us first assume that $f(z_0) \neq 0, z_0 \in D$. Since f is analytic in D, it can have only a finite number of zeros in D. We may then choose a positive r such that $f(z) \neq 0$ (in and) on the circle

$$\Omega(z_0, r) = \{|z - z_0| \leq r\}.$$

Let us set

$$\epsilon = min_{z \in \Omega(z_0, r)}|f(z)|.$$

Since the $f_n(z)$ converge to $f(z)$ uniformly in D, we can find a positive integer $N = N(r)$ such that $|f_n(z) - f(z)| < \epsilon$ for all z in $\Omega(z_0, r)$ and all $n > N$. Consequently,

$$|f_n(z) - f(z)| < |f(z)|$$

on $\partial\Omega(z_0, r) = \{|z - z_0| = r\}$ and by Rouché's theorem, the sum function

$$f_n(z) = [f_n(z) - f(z)] + f(z)$$

has as many zeros in $\Omega(z_0, r)$ as does $f(z)$. Since, therefore, $f(z) \neq 0$ on $\Omega(z_0, r)$, a point z_0 with $f(z_0) \neq 0$ could not be a limit point of the zeros of f.

Conversely, if we assume that z_0 is an m-fold zero of $f(z)$, then we may again choose a positive r so that $f(z) \neq 0, z \in \partial\Omega(z_0, r)$. Reasoning as in the previous paragraph, we now conclude from Rouché's theorem that each $f_n(z), n > N$, has precisely m zeros in $\Omega(z_0, r)$. Q. E. D.

The theorem, whose proof we have now completed, will provide our principal means of passing from certain theorems on the zeros of polynomials to the corresponding theorems on the zeros of entire functions and other analytic functions.

3.2 The Caratheodory inequalities

Let $f(z)$ be holomorphic on a neighborhood of the circle $|z| \leq R$. Put

$$M_f(r) := \max_{|z| \leq r} |f(z)|$$

and

$$A_f(r) := \max_{|z| \leq r} Re\, f(z) \quad (r < R).$$

From the maximum principle for harmonic functions, it follows that $A_f(r)$ is a monotone increasing function of r. Also we clearly have

$$|A_f(r)| \leq M_f(r).$$

For $R > r$, there is an inequality between $A_f(R)$ and $M_f(r)$, which is a sort of converse to the inequality above.

Theorem 3.2.1 *If $f(z)$ is any function holomorphic in the circle $|z| \leq R$, then*

$$M_f(r) \leq [A_f(R) - Re\, f(0)] \frac{2r}{R - r} + |f(0)| \quad (r < R). \qquad (2.1)$$

This inequality is called *the Caratheodory inequality for the circle.*
Proof: Our starting point is the formula of Schwarz,

$$f(z) = \frac{1}{2\pi} \int_{-\pi}^{\pi} u(Re^{i\theta}) \frac{Re^{i\theta} + z}{Re^{i\theta} - z} d\theta + iv(0)$$

$$(f(z) = u(z) + iv(z)), \qquad (2.2)$$

which represents a function holomorphic in the the circle $|z| \leq R$ in terms of the boundary values of its real part, see [Titchmarsh, 1939]. Let us add to the right side of (2.2) the quantity

$$u(0) - \frac{1}{2\pi} \int_{-\pi}^{\pi} u(Re^{i\theta}) d\theta,$$

which is equal to zero. We obtain the equation

$$f(z) = \frac{1}{\pi} \int_{-\pi}^{\pi} u(Re^{i\theta}) \frac{z}{Re^{i\theta} - z} d\theta + f(0).$$

In particular, for $f(z) \equiv 1$, we obtain

$$\frac{1}{\pi} \int_{-\pi}^{\pi} \frac{z}{Re^{i\theta} - z} d\theta = 0.$$

From the last two equations, we obtain the formula

$$-f(z) = \frac{1}{\pi} \int_{-\pi}^{\pi} [A_f(R) - u(Re^{i\theta})] \frac{z}{Re^{i\theta} - z} d\theta - f(0).$$

Since the quantity $A_f(R) - u(Re^{i\theta})$ is non-negative we obtain the inequality

$$|f(z)| \leq \frac{2r}{R-r}[A_f(R) - u(0)] + |f(0)|.$$

Clearly, it is possible to replace $|f(z)|$ by $M_f(r)$ and obtain (2.1). Q. E. D.

If the function $w = f(z)$ takes values lying in the left half-plane ($Re\ w < 0$) when z is in the circle $|z| < R$, then $A_f(R) < 0$ and

$$|f(z)| \leq -\frac{2r}{R-r}u(0) + |f(0)| \quad (|z| \leq r < R).$$

By mapping the circle $|z| < R$ onto the half-plane one can obtain from this inequality the corresponding inequality for functions analytic in a half-plane.

Theorem 3.2.2 *A function $f(z)$ that is holomorphic in the half-plane $Im\ z > 0$ and takes on it values in the upper half-plane $Im\ f(z) > 0$ satisfies, for $Im\ z > 0$ and $|z| > 1$, the inequalities*

$$\frac{1}{5}|f(i)|\frac{\sin\theta}{r} < |f(z)| < 5|f(i)|\frac{r}{\sin\theta} \tag{2.3}$$

$$(z = re^{\theta},\ 0 < \theta < \pi).$$

This inequality is called *the Caratheodory inequality for the half-plane.*
Proof: Map the upper half-plane $Im\ z > 0$ onto the unit circle

$$u = \frac{z-i}{z+i} \quad \text{or} \quad z = -i\frac{u+1}{u-1}.$$

The function

$$F(u) = if\left(-i\frac{u+1}{u-1}\right)$$

is defined in the unit circle and satisfies $Re\ F(u) \leq 0$. Consequently, $A_F(R) \leq 0$, and from Caratheodory's inequality for the circle $|u| \leq 1$ we have

$$|f(z)| \leq |f(i)| - Re\ (if(i))\frac{2|z-i|}{|z+i| - |z-i|} \quad (Im\ z > 0).$$

Note that for $|z| \geq 1$ we have

$$|z-i| \leq 2r,\ |z+i| \leq 2r \text{ and } |z+i|^2 - |z-i|^2 = 4r\sin\theta.$$

It is easy to obtain the last equality, if we consider the triangles with vertices at $0, z, i$ and at $0, z, -i$.

We thus have

$$|f(z)| < |f(i)|\left(1 + \frac{4r}{\sin\theta}\right) < 5|f(i)|\frac{r}{\sin\theta}. \qquad (2.4)$$

To obtain the left side of the required inequality, note that the function $f(z)$ does not vanish in the open upper half-plane. Writing the inequality (2.4) for the function $1/f(z)$ we obtain the left inequality in (2.3). As claimed. Q. E. D.

The Caratheodory theorem enables us to estimate from below the modulus of the holomorphic function $f(z)$ having no zeros in the circle $|z| < R$. Indeed, assume that $f(0) = 1$, and write inequality (2.1) for the holomorphic function $ln\ f(z)$. We obtain the inequality

$$|ln\ f(z)| \le ln\ M_f(R)\frac{2r}{R-r} \quad (0 < r < R,\ |z| \le r)$$

from which follows

$$ln\ \frac{1}{|f(z)|} \le ln\ M_f(R)\frac{2r}{R-r}$$

and

$$ln\ |f(z)| \ge -ln\ M_f(R)\frac{2r}{R-r}.$$

Thus we obtain the following theorem.

Theorem 3.2.3 *If the function $f(z)$ is holomorphic in the circle $|z| \le R$ and has no zeros in this circle, and if $f(0) = 1$, then its modulus in the circle $|z| \le r < R$ satisfies the inequality*

$$ln\ |f(z)| \ge -ln\ M_f(R)\frac{2r}{R-r}.$$

If the function $f(z)$ has zeros, such an estimate is, of course, impossible. However, one can give an analogous estimate in the domain obtained from the original circle by removing certain small circles containing the zeros.

3.3   Jensen's theorem

Denote by $\nu_f(t)$ the number of the zeros of $f(z)$ in the circle $|z| < t$. The function $\nu_f(t)$ will be called *the counting function* of the zeros of f (the counting function of f).

Theorem 3.3.1 *Let $f(z)$ be holomorphic in a circle of radius R with center at the origin, and $f(0) \ne 0$. Then*

$$\int_0^R \frac{\nu_f(t)}{t}dt = \frac{1}{2\pi}\int_0^{2\pi} ln\ |f(Re^{i\theta})|d\theta - ln\ |f(0)|. \qquad (3.1)$$

Note that if the function has no zeros in the circle $|z| < R$, then $\nu_f(t) = 0$ and equation (3.1) expresses a well-known property of harmonic functions. If there are zeros, then it follows from (3.1) that

$$ln\,|f(0)| < \frac{1}{2\pi}\int_0^{2\pi} ln\,|f(Re^{i\theta})|d\theta.$$

Proof: If the function has no zeros for $|z| = t$, then by the argument principle we have

$$\nu_f(t) = \frac{1}{2\pi}\int_0^{2\pi}\frac{d\,[arg(f(te^{i\theta}))]}{d\theta}d\theta.$$

Using the Cauchy-Riemann equations

$$\frac{\partial[arg\,f(te^{i\theta})]}{t\,\partial\theta} = \frac{\partial\,ln\,|f(te^{i\theta})|}{\partial t},$$

we have

$$\frac{\nu_f(t)}{t} = \frac{1}{2\pi}\int_0^{2\pi}\frac{\partial\,ln\,|f(te^{i\theta})|}{\partial t}d\theta$$

for all except a finite set of values of t in the interval $0 < t < R$. Integrating both sides of this equation from 0 to R we obtain (3.1). Here, of course, we are using the continuity of the function

$$\int_0^{2\pi} ln\,|f(te^{i\theta})|d\theta.$$

It is easy to establish this continuity if we represent the function $ln\,|f(z)|$ in the form

$$ln\,|f_1(z)| + \sum_{|z_n|\le R} ln\,|z - z_n|,$$

where z_k are the zeros of the function $f(z)$, and $f_1(z)$ has no zeros in the circle $|z| \le R$. The first term is continuous, and

$$\int_0^{2\pi} ln\,|Re^{i\theta} - z_k|d\theta = \begin{cases} 2\pi\,ln\,R & \text{if } R > |z_k|, \\ 2\pi\,ln\,|z_k| & \text{if } R < |z_k| \end{cases}.$$

So that this is also a continuous function. Thus the theorem is proved.
    Q. E. D.

Jensen's formula can be generalized to include the case when $f(z)$ has a zero of order λ at the origin. Then, applying formula (3.1) to the function $f(z)z^{-\lambda}$, we obtain

$$\lambda\,ln\,R + \int_0^R \frac{\nu_f(t)}{t}dt = \frac{1}{2\pi}\int_0^{2\pi} ln\,|f(Re^{i\theta})|d\theta - ln\,\frac{|f^{(\lambda)}(0)|}{\lambda!}.$$

Here $\nu_f(t)$ is defined to be the number of zeros of $f(z)$, different from the origin, in the circle $|z| < t$. The following lemma gives an important estimate for the number of zeros of $f(z)$ in a circle.

Lemma 3.3.2 *If $f(z)$ is holomorphic in the circle*

$$|z| \le er$$

and if $|f(0)| = 1$, then

$$\nu_f(r) \le \ln M_f(er). \tag{3.2}$$

Proof: From Jensen's formula and the monotonicity of $\nu_f(t)$ we have at once

$$\nu_f(r) \le \int_r^{er} \frac{\nu_f(t)}{t} dt \le \frac{1}{2\pi} \int_0^{2\pi} \ln |f(ere^{i\theta})| d\theta \le \ln M_f(er). \tag{3.3}$$

As claimed. Q. E. D.

Inequality (3.2) is called the *Jensen inequality*.

Let us investigate when the equality sign can hold in (3.2). For this it is necessary and sufficient that equality hold everywhere in (3.3). The equation

$$\nu_f(r) = \int_r^{er} \frac{\nu_f(t)}{t} dt$$

says that the function $f(z)$ has no zeros in the annulus $r < |z| < er$. The equation

$$\int_r^{er} \frac{\nu_f(t)}{t} dt = \frac{1}{2\pi} \int_0^{2\pi} \ln |f(ere^{i\theta})| d\theta \tag{3.4}$$

says, by (3.1) since $|f(0)| = 1$, that $f(z)$ has no zeros in the circle $|z| < r$. Finally, the equation

$$\frac{1}{2\pi} \int_0^{2\pi} \ln |f(ere^{i\theta})| d\theta = \ln M_f(er)$$

says that on the whole circumference $|z| = er$,

$$|f(ere^{i\theta})| = M_f(er).$$

All these conditions are fulfilled for functions of the form

$$\phi(z) = e^{n+i\alpha} \prod_{k=1}^n \frac{er(z - z_k)}{e^2 r^2 - \bar{z}_k z} \tag{3.5}$$

$$(|z_k| = r, \alpha \text{ is real }).$$

Indeed, each factor in this product maps the circle $|z| \le er$ onto the unit circle, and therefore has modulus one everywhere on the circumference $|z| = er$. Thus, on the circumference $|z| = er$, we have $|\phi(z)| = e^n$. In addition, all the roots of $\phi(z)$ are on the circle $|z| = r$, and $|\phi(0)| = 1$. It is not difficult to show that (3.5) is the general form of the functions for which equality holds in (3.2).

3.4   Lower bounds for moduli of holomorphic functions

At the beginning of this section, we mention an important theorem of Cartan.

Theorem 3.4.1 *Let*

$$P(z) = \prod_{k=1}^{n} (z - a_k) \quad (n < \infty, z \in \mathbf{C}),$$

where $a_1, ..., a_n$ are complex numbers. Then for any number $H > 0$, the inequality

$$|P(z)| > (H/e)^n$$

holds outside of at most n circles, the sum of whose radii is at most $2H$.

Proof: We divide the proof into several stages.

1. We choose the quantity H/n as the unit of measurement and show that there are closed circles in the complex plane having radius equal to the number of points $\{a_k\}$ contained within the circle. Indeed, form the smallest convex polygon containing all the points $\{a_k\}$, and choose any vertex a_j of this polygon. Clearly, there are circles of arbitrary radius that contain this point but do not contain any other of the points $\{a_k\}$. In particular, the radius can be chosen to equal the multiplicity of the point a_j.

2. From among all circles with radius equal to the number of points $\{a_k\}$ lying inside the circle choose one with the largest radius, $\lambda_1 H/n$, and call it C_1. Note that no circle in the plane with radius greater than or equal to $\lambda_1 H/n$ can contain more points of the set $\{a_k\}$ than the number of units of measurement in its radius. Indeed, suppose a circle of radius $\lambda H/n$, with $\lambda \geq \lambda_1$ contains $\lambda' > \lambda$ points of the set $\{a_k\}$. The concentric circle of radius $\lambda' H/n$ either contains λ' points or $\lambda'' > \lambda'$ points. In the second case, we consider the concentric circle of radius $\lambda'' H/n$ and so forth. Since the set $\{a_k\}$ is finite, we eventually come to a circle of radius $\lambda H/n$, larger than $\lambda_1 H/n$, that contains λ points. This is impossible since C_1 is the largest circle having this property. The points of the set $\{a_k\}$ that lie inside C_1 will be said to be of rank λ_1.

3. We remove the points of rank λ_1, and for the remaining $n - \lambda_1$ points, we construct the largest circle C_2 that contains the same number of points as there are units in its radius. Let its radius be $\lambda_2 H/n$. We shall show that $\lambda_2 < \lambda_1$.

Indeed, if this were not the case then the circle C_2 would have a radius larger than $\lambda_1 H/n$ and would contain at least as many of the original n points as there are units in its radius. But this contradicts the result of point 2. The points in C_2 will be said to be of rank λ_2. Now remove these points too, and for the remaining $n - \lambda_1 - \lambda_2$ points we find the largest circle C_3 containing the same number of points as there are units in its radius. Suppose that its radius is $\lambda_3 H/n$. Clearly, $\lambda_3 \leq \lambda_2$. The points in C_3 will be said to be of rank λ_3, and so forth. We thus obtain a sequence of circles

$$C_1, C_2, ..., C_p$$

with radii that contain $\lambda_1, \lambda_2, ..., \lambda_p$ units of measurement, respectively, where

$$\lambda_1 \geq \lambda_2 \geq ... \geq \lambda_p$$

and

$$\frac{H}{n}(\lambda_1 + \lambda_2 + ... + \lambda_p) = H.$$

4. We now form circles

$$\Gamma_1, \Gamma_2, ..., \Gamma_p$$

concentric with the circles $C_1, C_2, ..., C_p$ but with radii twice as large. Let z_0 be an arbitrary point lying outside all of the new circles. Take the circle

$$C_{z_0} = \{z \in \mathbf{C} : |z - z_0| \leq \lambda H/n\}$$

where λ_0 is some natural number. This circle intersects any of the circles C_j, that have a radius larger than or equal to λ. Thus, this circle can only contain points whose rank is less than λ. From the definition of rank it follows that after removing all points of rank greater than or equal to λ, no circle of radius greater than or equal to $\lambda H/n$ can contain as many of the remaining points as there are units of measurement in its radius. It follows from point 2 that the circle C_{z_0} can contain at most $\lambda - 1$ points.

5. Enumerating the points $\{a_k\}$ in the order of increasing distance from z_0 we have

$$|z - a_k| > k\frac{H}{n}$$

and

$$|z - a_1||z - a_2|...|z - a_n| > n!(\frac{H}{n})^n > (\frac{H}{e})^n.$$

As claimed. Q. E. D.

By the Cartan theorem, we will proof the following result.

Theorem 3.4.2 *Let $f(z)$ be holomorphic in the circle $|z| \leq 2eR$ $(R > 0)$ with $f(0) = 1$, and let η be an arbitrary positive number not exceeding $3e/2$. Then inside the circle $|z| \leq R$, but outside of a family of excluded circles the sum of whose radii is not greater than $4\eta R$, we have*

$$ln\, |f(z)| > -H(\eta)\, ln\, M_f(2eR) \qquad (4.1)$$

for

$$H(\eta) = 2 + ln\, \frac{3e}{2\eta}.$$

Proof: Construct the function

$$\phi(z) = \frac{(-2R)^n}{z_1 z_2 \cdots z_n} \prod_{k=1}^{n} \frac{2R(z - z_k)}{(2R)^2 - \overline{z}_k z},$$

where $z_1, z_2, ..., z_n$ are the roots of $f(z)$ in the circle $|z| \leq 2R$. We have

$$\phi(0) = 1 \text{ and } |\phi(2Re^{i\theta})| = \frac{(2R)^n}{|z_1 z_2 ... z_n|}.$$

The function

$$\psi(z) = \frac{f(z)}{\phi(z)}$$

has no roots in the circle $|z| \leq 2R$, and therefore by the corollary to the Caratheodory theorem for the circle (Theorem 3.2.3) we have for $|z| \leq R$,

$$\ln |\psi(z)| > -\ln M_f(2eR) + 2 \ln \frac{(2R)^n}{|z_1 z_2 ... z_n|}$$

$$\geq -\ln M_f(2eR). \qquad (4.2)$$

Now we estimate $|\phi(z)|$ from below. For $|z| \leq R$,

$$\prod_{k=1}^{n} |(2R)^2 - \overline{z}_k z| < (6R^2)^n.$$

By applying the Cartan theorem (see the previous theorem) to the polynomial in the numerator of $\phi(z)$, we see that, outside of a family of circles with the sum of the radii equal to $4\eta R$, we have

$$\left| \prod_{k=1}^{n} 2R(z - z_k) \right| > \left(\frac{2\eta R}{e}\right)^n (2R)^n$$

and consequently,

$$|\phi(z)| > \left(\frac{2\eta R}{e}\right)^n \frac{1}{(6R^2)^n} \frac{(2R)^{2n}}{|z_1 z_2 ... z_n|} \geq \left(\frac{2\eta}{3e}\right)^n.$$

By the Jensen inequality,

$$n = \nu_f(2R) \leq \ln M_f(2eR),$$

and this leads to the estimate

$$\ln |\phi(z)| > \ln\left(\frac{2\eta}{3e}\right) \ln M_f(2eR).$$

Combining this with the estimate (4.2) we see that in the circle $|z| < R$, but outside of a family of circles, the sum of whose radii does not exceed $4\eta R$, we have the required result. Q. E. D.

3.5   Order and type of an entire function

Recall that an entire function is a function of a complex variable holomorphic in the entire plane and consequently represented by an everywhere convergent power series

$$f(z) = \sum_{k=0}^{\infty} c_k z^k \quad (z \in \mathbf{C}). \tag{5.1}$$

Again use the function

$$M_f(r) := \max_{|z|=r} |f(z)|$$

(sometimes we shall simply write $M(r)$). It is easily seen that $M_f(r)$ is continuous. In fact, let $r_1 < r_2$ and $|f(r_2 e^{i\theta_0})| = M(r_2)$. Then

$$0 < M(r_2) - M(r_1) \le |f(r_2 e^{i\theta_0}) - f(r_1 e^{i\theta_0})| < \epsilon$$

for $r_2 - r_1 < \delta_\epsilon$. It follows from the maximum modulus principle that, as r increases, $M_f(r)$ grows monotonically. The rate of growth of the function $M_f(r)$ is an important characteristic of the behavior of the entire function. Let us show that for an entire function that is not a polynomial $M_f(r)$ grows faster than any positive power of r.

As usual, put

$$\underline{\lim}_{r \to \infty} \phi(r) = \lim_{r \to \infty} \inf_{t \ge r} \phi(t)$$

and

$$\overline{\lim}_{r \to \infty} \phi(r) = \lim_{r \to \infty} \sup_{t \ge r} \phi(t).$$

Theorem 3.5.1 *If there exists a positive integer* n *such that*

$$\underline{\lim}_{r \to \infty} \frac{M_f(r)}{r^n} < \infty,$$

then $f(z)$ *is a polynomial of degree at most* n.

Proof: If

$$f(z) = \sum_{k=0}^{\infty} c_k z^k \quad (z \in \mathbf{C})$$

and

$$P_n(z) = \sum_{k=0}^{n} c_k z^k \quad (z \in \mathbf{C})$$

then the function

$$\phi(z) = [f(z) - P_n(z)]z^{-n-1}$$

is entire and tends uniformly to zero on some sequence of circles $|z| = r_n$ ($r_n \rightarrow \infty$). It follows from the maximum principle that $\phi(z) = 0$, that is,

$$f(z) \equiv P_n(z).$$

As claimed. Q. E. D.

Thus, in order to estimate the growth of transcendental entire functions, one must choose comparison functions that grow more rapidly then powers of r. We choose comparison functions of the form

$$e^{r^k},$$

where $k > 0$.

An entire function $f(z)$ is said to be *a function of finite order* if there exists a positive constant k, such that the inequality

$$M_f(r) < e^{r^k}$$

is valid for all sufficiently large values of r ($r > r_0(k)$). The greatest lower bound of such numbers k is called *the order* of the entire function $f(z)$.

It follows from this definition that if ρ is the order of the entire function $f(z)$, and if ϵ is an arbitrary positive number, then

$$e^{r^{\rho-\epsilon}} < M_f(r) < e^{r^{\rho+\epsilon}}, \tag{5.2}$$

where the inequality on the right is satisfied for all sufficiently large values of r, and the inequality on the left holds for some sequence $\{r_n\}$ of values of r, tending to infinity. It is easy to verify that condition (5.2) is equivalent to the equation

$$\rho(f) := \overline{\lim}_{r\to\infty} \frac{\ln \ln M_f(r)}{\ln r},$$

which could therefore be taken as the definition of the order of the function.

An inequality that holds for all sufficiently large values of r will be called *an asymptotic inequality*. For functions of a given order a more precise characterization of the growth is given by the type of the function. By *the type of an entire function $f(z)$ of order ρ*, we mean the greatest lower bound of positive numbers A for which asymptotically $M_f(r) < e^{Ar^\rho}$. Just as with the definition of order it is easy to verify that the type σ a of a function $f(z)$ of order ρ is given by the equation

$$\sigma = \overline{\lim}_{r\to\infty} \frac{\ln M_f(r)}{r^\rho}.$$

If $\sigma = 0$, then function $f(z)$ is said to be *of minimal type*, if $0 < \sigma < \infty$ of *normal type*, and if $\sigma = \infty$ of *maximal type*.

An example of an entire function of order n and type σ is the function

$$e^{\sigma z^n}$$

for integral n.

It is easy to verify also that $\sin z$ is an entire function of order one and normal type $\sigma = 1$.

The function $\cos \sqrt{z}$ has order $\rho = 1/2$ and type $\sigma = 1$. The function

$$e^{e^z}$$

is an example of an entire function of infinite order.

We shall say that the function $f_2(z)$ is of larger growth than the function $f_1(z)$ if the order of $f_2(z)$ is greater than the order of $f_1(z)$, or if the orders are equal and the type of $f_2(z)$ is larger than the type of $f_1(z)$. It is easy to see that the order of the sum of two functions is not greater than the larger of the orders of the two summands, and if the orders of the summands and of the sum are all equal, then the type of the sum is not greater than the larger of the types of the two summands. In addition, if one of the two functions is of larger growth than the other, then the sum has the same order and type as the function of larger growth.

3.6 Taylor coefficients of an entire function

Let an entire function $f(z)$ have the power series

$$f(z) = \sum_{k=0}^{\infty} c_k z^k. \tag{6.1}$$

The radius of convergence is infinite and therefore

$$\lim_{r \to \infty} \sqrt[n]{|c_n|} = 0. \tag{6.2}$$

The following theorem enables one to determine the order and type of an entire function by the rate of decrease of its Taylor coefficients.

Theorem 3.6.1 *The order $\rho = \rho(f)$ and type $\sigma = \sigma(f)$ of an entire function are expressed in terms of its Taylor coefficients by the following equations*

$$\rho = \overline{\lim}_{n \to \infty} \frac{n \ln n}{\ln \frac{1}{|c_n|}} \tag{6.3}$$

and

$$(\sigma e \rho)^{\frac{1}{\rho}} = \overline{\lim}_{n \to \infty} (n^{\frac{1}{\rho}} \sqrt[n]{|c_n|}). \tag{6.4}$$

Proof: By the well-known inequality for the coefficients of a power series

$$|c_n| \leq \frac{M_f(r)}{r^n}. \tag{6.5}$$

If $f(z)$ is of finite order, then asymptotically

$$M_f(r) < e^{Ar^k} \tag{6.6}$$

and thus

$$|c_n| \leq \frac{e^{Ar^k}}{r^n}.$$

Employing the usual method for finding extrema it is easy to see that the function on the right side of this inequality takes its largest value in the range $r > 0$ for

$$r = \left(\frac{n}{Ak}\right)^{\frac{1}{k}}$$

and therefore asymptotically

$$|c_n| \leq \left(\frac{eAk}{n}\right)^{\frac{n}{k}}. \tag{6.7}$$

Conversely, assume that (6.7) holds for all indices n greater than $n_0(k, A)$, and let us estimate $M_f(r)$. For $n > m_r = [2^k eAkr^k]$, where $[x]$ is the integer part of $x > 0$, and all sufficiently large values of r we have by (6.7)

$$|c_n z^n| \leq 2^{-n}$$

and therefore

$$|f(z)| < \sum_{k=0}^{m_r} |c_n| r^n + 2^{-m_r}.$$

   Introducing the notation

$$\mu(r) := \max_n |c_n| r^n,$$

we have

$$M_f(r) \leq (1 + 2^k eAkr^k)\mu(r) + 2^{-m_r}. \tag{6.8}$$

If $f(z)$ is not a polynomial, then $M_f(r)$ and $\mu(r)$ grow faster than any power of r, and therefore the index of the largest term in the series (6.1) increases without bound as r grows. It follows from (6.7) that asymptotically

$$\mu(r) \leq \max_n \left(\frac{eAk}{n}\right)^{\frac{n}{k}} r^n.$$

The maximum of the right side is attained for

$$n = Akr^k$$

and therefore asymptotically

$$\mu(r) \le e^{Ar^k}.$$

From (6.8), we see that asymptotically

$$M_f(r) < (2 + 2^k eAkr^k)e^{Ar^k}. \tag{6.9}$$

Thus (6.7) follows from (6.6), and (6.9) follows from (6.7). This shows that the order ρ of an entire function $f(z)$ is equal to the greatest lower bound of numbers k for which the asymptotic inequality (6.7) holds, and the type is equal to the greatest lower bound of numbers A for which the asymptotic inequality (6.7) holds with $k = \rho$. From this both assertions of the theorem follow at once. Q. E. D.

We note an interesting corollary that follows at once from formulas (6.5) and (6.8). The maximal term $\mu(r)$ in the power series expansion of an entire function of finite order satisfies the asymptotic inequalities

$$\mu(r) \le M_f(r) < \mu(r)r^{\rho+\epsilon},$$

where ρ is the order of the function $f(z)$ and ϵ is an arbitrary positive number. With the aid of Theorem 3.6.1, one can easily construct entire functions of arbitrary order and type. To this end, consider the entire function

$$f(z) = \sum_{k=0}^{\infty} \frac{(A^\alpha z)^k}{\Gamma(\alpha k + 1)},$$

where $A > 0$ and $\alpha > 0$, and Γ is the Euler Gamma function. By Stirling's formula, we easily have from (6.3) and (6.4) the equalities

$$\rho = \frac{1}{\alpha} \text{ and } \sigma = A.$$

3.7   The theorem of Weierstrass

It is well known that every polynomial can be written as the product of linear factors. The analogue of this assertion for entire functions is the theorem of Weierstrass on the representation of entire functions by infinite products. This representation is important for the investigation of the basic question of the theory of entire functions - the question of the relation between the growth of the entire function and the distribution of its zeros in the complex plane.

Let $z_1, z_2, \ldots$ be a sequence of complex numbers, none of which is zero,

with the point at infinity as the only limit point. We shall construct an entire
function whose set of zeros is precisely this sequence.

We can assume that these points have been arranged in order of increasing
moduli (if several different points have the same modulus, then we take them
in any order). Choose a sequence of natural numbers p_n, such that the series

$$\sum_{k=1}^{\infty} |\frac{z}{z_k}|^{p_k} \tag{7.1}$$

converges uniformly in each bounded domain. Such a choice is possible since,
for $|z| < R$ $(R > 0)$ the inequality

$$|\frac{z}{z_n}| < q < 1$$

is satisfied for all sufficiently large values of n, and thus, for example, one can
choose $p_n = n$. We form the infinite product

$$\Pi(z) = \prod_{k=1}^{\infty} G(\frac{z}{z_k}, p_k) \tag{7.2}$$

in which

$$G(u, p) = (1 - u)e^{u + \frac{u^2}{2} + ... + \frac{u^p}{p}}, \, G(u, 0) = 1 - u$$

are primary factors. Let us show that the product (7.2) converges uniformly
on each closed bounded set that contains none of the points z_n, and therefore
defines an entire function that vanishes at the points z_n and only at them. To
this end, we estimate the quantity $|ln \, G(u, p)|$ for $|u| < q < 1$ and $|arg \, (1 - u)| < \pi$. From the expansion

$$ln \, G(u, p) = \frac{u^{p+1}}{p + 1} - \frac{u^{p+2}}{p + 2} - ...,$$

which is valid under the conditions indicated, we obtain

$$|ln \, G(u, p)| < \frac{|u|^{p+1}}{1 - |u|} \le \frac{|u|^{p+1}}{1 - q}.$$

It follows from this inequality that for $|z| < R$ and $n > n(q, R)$ we will have

$$|ln \, G(\frac{z}{z_n}, p_n)| < \frac{1}{1 - q} |\frac{z}{z_n}|^{p_n+1}$$

and this implies, because of the uniform convergence of the series (7.1) in the
circle $|z| < R$, that the series

$$\sum_{n=1}^{\infty} ln \, G(\frac{z}{z_n}, p_n)$$

and therefore also the product (7.2), converges uniformly in each closed subset
of this circle that contains none of the points z_n.

Theorem 3.7.1 *Every entire function $f(z)$ can be represented in the form*

$$f(z) = z^m e^{g(z)} \prod_{k=1}^{\omega} G\left(\frac{z}{z_k}, p_n\right) \quad (\omega \leq \infty), \qquad (7.3)$$

where $g(z)$ is an entire function, z_n are the nonzero roots of $f(z)$, and m is the order of the zero of $f(z)$ at the origin.

Proof: The nonzero roots z_n of the entire function $f(z)$ $(n = 1, 2, ..., \omega \leq \infty)$ have no finite limit point. We form the product

$$\Pi(z) = \prod_{n=1}^{\omega} G\left(\frac{z}{z_n}, p_n\right) \qquad (7.4)$$

in which $p_n = 0$, if ω is finite. The function

$$\phi(z) = \frac{f(z)}{\Pi(z) z^m}$$

is entire and has no zeros. Hence, the function $g(z) = \ln \phi(z)$ is also entire, and we obtain the desired representation. Q. E. D.

In the representation (7.3), the sequence of numbers p_n is not uniquely determined, and therefore the function $g(z)$ is not determined uniquely either. The representation of the function $f(z)$ is considerably simpler if the numbers z_n satisfy the following supplementary condition: the series

$$\sum_{k=1}^{\infty} \frac{1}{|z_k|^{\lambda}} \qquad (7.5)$$

converges for some positive λ. In this case let p denote the smallest integer for which the series

$$\sum_{k=1}^{\infty} \frac{1}{|z_k|^{p+1}}$$

converges. Clearly, $0 \leq p < \lambda$. The series (7.1) will converge uniformly if we put all $p_n = p$. The uniformly convergent infinite product

$$\Pi(z) = \prod_{n=1}^{\infty} G\left(\frac{z}{z_n}, p\right) \qquad (7.6)$$

is called *a canonical product*, and the number p is called *the genus of the canonical product*.

In the case just considered, it is customary in the representation (7.3) to choose the canonical product for the infinite product. Then the function $g(z)$ is uniquely determined. If $g(z)$ is a polynomial, $f(z)$ is said to be an entire function of finite genus. The larger of the numbers p and q, where q is the degree of the polynomial $g(z)$, is called *the genus* of the entire function $f(z)$. If $g(z)$ is not a polynomial, or if the series (7.5) diverges for all values of λ, then the genus is said to be infinite.

3.8 Density of zeros

The representation of an entire function as an infinite product makes it possible to establish a very important dependence between the growth of the function and the density of distribution of its zeros. As a measure of the density of the sequence of points z_n, having no finite limit point, we introduce the convergence exponent. By *the convergence exponent of the sequence*

$$z_1, z_2, \ldots \quad (z_n \neq 0, \ \lim_{n \to \infty} z_n = \infty),$$

we mean the greatest lower bound of numbers λ for which the series

$$\sum_{k=1}^{\infty} \frac{1}{|z_k|^{\lambda}} \tag{8.1}$$

converges. Clearly, the more rapidly the sequence of numbers $|z_k|$ increases, the smaller will be the convergence exponent. In particular, the convergence exponent can be zero. If the series (8.1) diverges for all $\lambda > 0$, then the convergence exponent is said to be infinite. If λ is equal to the convergence exponent, the series (8.1) may or may not converge, depending on the sequence $\{z_n\}$. For example, the sequences $z_n = n^{1/\rho}$ and $z_n = (n \ ln^2 n)^{1/\rho}$ have the same convergence exponent ρ, but in the first case, the series (8.1) diverges when $\lambda = \rho$, while in the second case, it converges. The sequences $\{e^n\}$ and $\{ln \ n\}$ have convergence exponents 0 and ∞, respectively. Let us recall the following obvious relation between the convergence exponent ρ_1 of the sequence $\{z_n\}$ and the genus p of the corresponding canonical product (7.6):

$$p \leq \rho_1 \leq p + 1.$$

If ρ_1 is an integer, then $p = \rho_1$ when the series (8.1) diverges for $\lambda = \rho_1$ while $\rho_1 = p + 1$ means that the series converges.

Define the function $n(r)$ as the number of points of the sequence z_k in the circle $|z| < r$. In other words, $n(r)$ is *the counting function of the numbers* z_k. The growth of the function $n(r)$ gives us a more precise description of the density of the sequence z_n than the convergence exponent. By the order of this monotone function, we mean the number

$$\rho_1 = \overline{\lim}_{r \to \infty} \frac{ln \ n(r)}{ln \ r}, \tag{8.2}$$

and by the *upper density of the sequence* $\{z_n\}$, we mean the number

$$\Delta = \overline{\lim}_{r \to \infty} \frac{n(r)}{r^{\rho_1}}. \tag{8.3}$$

If the limit exists, then Δ is simply called the *density*.

Lemma 3.8.1 *The convergence exponent of the sequence*

$$z_1, z_2, \dots \ (\lim_{n \to \infty} |z_n| = \infty)$$

is equal to the order ρ_1 of the corresponding function $n(r)$.

Proof: The series (8.1) can be rewritten as a Stieltjes integral in the following form:

$$\int_0^\infty \frac{dn(t)}{t^\lambda}.$$

Use the fact that $n(t) = 0$ for $0 \le t < |z_1|$. Then, integrating by parts, we have

$$\int_0^r \frac{dn(t)}{t^\lambda} = \frac{n(r)}{r^\lambda} + \lambda \int_0^r \frac{n(t)dt}{t^{\lambda+1}}. \tag{8.4}$$

If the series (8.1) converges, then both the positive terms on the right side of (8.4) are bounded, and since the second term is monotone increasing, the integral

$$\int_0^\infty \frac{n(t)dt}{t^{\lambda+1}} \tag{8.5}$$

converges. From this convergence, it follows that for arbitrary $\epsilon > 0$ and $r > r_0(\epsilon)$,

$$\epsilon > \lambda \int_r^\infty \frac{n(t)dt}{t^{\lambda+1}} \ge \lambda n(r) \int_r^\infty \frac{dt}{t^{\lambda+1}} = n(r)\frac{1}{r^\lambda}$$

or

$$\lim_{r \to \infty} \frac{n(r)}{r^\lambda} = 0.$$

Thus, the order of the function $n(r)$ is not greater than the convergence exponent of the sequence. Also, from the above reasoning, we see that conversely the convergence of the integral (8.5) implies the convergence of the series (8.1). Let now ρ_1 be the order of the function $n(r)$. Then, for $\epsilon > 0$, we have asymptotically

$$n(r) < r^{\rho_1+\epsilon/2}.$$

Therefore, for $\lambda = \rho_1 + \epsilon$, the integral (8.5) converges, and therefore the series (8.1) also converges. Thus, the convergence exponent is not greater than the order of the function $n(r)$. The lemma is proved. Q. E. D.

3.9   An estimate for canonical products in terms of counting functions

Now we are going to estimate the canonical product. To this end, we recall that by Lemma 2.14.2 for any integer $p \geq 1$ and all complex numbers u

$$ln \ |G(u,p)| \leq \beta_{p+1} \frac{|u|^{p+1}}{1+|u|}, \tag{9.1}$$

where β_p is a constant satisfying

$$\beta_p \leq 3e(2 + ln \ (p-1)).$$

In addition,

$$ln \ |G(u,0)| \leq ln \ (1+|u|).$$

Lemma 3.9.1 *If the series*

$$\sum_{k=1}^{\infty} \frac{1}{|z_k|^{p+1}} \tag{9.2}$$

converges, then the infinite product

$$\Pi(z) = \prod_{k=1}^{\infty} G(\frac{z}{z_k}, p)$$

satisfies in the entire complex plane the inequality

$$|ln \ \Pi(z)| < k_p r^p \ (\int_0^r \frac{dn(t)}{t^{p+1}} + r \int_r^{\infty} \frac{n(t)dt}{t^{p+2}}) \quad (r = |z|),$$

in which

$$k_p = 3e(p+1)(2 + ln \ p) \ for \ an \ integer \ p \geq 1$$

and $k_0 = 1$.

Proof: From inequality (9.1) with $p > 0$, we have

$$ln \ |\Pi(z)| \leq \beta_{p+1} \sum_{k=1}^{\infty} \frac{r^{p+1}}{|z_k|^p(|z_k|+r)} = \beta_{p+1}r^{p+1} \int_0^{\infty} \frac{dn(t)}{t^p(t+r)}.$$

As was proved in the proof of Lemma 3.8.1, it follows from the convergence of the series (9.2) that

$$\lim_{r \to \infty} \frac{n(r)}{r^{p+1}} = 0.$$

Integrating by parts and taking into account that $n(t) = 0, t < |z_1|$, we have

$$ln \ |\Pi(z)| \leq \beta_{p+1}r^{p+1} \int_0^{\infty} \frac{pr + (p+1)t}{t^{p+1}(t+r)^2} n(t)dt$$

$$< \beta_{p+1} r^{p+1}(p+1) \int_0^\infty \frac{1}{t^{p+1}(t+r)} n(t)dt$$

or

$$ln\,|\Pi(z)| \le \beta_{p+1}(p+1)r^p \Big(\int_0^r \frac{n(t)}{t^{p+1}} \, dt + r \int_r^\infty \frac{n(t)}{t^{p+1}} \, dt \Big).$$

The case $p = 0$ is verified in a similar manner. Q. E. D.

3.10 The convergence exponent of zeros

Theorem 3.10.1 *(Borel). The order ρ of the canonical product*

$$\Pi(z) = \prod_{n=1}^\infty G\Big(\frac{z}{z_n}, p\Big)$$

does not exceed the convergence exponent ρ_1 of the sequence $\{z_n\}$.

Proof: The number p is the smallest integer for which the series (9.2) converges. Thus, for the convergence exponent ρ_1, we have the inequality

$$p \le \rho_1 < p+1.$$

Suppose that $\rho_1 < p+1$ and $\rho_1 < \lambda < p+1$. Then by (8.2) there exists a constant C_λ such that for all values $t > 0$

$$n(t) < C_\lambda t^\lambda.$$

From Lemma 3.9.1, we have

$$ln\,|\Pi(z)| \le C_\lambda k_p \Big(r^p \int_0^r t^{\lambda-p-1}dt + r \int_r^\infty t^{\lambda-p-2}dt \Big) = C_\lambda B_\lambda r^p,$$

where

$$B_\lambda = k_p \Big[\frac{1}{\lambda - p} + \frac{1}{p+1-\lambda} \Big],$$

that is, the order of $\Pi(z)$ does not exceed λ, and therefore, it does not exceed ρ_1. If $\rho_1 = p+1$, then, as was shown in the proof of Lemma 3.8.1,

$$\lim_{r\to\infty} \frac{n(r)}{r^{p+1}} = 0,$$

and the integral

$$\int_0^\infty \frac{n(t)}{t^{p+2}} \, dt$$

converges. Using this, we have from Lemma 3.9.1 that asymptotically

$$|ln \, \Pi(z)| < \epsilon r^{p+1},$$

that is, $\Pi(z)$ is at most of order $p+1$ and minimal type. Q. E. D.

Remark 3.10.2 *If ρ_1 is not an integer and if the upper density of the sequence $\{z_n\}$ is finite, then $\Pi(z)$ is at most of order ρ_1 and normal type; if ρ_1 is not an integer and if the density of the sequence $\{z_n\}$ is zero, then $\Pi(z)$ is at most of order ρ_1 and minimal type.*

For details, see [Levin, 1980, p. 13].

Theorem 3.10.3 *The convergence exponent $\tilde{\rho}_1$ of the zeros of an arbitrary entire function f does not exceed its order $\rho(f)$.*

Proof: Without loss of generality, we may assume that $f(0) = 1$. Otherwise in place of $f(z)$ one considers

$$f_1(z) = \lambda! z^{-\lambda} \frac{f(z)}{f^{(\lambda)}(0)},$$

which has the same order and convergence exponent as $f(z)$. From the Jensen inequality, it follows that

$$\tilde{\rho}_1 = \overline{\lim}_{r \to \infty} \frac{\ln \nu_f(r)}{\ln r} \leq \overline{\lim}_{r \to \infty} \frac{\ln \ln M_f(er)}{\ln er} = \rho(f).$$

The theorem is proved. Q. E. D.

Note that from the Jensen inequality one can also obtain the following relations

$$\Delta_f := \overline{\lim}_{r \to \infty} \frac{\nu_f(r)}{r^{\tilde{\rho}_1}} \leq \overline{\lim}_{r \to \infty} e^{\tilde{\rho}_1} \frac{ln M_f(er)}{(er)^{\tilde{\rho}_1}},$$

and if $\tilde{\rho}_1 = \rho(f)$, we will have

$$\Delta_f \leq e^{\rho(f)} \sigma_f,$$

where σ_f is the type of f. Comparing this with Theorem 3.10.1 and Remark 3.10.2, we have the following theorem.

Theorem 3.10.4 *For canonical products the convergence exponent of the zeros is equal to the order of the function. In addition, if the convergence exponent is not an integer, then the canonical product is of maximal, minimal, or normal type according to whether the upper density of the set of zeros*

$$\Delta_f = \overline{\lim}_{r \to \infty} \frac{\nu_f(r)}{r^\rho}$$

is equal to infinity, to zero, or to a number different from zero and infinity.

3.11 Hadamard's theorem

The theorems of the preceding sections enable us to refine considerably the theorem on the representation of an entire function as an infinite product. This refinement, which is due to Hadamard, concerns the representation of entire functions of finite order, and is one of the classical theorems of the theory of entire functions.

Theorem 3.11.1 *The entire function $f(z)$ of finite order $\rho(f) = \rho$ can be represented in the form*

$$f(z) = z^m e^{P(z)} \prod_{k=1}^{\omega} G\left(\frac{z}{z_k}, p\right) \ (\omega \leq \infty),\tag{11.1}$$

where z_n are the nonzero roots of $f(z)$, $p \leq \rho$, $P(z)$ is a polynomial whose degree q does not exceed ρ, and m is the multiplicity of the zero at the origin.

Proof: The convergence exponent $\tilde{\rho}_1$ of the zeros of the entire function does not exceed its order ρ, and the integer p in the associated canonical product

$$\Pi(z) = \prod_{k=1}^{\infty} G\left(\frac{z}{z_k}, p\right)$$

does not exceed $\tilde{\rho}_1$. Consequently, $p \leq \rho$.

It remains to show that the function $P(z)$ in (11.1) is a polynomial of degree q not exceeding ρ.

To this end, we recall that the order of the canonical product is equal to the convergence exponent. Furthermore, it is not hard to check that the entire function

$$\psi(z) = \frac{f(z)}{\Pi(z)z^m}$$

has order at most ρ, cf. the corollary to Theorem 12 from the book [Levin, 1980, p. 24]. Therefore, ψ satisfies the asymptotic inequality

$$ln\ \psi(z) < r^{\rho+\epsilon}.$$

In addition, the function $\psi(z)$ has no zeros. Thus,

$$P(z) = ln\ \psi(z)$$

is an entire function, and we have asymptotically

$$A_P(r) < r^{\rho+\epsilon}.$$

By the Caratheodory inequality, it follows that asymptotically

$$M_P(r) < r^{\rho+2\epsilon}.$$

By Theorem 3.5.1, the function $P(z)$ is a polynomial of degree at most ρ. The theorem is proved. Q. E. D.

For functions of nonintegral order, the degree of the polynomial in the exponential factor is less than ρ. Therefore, the order of the function coincides with the order of the canonical product, and therefore with the convergence exponent. Thus, in this case

$$p < \rho < p + 1.$$

Example 3.11.2 *The function*

$$f(z) = \frac{\sin \pi \sqrt{z}}{\pi \sqrt{z}}$$

is entire, of order one-half, and has the zeros $1^2, 2^2, \dots$. By Hadamard's theorem, it has the representation

$$f(z) = c \prod_{k=1}^{\infty} \left(1 - \frac{z}{k^2}\right).$$

From the equation $f(0) = 1$ we have $c = 1$. Replacing z by z^2 we obtain

$$\sin \pi z = \pi z \prod_{k=1}^{\infty} \left(1 - \frac{z^2}{k^2}\right)$$

from which we also have

$$\sin \pi z = \pi z \prod_{k=-\infty, k \neq 0}^{\infty} \left(1 - \frac{z}{k}\right) e^{z/k}.$$

The function

$$f(z) = \pi z \prod_{k=1}^{\infty} \left(1 - \frac{z}{n \, ln^2 n}\right)$$

is entire of order one.

Functions of nonintegral order have an infinite set of zeros, and the convergence exponent of the set of zeros is equal to the order of the function, cf. [Levin, 1980, p. 26].

Let us mention the following useful result.

Theorem 3.11.3 *If the order ρ of the entire function $f(z)$ is not an integer, and if Δ_f is the upper density of the set of its zeros, then for $\Delta_f = 0$ the function is of minimal type, for $\Delta_f \neq 0, \infty$, the function is of normal type, while for $\Delta_f = \infty$, it is of maximal type.*

For the proof, see [Levin, 1980, p. 27].

The conclusion of this theorem is no longer valid when the order of the entire function is equal to an integer. Indeed, in this case, the order can be determined by the exponential factor, and the convergence exponent may be less than the order of the function. In particular, the function need not have any zeros at all. But this is not the only peculiarity of functions of integral order. It turns out that in this case the type of the canonical product is not determined just by the upper density of the zeros but depends also on the distribution of the arguments of the zeros. Two examples that illustrate this are the functions

$$sin\,\frac{\pi z}{2} = \frac{\pi z}{2} \prod_{k=1}^{\infty} \left(1 - \frac{z^2}{(2k)^2}\right)$$

and

$$\phi(z) = \frac{1}{\Gamma(z)} = ze^{\gamma z} \prod_{k=1}^{\infty} \left(1 + \frac{z}{k}\right)e^{-\frac{z}{k}} \quad (\gamma = const).$$

For both functions, we have

$$\nu_f(r) \approx r,$$

that is, the density of the zeros is the same. However, $sin(\pi z/2)$ is of order one and normal type, while $\phi(z)$ is of order one and maximal type. This is easily seen for the latter function by using Stirling's formula

$$-ln\,\Gamma(z) = \left(z - \frac{1}{2}\right) ln\,z - z + \frac{1}{2}\,ln\,2\pi + 0\left(\frac{1}{r}\right)$$

from which it follows that

$$ln\,M_\phi(r) \approx r\,ln\,r.$$

For the further development of the theory of the finite-order canonical products, see the excellent books [Levin, 1980] and [Levin, 1996].

3.12 The Borel transform

Entire functions of at most the first order and normal type are called *entire functions of exponential type*, and the exponential type of the function is the quantity

$$\sigma = \overline{\lim}_{r\to\infty} \frac{ln\,M_f(r)}{r}.$$

That is, an entire function $f(z)$ is of exponential type if there exist constants M and α, such that

$$|f(z)| \le Me^{\alpha|z|} \quad (z \in \mathbf{C}).$$

Thus functions of order less than one, or of order one and minimal type, are said to be of exponential type zero. Functions of exponential type occur frequently in the applications, especially in harmonic analysis and in boundary value problems in differential equations. To each entire function of exponential type

$$f(z) = \sum_{k=0}^{\infty} \frac{a_k}{k!} z^k, \tag{12.1}$$

there corresponds the function

$$\phi(z) = \sum_{k=0}^{\infty} \frac{a_k}{z^{k+1}} \tag{12.2}$$

called *the Borel transform of the function* $f(z)$. If σ is the exponential type of function $f(z)$, then

$$\sigma = \overline{\lim} \sqrt[n]{|a_n|}$$

and thus the series (12.2) represents a function holomorphic in the domain $|z| > \sigma$.

The smallest convex domain I_f containing all the singularities of $\phi(z)$ is called *the conjugate diagram of the function* $f(z)$. It clearly lies inside the circle $|z| \leq \sigma$.

Let us point two integral formulas. The first represents $f(z)$ in terms of its Borel transform

$$f(z) = \frac{1}{2\pi i} \int_C e^{zw} \phi(w) dw, \tag{12.3}$$

where C is any contour containing the conjugate diagram. Indeed, the integral on the right is unchanged if we replace the contour C by any contour containing the disc $|z| \leq \sigma$. But for such a contour $\phi(z)$ can be represented by the series (12.2) and integrating termwise we obtain (12.3). The second formula gives an integral representation for $\phi(z)$,

$$\phi(z) = \int_0^{\infty} e^{-zs} f(s) ds, \tag{12.4}$$

where the integration is along the ray $s = te^{-i\theta}, t > 0, \theta \in [0, 2\pi]$.

It is easy to verify that this integral converges absolutely and uniformly in the domain

$$Re \, ze^{-i\theta} > h_f(-\theta) + \epsilon \quad (\epsilon > 0), \tag{12.5}$$

where $h_f(\theta)$ is *the indicator function* of $f(z)$. That is,

$$h_f(\theta) = \overline{\lim}_{r \to \infty} \frac{\ln |f(re^{i\theta})|}{r}.$$

The convergence of the integral follows at once from the asymptotic inequalities

$$|e^{-tze^{-i\theta}}| \leq e^{-[h_f(-\theta)+\epsilon]t}$$

and

$$|f(te^{-i\theta})| \leq e^{[h_f(-\theta)+\frac{\epsilon}{2}]t}.$$

To prove (12.4), it is sufficient to show that the equation is valid in a part of the domain (12.5), for example, for

$$Re \, ze^{-i\theta} > 3\sigma.$$

We show that for these values of z the series (12.1) can simply be substituted into (12.4) and integrated termwise. Indeed, in that half-plane

$$|e^{-tze^{-i\theta}}| \leq e^{-3\sigma t}.$$

However, from the inequality

$$\frac{|a_n|}{n!} < \frac{M_f(r)}{r^n},$$

it follows that the remainder

$$R_n(t) = \sum_{k=n+1}^{\infty} \frac{|a_k|}{k!} t^k$$

of the series (12.1) satisfies the inequality

$$|R_n(t)| \leq \frac{M_f(r)}{1 - \frac{t}{r}} \left(\frac{t}{r}\right)^{n+1}$$

and if we put $r = 2t$, then

$$|R_n(t)| \leq \frac{1}{2^n} e^{2(\sigma+\epsilon)t}.$$

Thus, the series

$$\sum_{k=0}^{\infty} e^{-zs} \frac{a_k}{k!} s^k$$

converges on the whole ray $t > 0, s = te^{-i\theta}$. Integrating termwise, we obtain equation (12.4).

Note that, if $\theta = 0$, then the integral form of the Borel transform is just the Laplace transform.

3.13 Comments to Chapter 3

As it was already mentioned, Chapter 3 contains some fundamental notions and theorems of the theory of analytic functions, in particular, entire functions.

The proof of the Rouché theorem (Theorem 3.1.1) can be also found in many textbooks on the theory of analytic functions, cf. [Greene and Krantz, 2006], [Priestley, 2003], and [Wong, 2008].

In the proof of the Hurwitz theorem (Theorem 3.1.2), we have followed the book [Marden, 1966, Sections 3.2]. The rest of Chapter 3 is based on the book [Levin, 1980].

Chapter 4

Polynomials

In this chapter, we present various inequalities for the zeros of polynomials. The material of this chapter is basic for the next chapters, where the corresponding results are derived for entire functions via limits of sequences of polynomials.

4.1 Some classical theorems

4.1.1 The Cauchy theorem

For an integer $n \geq 2$, let us consider the polynomial

$$P(\lambda) = \sum_{k=0}^{n} c_k \lambda^{n-k} \quad (c_0 = 1) \tag{1.1}$$

with complex in general coefficients. Clearly, P is the characteristic polynomial of the matrix

$$A_n = \begin{pmatrix} -c_1 & -c_2 & \cdots & -c_{n-1} & -c_n \\ 1 & 0 & \cdots & 0 & 0 \\ 0 & 1 & \cdots & 0 & 0 \\ \cdot & \cdot & \cdots & \cdot & \cdot \\ 0 & 0 & \cdots & 1 & 0 \end{pmatrix}. \tag{1.2}$$

So

$$P(\lambda) = det\ (\lambda I - A_n)\ (\lambda \in \mathbf{C}),$$

where I is the unit $n \times n$-matrix, and

$$\lambda_k(A_n) = z_k(P)\ (k = 1, ..., n), \tag{1.3}$$

where $z_k(P)$ are the zeros of P and $\lambda_k(A_n)$ $(k = 1, ..., n)$ are the eigenvalues of A_n with their multiplicities.

Let us recall the classical result of Cauchy.

Theorem 4.1.1 *Let $P(z)$ be defined by (1.1) and $\tilde{r}$ be the unique positive root of the equation*

$$z^n + |c_1|z^{n-1} + ... + |c_{n-1}|z + |c_n| = 0. \tag{1.4}$$

Then all the zeros of $P(z)$ lie in the circle $|z| \le \tilde{r}$.

Proof: This result follows at once from (1.3) and Theorem 1.13.3. Q. E. D.

For another proof of this theorem, see [Milovanović et al., 1994, p. 244].

A. Cohn, cf. [Milovanović et al., 1994, p. 245] has proved that at least one of the zeros $z(P)$ of $P(z)$ satisfies the inequality

$$|z(P)| \ge \tilde{r}(\sqrt[n]{2} - 1),$$

where $\tilde{r}$ is the unique positive root of the equation (1.4).

L. Berwald has proved the following result: let $\tilde{r}$ be the unique positive root of equation (1.4). Then

$$\tilde{r}(\sqrt[n]{2} - 1) \le \frac{1}{n} \sum_{k=1}^{n} |z_k(P)| \le \tilde{r}$$

with equality in the second inequality if and only if $z_1(P) = ... = z_n(P)$, cf. [Milovanović et al., 1994, p. 245].

Note also that from the equality (1.3) and Lemma 1.1.3, we have the following well-known result.

Lemma 4.1.2 *Let P be defined (1.1). Then*

$$\max_{k=1,...,n} |z_k(P)| \leq \max_{k=1,...,n} |c_k| + 1$$

and

$$\max_{k=1,...,n} |z_k(P)| \leq \max\{1, \sum_{k=1}^{n} |c_k|\}.$$

Additional bounds for $max_k|z_k(P)|$ are presented in Section 1.6.

4.1.2 The Viéte and Varing formulas

Again, let $z_k = z_k(P)$ be the roots of the polynomial $P(z)$ defined by (1.1). Put

$$\sigma_1 := \sum_{k=1}^{n} z_k,$$

$$\sigma_2 := \sum_{1 \leq j < k \leq n} z_j z_k,$$

$$\sigma_3 := \sum_{1 \leq i < j < k \leq n} z_i z_j z_k$$

$$\cdots\cdots\cdots\cdots\cdots\cdots$$

$$\sigma_n := \prod_{k=1}^{n} z_k.$$

Let us recall the *Viéte formula*, cf. [Mishina and Proskuryakov, 1965, p. 187]:

$$\sigma_k = (-1)^k c_k.$$

Now put

$$\tilde{s}_m(P) := \sum_{k=1}^{n} z_k^m, \quad m = 1, 2, ..., n.$$

The following formula (the first Varing formula) is valid:

$$\tilde{s}_m(P) = m \sum (-1)^{2t_1 + 3t_2 + ... + (n+1)t_n} \frac{(t_1 + t_2 + ... + t_n - 1)! \sigma_1^{t_1}...\sigma_n^{t_n}}{t_1! t_2!...t_n!}$$

$$(m = 1, 2, ...),$$

where the sum is taken over all combinations of non-negative integer numbers t_k satisfying

$$t_1 + 2t_2 + ... + nt_n = m$$

cf. [Mishina and Proskuryakov, 1965, p. 245]. For additional relations pn $\tilde{s}_m(P)$ and σ_k, see [Milovanović et al., 1994, pp. 53 - 57].

4.1.3   The Eneström-Kakeya theorem

For polynomials with positive coefficients, Eneström and Kakeya have proved the following theorem.

Theorem 4.1.3 *Let $f(z) = a_0 + a_1 z + ... + a_n z^n$ be any real polynomial whose coefficients satisfy*

$$a_n \geq a_{n-1} \geq ... \geq a_1 \geq a_0 > 0.$$

Then f has no zeros for $|z| > 1$.

For the proof, see [Milovanović et al., 1994, p. 272].

Let us also point to the following result [Marden, 1985, p. 137].

Theorem 4.1.4 *All the zeros of the polynomial $a_0 + a_1 z + ... + a_n z^n$ having real positive coefficients a_k, lie in the ring $\rho_1 \leq |z| \leq \rho_2$, where*

$$\rho_1 = min_k(a_k/a_{k+1}) \ and \ \rho_2 = max_k(a_k/a_{k+1})$$

for $k = 0, ..., n - 1$.

4.1.4   The Gauss-Lucas theorem

In this subsection, we consider the location of the critical points of a polynomial, that is, the zeros of its derivative. Let the polynomial $P(z)$ defined by (1.1) have m different zeros $z_1, ..., z_m$, and their multiplicities are $n_1, ..., n_m$, respectively. Then, we have

$$P(z) = \prod_{k=1}^{m} (z - z_k)^{n_k}, \quad \sum_{k=1}^{m} n_k = n.$$

Since

$$P'(z) = P(z)\frac{P'(z)}{P(z)} = P(z)F(z),$$

where

$$F(z) = \frac{d}{dz} ln \, P(z) = \sum_{k=1}^{m} \frac{n_k}{z - z_k},$$

the zeros of P' can be separated into two classes. First, there are the points z_k for which $n_k > 1$ as zeros of P', with multiplicities $n_k - 1$. Their total multiplicity is

$$\sum_{k=1}^{m}(n_k - 1) = n - m.$$

Second, there are else zeros of P', which are the zeros of the logarithmic derivative. Evidently, if, we know the location of the zeros of the polynomial (1.1), then, we know a priori the location of the first class of zeros of P'.

However, the location of the second class of zeros of P', that is, the zeros of function F, remains as a problem. Some physical, geometric, and function-theoretic interpretation of zeros of F can be found in [Marden, 1985]. In a special case, we have the answer to the previous question. Namely, a particular corollary of Rolle's theorem says that any interval I_0 of the real line, which contains all the zeros of a polynomial P, also contains all the zeros of P'. This can be generalized in the sense that I_0 can be replaced by a line-segment in the complex plane. In a general case, we have (see [Marden, 1985, p. 22]):

Theorem 4.1.5 *All the critical points of a non-constant polynomial P lie in the convex hull D of the set of zeros of P. If the zeros of P are not collinear, no critical point of P lies on the boundary ∂D of D unless it is a multiple zero of P.*

This is a well-known result, called the Gauss-Lucas theorem, which was presented in a note of Gauss dated 1836, but, it was stated explicitly and proved by Lucas in 1874. From the Gauss-Lucas theorem follows:

Theorem 4.1.6 *Any circle C that encloses all the zeros of a polynomial $P(z)$ also encloses all the zeros of its derivative $P'(z)$.*

Indeed, if D is the smallest convex polygon enclosing the zeros of $P(z)$, then D lies in C and therefore by the Gauss-Lucas theorem all the zeros of $P'(z)$ being in D, also lie in C. It can be proved that the previous two theorems are equivalent, cf. [Marden, 1985, p. 23].

Let us mention also the following well-known result, [Milovanović et al., 1994, p. 181].

Theorem 4.1.7 (S. Bernstein) *If P and Q are polynomials satisfying $|P(z)| \leq |Q(z)|$, $Q(z) \neq 0$, for any z in the upper half-plane or on the real axis, then we have $|P'(z)| \leq |Q'(z)|$ for those values of z.*

4.1.5 Self-inversive polynomials

Let

$$f(z) = \sum_{k=0}^{n} a_k z^k = a_n \prod_{k=0}^{n} (z - w_k)$$

be a polynomial with zeros $w_1, ..., w_n$. Define the polynomial $f^*(z)$ by

$$f^*(z) = z^n \overline{f}(1/z) = \sum_{k=0}^{n} \overline{a}_k z^{n-k} = \overline{a}_0 \prod_{k=1}^{n} (z - w_k^*)$$

whose zeros $w_k^* = 1/\overline{w}_k$ are the inverses of the zeros w_k with respect to the unit circle $|z| = 1$. Any zero of f on the unit circle is also a zero of f^*. However, if f has no zeros on the unit circle, then f^* has also no zeros on $|z| = 1$. For polynomial f^*, we say that it is the *inverse* of f.

A polynomial $f(z)$ of degree n with zeros $w_1, ..., w_n$ is said to be self-inversive if

$$\{w_1, w_2, ..., w_n\} = \{1/\overline{w}_1, 1/\overline{w}_2, ..., 1/\overline{w}_n\}.$$

In the proof of the next theorem, we followed the book [Milovanović et al., 1994, p. 17].

Theorem 4.1.8 *Let $f(z)$ be a polynomial of degree n. If there exists $u \in \mathbf{C}$, $|u| = 1$, such that $z^n \overline{f}(1/z) = uf(z)$, then polynomial f is self-inversive.*

Proof: Indeed, we have

$$f^*(z) = z^n \overline{f}(1/z) = z^n \overline{f(\overline{z})} = z^n \overline{a_n} \prod_{k=1}^{n} \overline{\left(\frac{1}{z} - w_k\right)}.$$

I.e.

$$f^*(z) = \overline{a}_n \prod_{k=1}^{n} (1 - \overline{w}_k z) = (-1)^n \frac{\overline{a}_n}{w_1 w_2 ... w_n} \prod_{k=1}^{n} (z - w_k).$$

Using the Viéte formula

$$w_1 w_2 ... w_n = (-1)^n \frac{a_n}{a_0},$$

we obtain that

$$f^*(z) = \frac{\overline{a}_n}{a_0} f(z).$$

Since for points on the unit circle $|z| = 1$, we have

$$|f^*(z)| = |z^n \overline{f(1/\overline{z})}| = |f(z)|,$$

we conclude that it must be $|a_n| = |a_0|$. Therefore, the equality $\frac{\overline{a}_n}{a_0} f(z) = f^*(z)$ reduces to $f^*(z) = uf(z)$, where $u = \overline{a}_n/a_0$ and $|u| = 1$. Q. E. D.

The following very interesting result can be found in [Milovanović et al, 1994, p. 17].

Corollary 4.1.9 *If $f(z) = \sum_{k=0}^{n} a_k z^k$, $a_n \neq 0$, then the following statements are equivalent:*
a) f is self-inversive.
b) $\overline{a}_n f(z) = a_0 z^n \overline{f}(1/z)$ for each complex number z.
c) $\overline{a}_k = u a_{n-k}$, $k = 0, 1, ..., n$, where $|u| = 1$.

4.2 Equalities for real and imaginary parts of zeros

Again consider the polynomial $P(\lambda)$ defined by (1.1) and denote

$$\tau(P) := \sum_{k=1}^{n} |c_k|^2 + n - 1.$$

For a $t \in [0, 2\pi)$ put

$$\psi(P, t) := Re\,(c_1 e^{it})^2 - 2Re\,(c_2 e^{2it}) + \tau(P).$$

Theorem 4.2.1 *For any* $t \in [0, 2\pi)$, *we have*

$$\tau(P) - \sum_{k=1}^{n} |z_k(P)|^2 = \psi(P, t) - 2\sum_{k=1}^{n} (Re\,e^{it} z_k(P))^2 \geq 0.$$

Proof: Let A_n be the matrix defined by (1.2). Taking into account (1.3), by Theorem 1.9.1 with $A = ie^{it}A_n$, we have

$$N_2^2(A_n) - \sum_{k=1}^{n} |\lambda_k(A_n)|^2$$

$$= N_2^2(e^{it}A_n + e^{-it}A_n^*)/2 - \frac{1}{2}\sum_{k=1}^{n} |e^{it}\lambda_k(A_n) + e^{-it}\overline{\lambda}_k(A_n)|^2, \qquad (2.1)$$

where $N_2(A)$ is the Hilbert-Schmidt norm of an operator A. Moreover,

$$e^{it}A_n + A_n^* e^{-it} = \begin{pmatrix} -c_1 e^{it} - e^{-it}\overline{c}_1 & e^{-it} - c_2 e^{it} & \ldots & -c_{n-1}e^{it} & -c_n e^{it} \\ e^{it} - \overline{c}_2 e^{-it} & 0 & \ldots & 0 & 0 \\ -\overline{c}_3 e^{-it} & e^{it} & \ldots & 0 & 0 \\ \cdot & \cdot & \ldots & \cdot & \cdot \\ -\overline{c}_n e^{-it} & 0 & \ldots & e^{it} & 0 \end{pmatrix}.$$

Simple calculations show that

$$N_2^2(A_n) = \tau(P)$$

and

$$N_2^2(e^{it}A + e^{-it}A^*)/2 = |c_1 e^{it} + \overline{c}_1 e^{-it}|^2/2$$

$$+|1 - c_2 e^{2it}|^2 + \sum_{k=3}^{n} |c_k|^2 + n - 2.$$

So

$$N_2^2(e^{it}A + e^{-it}A^*)/2 = |c_1|^2 + (c_1^2 e^{2it} + \overline{c}_1^2 e^{-2it})/2 + 1 + |c_2|^2$$

$$-(c_2 e^{2it} + \bar{c}_2 e^{-2it})$$

$$+ \sum_{k=3}^{n} |c_k|^2 + n - 2 = \psi(P, t).$$

Hence (1.3) and (2.1) imply the required result. Q. E. D.

Note that
$$\psi(P, \pi/2) = -Re\ c_1^2 + 2Re\ c_2 + \tau(P)$$
and
$$\psi(P, 0) = Re\ c_1^2 - 2Re\ c_2 + \tau(P).$$

Now the previous theorem yields

Corollary 4.2.2 *The relations*

$$\tau(P) - \sum_{k=1}^{n} |z_k(P)|^2 = \psi(P, \pi/2) - 2\sum_{k=1}^{n} (Im\ z_k(P))^2$$

$$= \psi(P, 0) - 2\sum_{k=1}^{n} (Re\ z_k(P))^2 \geq 0$$

are valid.

Hence, it follows

Corollary 4.2.3 *The inequalities*

$$\sum_{k=1}^{n} |z_k(P)|^2 \leq \tau(P), \tag{2.2}$$

$$2\sum_{k=1}^{n} (Im\ z_k(P))^2 \leq \psi(P, \pi/2)$$

and

$$2\sum_{k=1}^{n} (Re\ z_k(P))^2 \leq \psi(P, 0)$$

hold.

Take into account that
$$Re\ z^2 = (Re\ z)^2 - (Im\ z)^2$$
and
$$Im\ z^2 = 2Re\ z\ Im\ z \quad (z \in \mathbf{C}).$$

In Section 3.4, we will prove that

$$\sum_{k=1}^{n} z_k^2(P) = c_1^2 - 2c_2.$$ (2.3)

Hence, it follows

$$\sum_{k=1}^{n} (Re\ z_k(P))^2 - (Im\ z_k(P))^2 = Re\ (c_1^2 - 2c_2)$$

and

$$2\sum_{k=1}^{n} (Re\ z_k(P))\ (Im\ z_k(P)) = Im\ (c_1^2 - 2c_2).$$

Furthermore, take into account that

$$|z|^2 + Im\ z^2 = (Re\ z + Im\ z)^2$$

and

$$|z|^2 - Im\ z^2 = (Re\ z - Im\ z)^2 \quad (z \in \mathbf{C}).$$

Therefore, relations (2.3) and (2.2) imply

Corollary 4.2.4 *We have*

$$\sum_{k=1}^{n} (Re\ z_k(P) + Im\ z_k(P))^2 \le \tau(P) + Im\ (c_1^2 - 2c_2)$$

and

$$\sum_{k=1}^{n} (Re\ z_k(P) - Im\ z_k(P))^2 \le \tau(P) - Im\ (c_1^2 - 2c_2).$$

Let

$$|P(e^{it})|_{L^2} = [\frac{1}{2\pi} \int_0^{2\pi} |P(e^{it})|^2 dt]^{1/2}.$$

Then because of the Parseval equality

$$\sum_{k=0}^{n} |c_k|^2 = |P(e^{it})|_{L^2}^2 \le \max_{|z|=1} |P(z)|^2.$$ (2.4)

Therefore,

$$\tau(P) = |P(e^{it})|_{L^2}^2 + n - 2 \le \max_{|z|=1} |P(z)|^2 + n - 2.$$

Moreover,

$$\psi(P, \pi/2) = -Re\ c_1^2 + 2Re\ c_2 + |P(e^{it})|_{L^2}^2 + n - 2$$

$$\leq \max_{|z|=1} |P(z)|^2 \ - Re\ c_1^2 + 2Re\ c_2 + n - 2$$

and

$$\psi(P,0) = Re\ c_1^2 - 2Re\ c_2 + n - 2 + |P(e^{it})|_{L^2}^2$$
$$\leq Re\ c_1^2 + 2Re\ c_2 + n - 2 + \max_{|z|=1} |P(z)|^2.$$

4.3   Partial sums of zeros and the counting function

Again consider the polynomial $P(\lambda) = \lambda^n + c_1 \lambda^{n-1} + ... + c_n$. Enumerate its zeros in the descending order: $|z_k(P)| \geq |z_{k+1}(P)|$ $(k = 1, ..., n - 1)$. Denote

$$\theta_P := [\sum_{k=1}^{n} |c_k|^2]^{1/2}.$$

Theorem 4.3.1 *The zeros of the polynomial P defined by (1.1) satisfy the inequalities*

$$\sum_{k=1}^{j} |z_k(P)| < \theta_P + j \quad (j = 1, ..., n - 1), \tag{3.1}$$

and

$$\sum_{k=1}^{n} |z_k(P)| < \theta_P + n - 1. \tag{3.2}$$

Proof: Again use the matrix A_n defined by (1.2). According to Lemma 1.1.1

$$\sum_{k=1}^{j} |\lambda_k(A_n)| = \sum_{k=1}^{j} |\lambda_k(A_n^*)| < \sum_{k=1}^{j} s_k(A_n^*) \quad (j = 1, ..., n), \tag{3.3}$$

where $s_k(A_n^*) = \sqrt{\lambda_k(AA^*)}, k = 1, 2, ...$ are the singular numbers of A_n^* ordered in the decreasing way. But $A_n = M + C$, where

$$C = \begin{pmatrix} -c_1 & -c_2 & -c_3 & ... & -c_n \\ 0 & 0 & 0 & ... & 0 \\ . & . & . & ... & . \\ 0 & 0 & 0 & ... & 0 \end{pmatrix}$$

and

$$M = \begin{pmatrix} 0 & 0 & 0 & ... & 0 & 0 \\ 1 & 0 & 0 & ... & 0 & 0 \\ 0 & 1 & 0 & ... & 0 & 0 \\ . & . & . & ... & . & . \\ 0 & 0 & 0 & ... & 1 & 0 \end{pmatrix}.$$

Clearly,

$$CC^* = \begin{pmatrix} \theta_P^2 & 0 & 0 & \dots & 0 \\ 0 & 0 & 0 & \dots & 0 \\ 0 & 0 & 0 & \dots & 0 \\ \cdot & \cdot & \cdot & \dots & \cdot \\ 0 & 0 & 0 & \dots & 0 \end{pmatrix}$$

and

$$MM^* = \begin{pmatrix} 0 & 0 & 0 & \dots & 0 & 0 \\ 0 & 1 & 0 & \dots & 0 & 0 \\ 0 & 0 & 1 & \dots & 0 & 0 \\ \cdot & \cdot & \cdot & \dots & \cdot & \cdot \\ 0 & 0 & 0 & \dots & 0 & 1 \end{pmatrix}.$$

Since the diagonal entries of diagonal matrices are the eigenvalues, we can write out

$$s_1(C^*) = \theta_P, \ s_k(C^*) = 0 \ (k = 2, ..., n).$$

In addition,

$$s_k(M^*) = 1 \ (k = 1, ..., n-1), \ s_n(M^*) = 0.$$

But according to Lemma 1.1.2

$$\sum_{k=1}^{j} s_k(A_n^*) = \sum_{k=1}^{j} s_k(M^* + C^*) \le \sum_{k=1}^{j} s_k(M^*) + \sum_{k=1}^{j} s_k(C^*).$$

So

$$\sum_{k=1}^{j} s_k(A_n^*) \le \theta_P + j \ (j = 1, ..., n-1)$$

and

$$\sum_{k=1}^{n} s_k(A_n^*) \le \theta_P + n - 1.$$

Now (1.3) and (3.3) yield (3.1) and (3.2).

The theorem is proved. Q. E. D.

Since the zeros of P are enumerated in the decreasing order, from Theorem 4.3.1 it follows that

$$j|z_j(P)| < \theta(P) + j \ (j = 1, ..., n).$$

Therefore, P has in the circle

$$\{z \in \mathbf{C} : |z| \le 1 + \frac{\theta(P)}{j}\} \ (j \le n)$$

no more than $n - j + 1$ zeros. Recall that the counting function $\nu_P(r)$ denotes the number of the zeros of P in $|z| \le r$. We thus get

Corollary 4.3.2 *The counting function $\nu_P(r)$ of the zeros of P satisfies the inequality*

$$\nu_P(r) \le n - j + 1 \quad (j = 1, 2, ..., n)$$

for any

$$r \le 1 + \frac{\theta(P)}{j}.$$

4.4   Sums of powers of zeros

In this section, we establish relations for sums of powers of zeros that supplement the Varing formula.

For any $m \le n$, due to Lemma 1.3.1,

$$Trace \, A_m^m = Trace \, A_n^m,$$

where

$$A_m = \begin{pmatrix} -c_1 & -c_2 & \cdots & -c_{m-1} & -c_m \\ 1 & 0 & \cdots & 0 & 0 \\ 0 & 1 & \cdots & 0 & 0 \\ \cdot & \cdot & \cdots & \cdot & \cdot \\ 0 & 0 & \cdots & 1 & 0 \end{pmatrix}.$$

We thus get

Theorem 4.4.1 *Let $P(\lambda)$ be defined by (1.1). Then for any $m \le n$ the equality*

$$\sum_{k=1}^{n} z_k^m(P) = Trace \, A_m^m$$

is valid.

In particular, from the previous theorem, it follows that

$$\sum_{k=1}^{n} z_k^2(P) = c_1^2 - 2c_2 \quad (n \ge 2)$$

and

$$\sum_{k=1}^{n} z_k^3(P) = -c_1^3 + 3c_1 c_2 - 3c_3 \quad (n \ge 3).$$

4.5   The Ostrowski type inequalities

Consider the polynomial

$$f(\lambda) = \sum_{k=0}^{n} a_k \lambda^k \quad (a_n \neq 0, a_0 \neq 0) \tag{5.1}$$

with complex coefficients. Again $z_k(f), k = 1, 2, ..., n$ are the zeros of f with multiplicities taken into account, and

$$\alpha_k(f) = Re \ z_k(f), \quad \omega_k(f) = Im \ z_k(f).$$

A. M. Ostrowski has established the following inequalities

$$\sum_{k=1}^{n} \frac{1}{\omega_k^2(f)} \geq 8Re \ \frac{a_1^2 - 2a_2a_0}{a_0^2}$$

and

$$\sum_{k=1}^{n} \frac{1}{\alpha_k^2(f)} \geq 8Re \ \frac{2a_2a_0 - a_1^2}{a_0^2}$$

(see [Milovanović et al., 1994, p. 113]). In the next section, we prove the following result.

Theorem 4.5.1 *The inequalities*

$$\sum_{k=1}^{n} \frac{1}{\alpha_k^2(f)} \geq Re \ \frac{a_1^2 - 2a_2a_0}{a_0^2} \tag{5.2}$$

and

$$\sum_{k=1}^{n} \frac{1}{\omega_k^2(f)} \geq Re \ \frac{2a_2a_0 - a_1^2}{a_0^2} \tag{5.3}$$

are true.

In some cases Theorem 4.5.1, improves Ostrowski's type inequalities. Indeed, consider the polynomial

$$f(\lambda) = 1 - 4\lambda + 4\lambda^2 = (1 - 2\lambda)^2.$$

Ostrowski's inequalities give us the relation $8 \geq -64$. At the same time, Theorem 4.5.1 yields the equality $8 = 8$. Combining Ostrowski's inequalities with Theorem 4.5.1, we can write out the inequalities

$$\sum_{k=1}^{n} \frac{1}{\alpha_k^2(f)} \geq \max \Big\{ Re \ \frac{a_1^2 - 2a_2a_0}{a_0^2}, 8Re \ \frac{2a_2a_0 - a_1^2}{a_0^2} \Big\}$$

and

$$\sum_{k=1}^{n} \frac{1}{\omega_k^2(f)} \geq \max \Big\{ Re \ \frac{2a_2a_0 - a_1^2}{a_0^2}, 8Re \ \frac{a_1^2 - 2a_2a_0}{a_0^2} \Big\}.$$

4.6   Proof of Theorem 4.5.1

For simplicity, put

$$\alpha_k(f) = \alpha_k \text{ and } \omega_k(f) = \omega_k.$$

Let $P(\lambda)$ be defined by (1.1). Clearly,

$$a_0 P(\lambda) = \lambda^n f(1/\lambda), \quad c_k = a_k/a_0.$$

So $z_k(P) = 1/z_k(f)$. P is the characteristic polynomial of the matrix A_n defined by (1.2) and (1.3) holds. Hence,

$$\sum_{k=1}^{n} z_k^2(P) = Trace\, A_n^2 = c_1^2 - 2c_2.$$

Thus,

$$\sum_{k=1}^{n} \frac{1}{z_k^2(f)} = \sum_{k=1}^{n} z_k^2(P) = \frac{a_1^2}{a_0^2} - \frac{2a_2}{a_0}. \tag{6.1}$$

Moreover

$$\sum_{k=1}^{n} \frac{1}{z_k^2(f)} = \sum_{k=1}^{n} \frac{\bar{z}_k^2(f)}{|z_k(f)|^4} = \sum_{k=1}^{n} \frac{\alpha_k^2 - \omega_k^2 - 2i\alpha_k\omega_k}{(\alpha_k^2 + \omega_k^2)^2}.$$

This and (6.1) imply

$$Re\,\left(\frac{a_1^2}{a_0^2} - \frac{2a_2}{a_0}\right) = \sum_{k=1}^{n} \frac{\alpha_k^2 - \omega_k^2}{(\alpha_k^2 + \omega_k^2)^2} \tag{6.2}$$

and

$$-\sum_{k=1}^{n} \frac{2\alpha_k\omega_k}{(\alpha_k^2 + \omega_k^2)^2} = Im\,\left(\frac{a_1^2}{a_0^2} - \frac{2a_2}{a_0}\right).$$

From (6.2), it follows

$$Re\,\left(\frac{a_1^2}{a_0^2} - \frac{2a_2}{a_0}\right) \leq \sum_{k=1}^{n} \frac{\alpha_k^2}{(\alpha_k^2 + \omega_k^2)^2} \leq \sum_{k=1}^{n} \frac{1}{\alpha_k^2}.$$

So (5.2) is proved. In addition, (6.2) implies

$$-Re\,\left(\frac{a_1^2}{a_0^2} - \frac{2a_2}{a_0}\right) = \sum_{k=1}^{n} \frac{\omega_k^2 - \alpha_k^2}{(\alpha_k^2 + \omega_k^2)^2} \leq \sum_{k=1}^{n} \frac{1}{\omega_k^2}.$$

So (5.3) is proved. As claimed. Q. E. D.

4.7 Higher powers of real parts of zeros

In this section, we particularly generalize the results proved in the previous section. Recall that matrix A_n is introduced by (1.2) and P is defined by (1.1). Put $\alpha_k(P) = Re \, z_k(P)$.

Lemma 4.7.1 *For any positive integer m, such that $2^m \leq n$, we have*

$$\sum_{k=1}^n \alpha_k^{2^m}(P) \geq Re \, Trace \, A_{2^m}^{2^m}.$$

Proof: Let $z = a + ib$, $a, b \in \mathbf{R}$. Let us check that

$$Re \, z^{2^m} \leq a^{2^m}. \tag{7.1}$$

Indeed,

$$z^{2^m} = |z|^{2^m} e^{2^m i\phi} = (a^2 + b^2)^{2^{m-1}} e^{2^m i\phi},$$

where ϕ is the argument of z. Hence

$$Re \, z^{2^m} = (a^2 + b^2)^{2^{m-1}} cos \, (2^m \phi).$$

So we should prove that

$$(a^2 + b^2)^{2^{m-1}} cos \, (2^m \phi) \leq (a^2 + b^2)^{2^{m-1}} cos^{2^m}(\phi).$$

Or that

$$cos \, (2^m \phi) \leq cos^{2^m}(\phi). \tag{7.2}$$

Take $m = 1$. Then

$$cos \, (2\phi) = cos^2(\phi) - sin^2(\phi) \leq cos^2(\phi).$$

So in this case (7.2) is valid. Assuming that (7.2) is correct for any m, let us prove that

$$cos \, (2^{m+1}\phi) \leq cos^{2^{m+1}}(\phi). \tag{7.3}$$

But

$$cos \, (2^{m+1}\phi) = cos^2(2^m \phi) - sin^2(2^m \phi) \leq (cos^{2^m} \phi)^2.$$

This proves inequality (7.3) and therefore inequality (7.1). Furthermore, we have by Theorem 1.3.1 and (1.3),

$$\sum_{k=1}^n z_k^{2^m}(P) = Trace \, A_{2^m}^{2^m}.$$

Hence, because of (7.1), we get the required result. Q. E. D.

4.8   The Gerschgorin type sets for polynomials

Let us apply theorems from Sections 1.11 - 1.13 to the $n \times n$-matrix A_n defined by (1.2) and thus obtain some results on the location of the zeros of the polynomial given by $P(z) = z^n + c_1 z^{n-1} + \ldots + c_n$, since P is the characteristic polynomial of A_n.

From Theorem 1.11.1 thus follows our next result.

Corollary 4.8.1 *The zeros of the polynomial P lie in the union of the disks*

$$|z| \leq 1, |z + a_1| \leq \sum_{k=2}^{n} |c_k|.$$

From Theorem 1.13.1, we get

Corollary 4.8.2 *If*

$$|c_1| > 1 + \sum_{k=2}^{n} |c_k|,$$

then one and only one zero of P lies in the disk

$$|z + c_1| \leq \sum_{k=2}^{n} |c_k|.$$

Moreover, Theorem 1.12.1 implies

Corollary 4.8.3 *Let $c_k \neq 0$ for at least one $k = 2, \ldots, n$. Then any zero $z(P)$ of P either satisfies the inequality $|z(P)| \leq 1$, or*

$$|z^2(P) + c_1 z(P) + c_2| = |z(P) - r_1||z(P) - r_2| \leq p_3(P),$$

where

$$p_3(P) = \sum_{k=3}^{n} |c_k|$$

and

$$r_{1,2} = -\frac{c_1}{2} \pm \sqrt{\frac{c_1^2}{4} - c_2}.$$

4.9   Perturbations of polynomials

For a finite integer $n \geq 2$, let us consider the polynomials

$$P(\lambda) = \sum_{k=0}^{n} c_k \lambda^{n-k} \text{ and } Q(\lambda) = \sum_{k=0}^{n} b_k \lambda^{n-k} \quad (c_0 = b_0 = 1; \ \lambda \in \mathbf{C}). \quad (9.1)$$

Denote

$$var_P(Q) := \max_j \min_k |z_k(P) - z_j(Q)|.$$

We will call $var_P(Q)$ the variation of the zeros of Q with respect to the zeros of P.

Recall that

$$\tau(P) = n - 1 + \sum_{k=1}^{n} |c_k|^2.$$

Put

$$g_0(P) = \left[\tau(P) - \sum_{k=1}^{n} |z_k(P)|^2\right]^{1/2}$$

and

$$q_n := \left[\sum_{k=1}^{n} |c_k - b_k|^2\right]^{1/2}.$$

Thanks to Theorem 4.4.1

$$\sum_{k=1}^{n} z_k^2(P) = c_1^2 - 2c_2.$$

Since

$$\sum_{k=1}^{n} |z_k(P)|^2 \geq |\sum_{k=1}^{n} z_k^2(P)|,$$

we get

$$g_0(P) \leq g_2(P) \leq \sqrt{\tau(P)}, \qquad (9.2)$$

where

$$g_2(P) := [\tau(P) - |c_1^2 - 2c_2|]^{1/2}.$$

Everywhere below one can replace $g_0(P)$ by the simple calculated quantity $g_2(P)$.

Theorem 4.9.1 *For any zero $z(Q)$ of Q, there is a zero $z(P)$ of P, such that*

$$|z(P) - z(Q)| \leq y(P, q_n),$$

where $y(P, q_n)$ is the unique positive (simple) root of the algebraic equation

$$1 = \frac{q_n}{y}\left[1 + \frac{1}{n-1}\left(1 + \frac{g_0^2(P)}{y^2}\right)\right]^{(n-1)/2}. \qquad (9.3)$$

That is, $var_P(Q) \leq y(P, q_n)$.

This theorem is proved in the next section.
    The previous theorem implies

Corollary 4.9.2 *All the zeros of Q lie in the union of the sets*

$$\{\lambda \in \mathbf{C} : \frac{q_n}{|z_k(P) - \lambda|}[1 + \frac{1}{n-1}(1 + \frac{g_0^2(P)}{|z_k(P) - \lambda|^2})]^{(n-1)/2} \geq 1\}$$

$$(k = 1, ..., n).$$

Note that the sets pointed in this corollary include the case $z_k(P) = z_k(Q)$ for some k.

Put in (9.3)

$$x = \frac{y}{g_0(P)}.$$

Then, we have the equation

$$1 = \frac{q_n}{x g_0(P)}[1 + \frac{1}{n-1}(1 + \frac{1}{x^2})]^{(n-1)/2}.$$

Denote

$$v_n := [1 + \frac{2}{n-1}]^{(n-1)/2} = [\frac{n+1}{n-1}]^{(n-1)/2}$$

and

$$\delta(P, q_n) = \begin{cases} q_n v_n & \text{if } q_n v_n \geq g_0(P), \\ g_0^{1-1/n}(P)[q_n v_n]^{1/n} & \text{if } q_n v_n \leq g_0(P). \end{cases}$$

Because of Lemma 1.6.1, we easily get the inequality

$$y(P, q_n) \leq \delta(P, q_n).$$

Now Theorem 4.9.1 yields

Corollary 4.9.3 *The inequality* $var_P(Q) \leq \delta(P, q_n)$ *is true.*

Recall that

$$|P(e^{it})|_{L^2} = [\frac{1}{2\pi}\int_0^{2\pi}|P(e^{it})|^2 dt]^{1/2}$$

and due to the Parseval equality,

$$\sum_{k=0}^{n}|c_k|^2 = |P(e^{it})|_{L^2}^2$$

and

$$q_n^2 = \sum_{k=1}^{n}|c_k - b_k|^2 = |P(e^{it}) - Q(e^{it})|_{L^2}^2.$$

Thus, according to (9.2),

$$g_0(P) \leq [n - 2 + |P(e^{it})|_{L^2}^2 - |c_1^2 - 2c_2|]^{1/2}$$

$$\leq [n - 2 + \max_{|z|=1}|P(z)|^2 - |c_1^2 - 2c_2|]^{1/2}.$$

In addition,

$$q_n \leq \max_{|z|=1}|P(z) - Q(z)|.$$

4.10 Proof of Theorem 4.9.1

In a Euclidean space $\mathbf{C}^n$ with the Euclidean norm $\|.\|$, introduce the operators (matrices) A_n defined by (1.2) and

$$B_n = \begin{pmatrix} -b_1 & -b_2 & \cdots & -b_{n-1} & -b_n \\ 1 & 0 & \cdots & 0 & 0 \\ 0 & 1 & \cdots & 0 & 0 \\ \cdot & \cdot & \cdots & \cdot & \cdot \\ 0 & 0 & \cdots & 1 & 0 \end{pmatrix}.$$

So

$$P(\lambda) = det\,(\lambda I - A_n) \text{ and } Q(\lambda) = det\,(I\lambda - B_n) \ (\lambda \in \mathbf{C}),$$

where I is the unit matrix. Hence,

$$\lambda_k(A_n) = z_k(P), \lambda_k(B_n) = z_k(Q), \tag{10.1}$$

where $\lambda_k(.)$ $(k = 1, ..., n)$ are the eigenvalues with their multiplicities. Moreover,

$$q_n = \|A_n - B_n\|.$$

Recall that for an $n \times n$-matrix A,

$$g(A) = (N_2^2(A) - \sum_{k=1}^n |\lambda_k(A)|^2)^{1/2}$$

(see Section 1.5). Because of the Lemma 1.7.3, for any $\lambda_k(B_n)$, there is a $\lambda_k(A_n)$, such that

$$|\lambda_k(A_n) - \lambda_k(B_n)| \le y(q_n), \tag{10.2}$$

where $y(q_n)$ is the unique positive (simple) zero of equation

$$\frac{q_n}{y}\left[1 + \frac{1}{n-1}\left(1 + \frac{g^2(A_n)}{y^2}\right)\right]^{(n-1)/2} = 1.$$

Take into account that

$$N_2^2(A_n) = \sum_{k=1}^n |c_k|^2 + n - 1$$

and thus

$$g(A_n) = g_0(P). \tag{10.3}$$

This equality, (10.1) and (10.2) prove the required result. Q. E. D.

4.11   Preservation of multiplicities

Again consider the polynomials P and Q defined by (9.1). Let $z_m(P)$ be a fixed zero of P of a multiplicity μ_m. Denote

$$\beta_m = \min_{z_k(P) \neq z_m(P)} |z_k(P) - z_m(P)|/2.$$

Recall that

$$\Omega(b, r) = \{b \in \mathbf{C} : |z - b| \leq r\}$$

for a $b \in \mathbf{C}$ and a positive r, and $g_0(P)$ and q_n are defined in the Section 4.9.

Theorem 4.11.1 *Let a zero $z_m(P)$ of P have a multiplicity μ_m. In addition, for a positive $r \leq \beta_m$, let*

$$\frac{q_n}{r}\left[1 + \frac{1}{n-1}\left(1 + \frac{g_0^2(P)}{r^2}\right)\right]^{(n-1)/2} < 1. \tag{11.1}$$

Then Q has in disk $\Omega(z_m(P), r)$ zeros whose total multiplicity is also equal to μ_m.

Proof: Again, we use the matrices A_n and B_n introduced in Sections 4.1 and 4.10 . Besides, (10.1) holds. As is it was shown in Section 4.10, $g(A_n) = g_0(P)$ and $q_n = \|A_n - B_n\|$. Now the required result is due to Corollary 1.8.3.
   Q. E. D.

Since the left-hand part of (11.1) decreases in $r > 0$, from the previous theorem, we easily have

Corollary 4.11.2 *Let a zero $z_m(P)$ of P have a multiplicity μ_m. In addition, let the unique positive zero $y(P, q_n)$ of the equation*

$$y^n = q_n y \left[1 + \frac{1}{n-1}\left(y^2 + g_0^2(P)\right)\right]^{(n-1)/2} \tag{11.2}$$

satisfy the inequality $y(P, q_n) \leq \beta_m$. Then Q has in disk $\Omega(z_m(P), y(P, q_n))$ zeros whose total multiplicity is also equal to μ_m.

Let $\delta(P, q_n)$ be defined as in Section 4.9. As it was above shown,

$$y(P, q_n) \leq \delta(P, q_n).$$

Now the previous corollary yields

Corollary 4.11.3 *Let a zero $z_m(P)$ of P have a multiplicity μ_m. In addition, let*

$$\delta(P, q_n) \leq \beta_m.$$

Then Q has in disk $\Omega(z_m(P), \delta(P, q_n))$ zeros whose total multiplicity is also equal to μ_m.

4.12 Distances between zeros and critical points

Let

$$P(z) = \sum_{k=0}^{n} c_k z^{n-k} \quad (c_0 = 1).$$

Clearly,

$$P'(z) = \sum_{k=0}^{n-1} (n-k)c_k z^{n-k-1}.$$

Take the polynomial

$$Q(z) = \frac{z}{n} P'(z) = \frac{1}{n} \sum_{k=0}^{n-1} (n-k)c_k z^{n-k}.$$

Then the nonzero roots of Q and of P' coincide with their multiplicities. If P' has at the origin a zero of a multiplicity n_0, then Q has at the origin the zero of the multiplicity $n_0 + 1$.

In the considered case $q_n = q_P$, where

$$q_P = \Big[\sum_{k=1}^{n} \Big(\frac{n-k}{n} - 1\Big)^2 |c_k|^2 \Big]^{1/2} = \frac{1}{n} \Big[\sum_{k=1}^{n} k^2 |c_k|^2 \Big]^{1/2}.$$

Define $g_0(P)$ and v_n as in Section 4.9. Thanks to Corollary 4.9.3, we get

Corollary 4.12.1 *For any nonzero root $z(P')$ of P', there is a root $z(P)$ of P, such that*

$$|z(P) - z(P')| \le \delta(P, q_P),$$

where

$$\delta(P, q_P) = \begin{cases} q_P v_n & \text{if } q_P v_n \ge g_0(P) \\ g_0^{1-1/n}(P)[q_P v_n]^{1/n} & \text{if } q_P v_n \le g_0(P) \end{cases}.$$

Moreover, thanks to Corollary 4.9.2, *all the nonzero roots of P' lie in the union of the sets*

$$\{\lambda \in \mathbf{C}: \frac{q_P}{|z_k(P) - \lambda|}\Big[1 + \frac{1}{n-1}\Big(1 + \frac{g_2^2(P)}{|z_k(P) - \lambda|^2}\Big)\Big]^{(n-1)/2} \ge 1\} \quad (k = 1, ..., n).$$

This result supplements the Gauss-Lukas theorem.

Clearly, the results of Section 4.9 enable us to establish location of the zeros of P if we know the zeros of P'.

4.13   Partial sums of imaginary parts of zeros

Again consider the polynomial

$$P(\lambda) = \sum_{k=0}^{n} c_k \lambda^{n-k} \quad (n \geq 3).$$

Let the zeros $z_k(P)$ of P with their multiplicities be ordered in the following way:

$$|Im\ z_k(P)| \geq |Im\ z_{k+1}(P)|.$$

Put

$$\eta(P) := \left[\sum_{k=3}^{n} |c_k|^2 \right]^{1/2},$$

$$v_1 := \frac{1}{2}(|\sigma_1\ Im\ c_1 + \sqrt{|Im\ c_1|^2 + |1 + c_2|^2}\ |),$$

$$v_2 := \frac{1}{2}(|\sigma_1\ Im\ c_1 - \sqrt{|Im\ c_1|^2 + |1 + c_2|^2}\ |), \quad v_k = 0 \ (k \geq 3),$$

where $\sigma_1 = +1$ if $Im\ c_1 \geq 0$ and $\sigma_1 = -1$ if $Im\ c_1 < 0$.

Lemma 4.13.1 *The zeros of P satisfy the inequalities*

$$\sum_{k=1}^{j} |Im\ z_k(P)| \leq j + (1 + j)\eta(P)/2 + \sum_{k=1}^{j} v_k \quad (j = 1, ..., n)$$

Proof: Recall that

$$z_k(P) = \lambda_k(A_n), \tag{13.1}$$

where $\lambda_k(A)$ $(k = 1, ..., n)$ mean the eigenvalues with their multiplicities of the $n \times n$-matrix A_n defined by (1.2). Denote $Im\ A_n = (A_n - A_n^*)/2i$. The asterisk means the adjointness. Because of Lemma 1.1.4

$$\sum_{k=1}^{j} |Im\ \lambda_k(A_n)| \leq \sum_{k=1}^{j} s_k(Im\ A_n) \quad (j = 1, ..., n). \tag{13.2}$$

Obviously, $Im\ A_n = B + (C - C^*)/2i + (D - D^*)/2i$, where B, C and D are the following $n \times n$-matrices:

$$B = \begin{pmatrix} -Im\ c_1 & i(c_2 + 1)/2 & 0 & ... & 0 & 0 \\ -i(\bar{c}_2 + 1)/2 & 0 & 0 & ... & 0 & 0 \\ 0 & 0 & 0 & ... & 0 & 0 \\ \cdot & \cdot & \cdot & ... & \cdot & \cdot \\ 0 & 0 & 0 & ... & 0 & 0 \end{pmatrix},$$

$$C = \begin{pmatrix} 0 & 0 & -c_3 & ... & -c_n \\ 0 & 0 & 0 & ... & 0 \\ . & . & . & ... & . \\ 0 & 0 & 0 & ... & 0 \end{pmatrix} \quad \text{and } D = \begin{pmatrix} 0 & 0 & 0 & ... & 0 & 0 \\ 0 & 0 & 0 & ... & 0 & 0 \\ 0 & 1 & 0 & ... & 0 & 0 \\ . & . & . & ... & . & . \\ 0 & 0 & 0 & ... & 1 & 0 \end{pmatrix}.$$

Obviously,

$$CC^* = \begin{pmatrix} \eta_\psi^2(P) & 0 & 0 & ... & 0 \\ 0 & 0 & 0 & ... & 0 \\ 0 & 0 & 0 & ... & 0 \\ . & . & . & ... & . \\ 0 & 0 & 0 & ... & 0 \end{pmatrix}$$

and

$$DD^* = \begin{pmatrix} 0 & 0 & 0 & ... & 0 & 0 \\ 0 & 0 & 0 & ... & 0 & 0 \\ 0 & 0 & 1 & ... & 0 & 0 \\ . & . & . & ... & . & . \\ 0 & 0 & 0 & ... & 0 & 1 \end{pmatrix}.$$

Since the diagonal entries of diagonal matrices are the eigenvalues, we can write out $s_1(C^*) = \eta_\psi(P)$, $s_k(C^*) = 0$ ($k = 2, ..., n$). Besides, $s_k(C) \leq s_1(C^*) = \|C^*\| = \eta_\psi(P)$ ($k = 1, ..., n$), and $s_k(B) = v_k$. In addition,

$$s_k(D) = 1 \ (k = 1, ..., n-2), \ s_{n-1}(D) = s_n(D) = 0,$$

and $s_k(D) = s_k(D^*)$. Take into account that

$$\sum_{k=1}^{j} s_k(Im \ A_n) = \sum_{k=1}^{j} s_k(B + (C - C^*)/2i + (D - D^*)/2i)$$

$$\leq \sum_{k=1}^{j} s_k(B) + s_k(C)/2 + s_k(C^*)/2 + s_k(D)$$

(see Lemma 1.1.2). So

$$\sum_{k=1}^{j} s_k(Im \ A_n) \leq \eta_\psi/2 + j\eta_\psi/2 + j + \sum_{k=1}^{j} v_k \ (j = 1, ..., n).$$

This proves the lemma. Q. E. D.

Furthermore, put

$$\eta_1(P) = |Im \ c_1| + \frac{1}{2}\Big(|1 + c_2| + \sum_{k=3}^{n} |c_k|\Big).$$

Take into account that

$$|Im \ z_j(P)| \leq \max_k |\lambda_k(Im \ A_n)| = r_s(Im \ A_n) \ (j = 1, ..., n).$$

Because of Lemma 1.1.3 this implies

Lemma 4.13.2 *The inequality*

$$\max_{j} |Im \ z_j(P)| \leq \max \left\{ \eta_1(P), \frac{1}{2}(|1+c_2|+1), 1 + \max_{k=3,...,n} \frac{|c_k|}{2} \right\}$$

is true.

4.14   Functions holomorphic on a circle

Consider the function

$$f(\lambda) = \sum_{k=0}^{\infty} c_k \lambda^k \quad (\lambda \in \mathbf{C}, \ c_0 = 1) \tag{14.1}$$

holomorphic on a circle $|z| \leq R, 1 < R < \infty$. So

$$\overline{\lim}_{k\to\infty} \sqrt[k]{|c_k|} \leq \frac{1}{R} < 1$$

and therefore

$$\theta(f) := \Big[\sum_{k=1}^{\infty} |c_k|^2 \Big]^{1/2} < \infty.$$

Everywhere in this section $\{z_k(f)\}_{k=1}^{l}$ is the set of all the zeros of f lying in $|z| < R$, taken with their *multiplicities* and enumerated in the increasing order:

$$|z_k(f)| \leq |z_{k+1}(f)| \quad (k = 1, ..., l-1).$$

Put

$$|f|_{L^2}^2 = \Big[\frac{1}{2\pi} \int_0^{2\pi} |f(e^{is})|^2 ds \Big]^{1/2}.$$

According to the Parseval equality, we have

$$\theta^2(f) := |f|_{L^2}^2 - 1 \leq \max_{|z|=1} |f(z)|^2 - 1.$$

Theorem 4.14.1 *Let the function* f *defined by (14.1) be holomorphic on* $|z| \leq R, 1 < R < \infty$. *Then the roots of* f *satisfy the inequalities*

$$\sum_{k=1}^{j} \frac{1}{|z_k(f)|} \leq \theta(f) + j \quad (j = 1, ..., l).$$

Proof: Consider the polynomial

$$f_n(z) = 1 + c_1 z + \ldots + c_n z^n \ (3 \le n < \infty)$$

with the zeros $z_k(f_n)$ ordered in the increasing way. Let

$$P(z) = \sum_{k=0}^{n} c_k z^{n-k}.$$

We have

$$\frac{1}{z_k(f_n)} \doteq z_k(P). \tag{14.2}$$

Now Theorem 4.3.1 yields

$$\sum_{k=1}^{j} \frac{1}{|z_k(f_n)|} \le j + \theta_P \le j + \theta(f) \ (j = 1, \ldots, l).$$

So for any fixed natural $j \le l$, by the Hurwitz theorem, we obtain

$$\sum_{k=1}^{j} |\frac{1}{z_k(f_n)}| \to \sum_{k=1}^{j} |\frac{1}{z_k(f)}|$$

as $n \to \infty$. This implies the required result. Q. E. D.

Now enumerate $z_k(f)$ $(|z_k| < R)$ in the following way:

$$|Im \frac{1}{z_k(f)}| \ge |Im \frac{1}{z_{k+1}(f)}|. \tag{14.3}$$

Put

$$\eta(f) := \left[\sum_{k=3}^{\infty} |c_k|^2 \right]^{1/2},$$

$$v_1 := \frac{1}{2}(|\sigma_1 \, Im \, c_1 + \sqrt{|Im \, c_1|^2 + |1 + c_2|^2} \,|),$$

$$v_2 := \frac{1}{2}(|\sigma_1 \, Im \, c_1 - \sqrt{|Im \, c_1|^2 + |1 + c_2|^2} \,|), \ v_k = 0 \ (k \ge 3),$$

where $\sigma_1 = +1$ if $Im \, c_1 \ge 0$ and $\sigma_1 = -1$ if $Im \, c_1 < 0$.

Theorem 4.14.2 *Let the function f defined by (14.1) be holomorphic on the circle $|z| \le R, 1 < R < \infty$. Then the zeros of f lying in that circle satisfy the inequalities*

$$\sum_{k=1}^{j} |Im \frac{1}{z_k(f)}| \le j + (1+j)\eta(P)/2 + \sum_{k=1}^{j} v_k + \ (j = 1, 2, \ldots, l_1),$$

where l_1 is the number of the nonreal roots of f in the circle $|z| \le R$.

The proof of this theorem is the same as the proof of the previous theorem with Theorem 4.13.1 instead of Theorem 4.3.1.

Furthermore, put

$$\theta_1(f) = \sum_{k=1}^{\infty} |c_k|$$

and

$$\eta_1(f) = |Im \ c_1| + \frac{1}{2}\left(|1 + c_2| + \sum_{k=3}^{\infty} |c_k|\right).$$

Taking into account (14.2), by the Hurwitz theorem and Lemmas 4.1.2 and 4.13.2, we have our next result.

Corollary 4.14.3 *Let the function f defined by (14.1) be holomorphic on $|z| \leq R, 1 < R < \infty$. Then the inequalities*

$$\inf_j |z_j(f)| \geq \frac{1}{\max\{\theta_1(f), 1\}}$$

and

$$\inf_j |z_j(f)| \geq \frac{1}{1 + \max_k |c_k|}$$

hold. In addition,

$$\sup_j \left|Im \ \frac{1}{z_j(f)}\right| \leq \max\left\{\eta_1(f), \frac{1}{2}(|1 + c_2| + 1), 1 + \max_{k=3,4,\dots} \frac{|c_k|}{2}\right\}.$$

4.15   Comments to Chapter 4

Section 4.1 contains the classical results on the zeros of polynomials. Sections 4.2 - 4.7 represent the results from the paper [Gil', 2007a]. Corollaries 4.8.1 and 4.8.2 are taken from the book [Marden, 1985]. Corollary 4.8.3 is probably new. The material of Sections 4.9 - 4.13 is adopted from [Gil', 2000c]. Theorems 4.14.1 and 4.14.2 are probably new.

About the recent very interesting results on the zeros of polynomials, see [Alzer, 2001], [Borwein and Erdelyi, 1995], [Chanane, 2003], [Dyakonov, 2000], [Schmieder and Szynal, 2002].

Chapter 5

Bounds for Zeros of Entire Functions

This is one of the main chapters of the book. Let f be an entire function whose zeros $z_1(f), z_2(f), \dots$ are taken with their multiplicities and enumerated in the increasing order. If f has $l < \infty$ finite zeros, we put

$$\frac{1}{z_k(f)} = 0 \ (k = l+1, l+2, \dots).$$

In this chapter, estimates for the sums

$$\sum_{k=1}^{j} \frac{1}{|z_k(f)|} \text{ and } \sum_{k=1}^{j} \left| Im \ \frac{1}{z_k(f)} \right| \ (j = 1, 2, \dots)$$

are derived. These estimates enable us to obtain inequalities for the counting function of the zeros of f, which supplement the Jensen inequality. We also establish relations between the series

$$\hat{s}_m(f) := \sum_{k=1}^{\infty} \frac{1}{z_k^m(f)} \ (m > \rho(f))$$

and the trace of some finite matrix. Here $\rho(f)$ is the order of f.

5.1   Partial sums of zeros

Consider the entire function

$$f(\lambda) = 1 + \sum_{k=1}^{\infty} c_k \lambda^k \quad (\lambda \in \mathbf{C}), \tag{1.1}$$

where c_k are complex numbers. Let

$$\psi_1 = 1 \text{ and } \psi_k, k = 2, 3, \dots$$

be positive numbers having the following property: the sequence

$$m_1 = 1, \ m_j := \frac{\psi_j}{\psi_{j-1}} \quad (j = 2, 3, \dots)$$

is nonincreasing and tends to zero. So

$$\psi_j := \prod_{k=1}^{j} m_k \quad (j = 1, 2, \dots).$$

Put

$$a_k = \frac{c_k}{\psi_k}, \ k = 1, 2, \dots \ .$$

Then

$$f(\lambda) = 1 + \sum_{k=1}^{\infty} a_k \psi_k \lambda^k \quad (\lambda \in \mathbf{C}). \tag{1.2}$$

We will call *(1.2) the ψ-representation of f.*
    Assume that

$$\alpha_\psi(f) := \overline{\lim}_{k \to \infty} \sqrt[k]{|a_k|} = \overline{\lim}_{k \to \infty} \sqrt[k]{\frac{|c_k|}{\psi_k}} < 1. \tag{1.3}$$

We have

$$|c_k| \leq const \ \psi_k \text{ and } \psi_{k+1}/\psi_k = m_{k+1} \to 0.$$

So f is really an entire function. For instance, the function

$$f(\lambda) = \sum_{k=0}^{\infty} \frac{a_k \lambda^k}{k!} \quad (a_0 = 1)$$

has the form (1.2) with $m_k = 1/k$ $(k = 1, 2, \dots)$. More general, the finite order function

$$f(\lambda) = \sum_{k=0}^{\infty} \frac{a_k \lambda^k}{(k!)^\gamma} \quad (a_0 = 1 , \gamma > 0)$$

can also be written as (1.2) with $m_k = 1/k^\gamma$ $(k = 1, 2, ...)$.

Now let us consider the function

$$f(\lambda) = 1 + \frac{\lambda}{2} + \sum_{k=2}^{\infty} \frac{\lambda^k}{2^k \, ln^k(1+k)}. \tag{1.4}$$

Take

$$a_k = 1/2^k, \psi_k = 1/(ln^k(1+k)) \ (k = 2, 3, ...).$$

Then for $k > 1$,

$$m_k = ln^{k-1}k/(ln^k \, (1+k)) \to 0, k \ \to \infty.$$

Clearly, the function defined by (1.4) is of infinite order. Under condition (1.3), the series

$$\theta_\psi(f) := \Big[\sum_{k=1}^{\infty} |a_k|^2\Big]^{1/2} = \Big[\sum_{k=1}^{\infty} \frac{|c_k|^2}{\psi_k^2}\Big]^{1/2}.$$

Theorem 5.1.1 *Let f be represented by (1.2) and condition (1.3) hold. Then*

$$\sum_{k=1}^{j} \frac{1}{|z_k(f)|} < \theta_\psi(f) + \sum_{k=1}^{j} m_{k+1} \tag{1.5}$$

This theorem is proved in the next section.

Furthermore, from (1.5) we get the following result.

Corollary 5.1.2 *Let f be represented by (1.2) and condition (1.3) hold. Let*

$$\phi(t) \ (0 \le t < \infty)$$

be a continuous convex scalar-valued function, such that $\phi(0) = 0$. Then

$$\sum_{k=1}^{j} \phi(|z_k(f)|^{-1}) < \phi(\theta_\psi(f) + m_2) + \sum_{k=2}^{j} \phi(m_{k+1}) \ (j = 2, 3, ...). \tag{1.6}$$

In particular, for any $p \ge 1$ and $j = 2, 3, ... ,$

$$\sum_{k=1}^{j} \frac{1}{|z_k(f)|^p} < (\theta_\psi(f) + m_2)^p + \sum_{k=2}^{j} m_{k+1}^p.$$

Indeed, this result is due to Lemma 1.2.1 and Theorem 5.1.1. Moreover, the previous corollary implies

Corollary 5.1.3 *Let f be defined by (1.2) and condition (1.3) hold. Then*

$$S_p(f) < (\theta_\psi(f) + m_2)^p + \sum_{k=2}^{\infty} m_{k+1}^p \ (p \ge 1), \tag{1.7}$$

provided the series

$$\sum_{k=1}^{\infty} m_{k+1}^p = \sum_{k=1}^{\infty} \frac{\psi_{k+1}^p}{\psi_k^p}$$

converges.

Furthermore, thanks to Lemma 1.2.2 and Theorem 5.1.1, we arrive at our next result.

Corollary 5.1.4 *Let f be defined by (1.2) and condition (1.3) hold. In addition, let a scalar-valued function $\Phi(t_1, t_2, ..., t_j)$ with an integer j be defined on the domain*

$$0 \le t_j \le t_{j-1}... \le t_2 \le t_1 < \infty \qquad (1.8)$$

and satisfy the condition

$$\frac{\partial \Phi}{\partial t_1} > \frac{\partial \Phi}{\partial t_2} > ... > \frac{\partial \Phi}{\partial t_j} > 0 \text{ for } t_1 > t_2 > ... > t_j \ge 0. \qquad (1.9)$$

Then

$$\Phi\left(\frac{1}{|z_1(f)|}, \frac{1}{|z_2(f)|}, ..., \frac{1}{|z_j(f)|}\right) < \Phi\left(\theta_\psi(f) + m_2, m_3, ..., m_{j+1}\right).$$

In particular, let $\{d_k\}_{k=1}^{\infty}$ be a decreasing sequence of positive numbers with $d_1 = 1$. Then the previous corollary yields the inequality

$$\sum_{k=1}^{j} \frac{d_k}{|z_k(f)|} < \theta_\psi(f) + \sum_{k=1}^{j} d_k m_{k+1} \quad (j = 1, 2, ...).$$

Recall that $\nu_f(r)$ denotes the counting function of the zeros of function f. Since $|z_j(f)| \le |z_{j+1}(f)|$, because of (1.7),

$$\frac{j}{|z_j(f)|} < \theta_\psi(f) + \sum_{k=1}^{j} m_{k+1} \quad (j = 1, 2, ...).$$

So $|z_j(f)| > w_j(f)$, where

$$w_j(f) := \frac{j}{\sum_{k=1}^{j} m_{k+1} + \theta_\psi(f)} \quad (j = 1, 2, ...).$$

Consequently, f has in the circle $|z| \le w_j$ no more than $j - 1$ zeros. We thus get

Corollary 5.1.5 *Let f be defined by (1.2). Then, under condition (1.3), the inequality $\nu_f(r) \le j - 1$ is valid for any $r \le w_j(f)$.*

5.2 Proof of Theorem 5.1.1

For an integer $n > 1$, let us consider the polynomial

$$P(\lambda) = \lambda^n + \sum_{k=1}^{n} a_k \psi_k \lambda^{n-k}, \tag{2.1}$$

where ψ_k are the same as in the previous section. Enumerate the zeros of P in the descending order. Denote

$$\theta_\psi(P) := \left[\sum_{k=1}^{n} |a_k|^2 \right]^{1/2}.$$

Introduce the $n \times n$-matrix

$$A_n = \begin{pmatrix} -a_1 & -a_2 & ... & -a_{n-1} & -a_n \\ m_2 & 0 & ... & 0 & 0 \\ 0 & m_3 & ... & 0 & 0 \\ . & . & ... & . & . \\ 0 & 0 & ... & m_n & 0 \end{pmatrix}. \tag{2.2}$$

Lemma 5.2.1 *Let P be defined by (2.1). Then*

$$\lambda_k(A_n) = z_k(P) \quad (k = 1, ..., n). \tag{2.3}$$

Proof: Clearly, P is the characteristic polynomial of the matrix

$$B = \begin{pmatrix} -a_1 & -a_2\psi_2 & ... & -a_{n-1}\psi_{n-1} & -a_n\psi_n \\ 1 & 0 & ... & 0 & 0 \\ 0 & 1 & ... & 0 & 0 \\ . & . & ... & . & . \\ 0 & 0 & ... & 1 & 0 \end{pmatrix}.$$

Let μ be an eigenvalue of B:

$$-a_1x_1 - a_2\psi_2x_2 - ... - a_{n-1}\psi_{n-1}x_{n-1} - a_n\psi_nx_n = \mu x_1,$$

$x_k = \mu x_{k+1} \quad (k = 1, ..., n - 1)$ for the eigenvector $(x_k)_{k=1}^{n} \in \mathbf{C}^n$.

Put $x_k = y_k/\psi_k$. Then

$$-a_1y_1 - a_2y_2... - a_{n-1}y_{n-1} - a_ny_n = \mu y_1$$

and

$$\frac{y_k}{\psi_k} = \mu \frac{y_{k+1}}{\psi_{k+1}} \quad (k = 1, ..., n - 1).$$

Or

$$m_{k+1}y_k = \frac{y_k \psi_{k+1}}{\psi_k} = \mu y_{k+1} \ (k = 1, ..., n-1).$$

These equalities are equivalent to the equality $A_n y = \mu y$ with $y = (y_k)$. This proves the lemma. Q. E. D.

Lemma 5.2.2 *The zeros of the polynomial P defined by (2.1) satisfy the inequalities*

$$\sum_{k=1}^{j} |z_k(P)| \leq \theta_\psi(P) + \sum_{k=1}^{j} m_{k+1} \ (j = 1, ..., n-1) \tag{2.4}$$

and

$$\sum_{k=1}^{n} |z_k(P)| \leq \theta_\psi(P) + \sum_{k=1}^{n-1} m_{k+1} \tag{2.5}$$

Proof: Because of Lemma 1.1.1,

$$\sum_{k=1}^{j} |\lambda_k(A_n^*)| < \sum_{k=1}^{j} s_k(A_n^*) \ (j = 1, ..., n), \tag{2.6}$$

where $s_k(A_n^*), k = 1, 2, ...$ are the singular numbers of A_n^* ordered in the decreasing way. But $A_n = M + C$, where

$$C = \begin{pmatrix} -a_1 & -a_2 & -a_3 & ... & -a_n \\ 0 & 0 & 0 & ... & 0 \\ . & . & . & ... & . \\ 0 & 0 & 0 & ... & 0 \end{pmatrix}$$

and

$$M = \begin{pmatrix} 0 & 0 & 0 & ... & 0 & 0 \\ m_2 & 0 & 0 & ... & 0 & 0 \\ 0 & m_3 & 0 & ... & 0 & 0 \\ . & . & . & ... & . & . \\ 0 & 0 & 0 & ... & m_n & 0 \end{pmatrix}.$$

CC^* and MM^* are diagonal matrices. Consequently, their diagonal entries are the eigenvalues. So

$$s_1(C^*) = \theta_\psi(P), \ s_k(C^*) = 0 \ (k = 2, ..., n), \ s_k(M^*) = m_{k+1} \ (k = 1, ..., n-1),$$

and $s_n(M^*) = 0$. Take into account that by Lemma 1.1.2,

$$\sum_{k=1}^{j} s_k(A_n^*) = \sum_{k=1}^{j} s_k(M^* + C^*) \leq \sum_{k=1}^{j} s_k(M^*) + \sum_{k=1}^{j} s_k(C^*).$$

So

$$\sum_{k=1}^{j} s_k(A_n^*) \le \theta_\psi(P) + \sum_{k=1}^{j} m_{k+1} \quad (j = 1, ..., n).$$

Now (2.3) and (2.6) yield (2.4). To prove inequality (2.5) take into account that $s_n(M) = 0$. Q. E. D.

Proof of Theorem 5.1.1: Consider the polynomial

$$f_n(\lambda) = 1 + \sum_{k=1}^{n} a_k \psi_k \lambda^k.$$

Clearly, $\lambda^n f_n(1/\lambda) = P(\lambda)$. So $z_k(P) = 1/z_k(f_n)$. Taking into account that the roots continuously depend on coefficients, have the required result, letting in the previous lemma $n \to \infty$. Q. E. D.

5.3   Functions represented in the root-factorial form

Consider the function defined by (1.2) with

$$\psi_k = \frac{1}{(k!)^\gamma} \quad (\gamma \in (0,1)).$$

The case $\gamma = 1$ is considered in Chapter 8. That is, under consideration,

$$a_k = c_k (k!)^\gamma$$

and

$$f(\lambda) = \sum_{k=0}^{\infty} \frac{a_k \lambda^k}{(k!)^\gamma} \quad (\lambda \in \mathbf{C}, \ a_0 = 1). \tag{3.1}$$

We will call (3.1) *the root-factorial representation of* f.
     Assume that

$$\alpha_\gamma(f) := \overline{\lim} \sqrt[k]{|c_k|(k!)^\gamma} = \overline{\lim} \sqrt[k]{|a_k|} < 1. \tag{3.2}$$

Since a_k depends on the choice of γ, $\alpha_\gamma(f)$ *in (3.2) also depends on* γ.
     Relations (3.1) and (3.2), and the Hólder inequality imply that function f has order $\rho(f) \le 1/\gamma$. Moreover, for any function f with $f(0) = 1$, whose order is $\rho(f) < \infty$, we can take $\gamma > 1/\rho(f)$, such that representation (3.1) holds with condition (3.2).

Under (3.1) and (3.2), we have $\theta_\psi(f) = \theta_\gamma(f)$, where

$$\theta_\gamma(f) := \left[\sum_{k=1}^\infty |c_k(k!)^\gamma|^2\right]^{1/2} = \left[\sum_{k=1}^\infty |a_k|^2\right]^{1/2}.$$

Enumerate the zeros of f in the increasing way: $|z_k(f)| \leq |z_{k+1}(f)|$ ($k = 1, 2, ...$).

Theorem 5.3.1 *Let f be defined by (3.1) and condition (3.2) hold. Then*

$$\sum_{k=1}^j \frac{1}{|z_k(f)|} < \theta_\gamma(f) + \sum_{k=1}^j \frac{1}{(k+1)^\gamma} \quad (j = 1, 2, ...).$$

Proof: The required result is due to Theorem 5.1.1 with $\psi_k = 1/(k!)^\gamma$.
Q. E. D.

Since $|z_k(f)| \leq |z_{k+1}(f)|$, the previous theorem implies

$$\frac{j}{|z_j(f)|} < \theta_\gamma(f) + \sum_{k=1}^j \frac{1}{(k+1)^\gamma},$$

and therefore

$$|z_j(f)| > \frac{j}{\theta_\gamma(f) + \sum_{k=1}^j \frac{1}{(k+1)^\gamma}} \quad (j = 1, 2, ...).$$

But

$$\sum_{k=1}^j (k+1)^{-\gamma} \leq \int_1^{j+1} \frac{dx}{x^\gamma} = \frac{(1+j)^{1-\gamma} - 1}{1 - \gamma} \quad (0 < \gamma < 1).$$

We thus, we get

Corollary 5.3.2 *Let f be defined by (3.1) with $\gamma \in (0,1)$ and condition (3.2) hold. Then with the notation*

$$v_j(f) := \frac{j(1-\gamma)}{\theta_\gamma(f)(1-\gamma) + (1+j)^{1-\gamma} - 1},$$

the inequality $|z_j(f)| > v_j(f)$ holds and thus $\nu_f(r) \leq j - 1$ for any

$$r \leq v_j(f) \quad (j = 1, 2, ...).$$

In addition, Corollaries 5.1.2 and 5.1.3 with the notations

$$\vartheta_1 = \theta_\gamma(f) + \frac{1}{2^\gamma} \text{ and } \vartheta_k = \frac{1}{(k+1)^\gamma} \quad (k = 2, 3, ...),$$

imply the following result.

Corollary 5.3.3 *Under (3.1) and (3.2), let $\phi(t)$ $(0 \leq t < \infty)$ be a continuous convex scalar-valued function, such that $\phi(0) = 0$. Then*

$$\sum_{k=1}^{j} \phi(|z_k(f)|^{-1}) < \sum_{k=1}^{j} \phi(\vartheta_k) \quad (j = 1, 2, ...).$$

In particular, for any $p \geq 1$ and $j = 2, 3, ...,$

$$\sum_{k=1}^{j} \frac{1}{|z_k(f)|^p} < \sum_{k=1}^{j} \vartheta_k^p = \left(\theta_\gamma(f) + \frac{1}{2^\gamma}\right)^p + \sum_{k=2}^{j} \frac{1}{(k+1)^{p\gamma}}.$$

Moreover,

$$S_p(f) < \left(\theta_\gamma(f) + \frac{1}{2^\gamma}\right)^p + \zeta(p\gamma) - \frac{1}{2^\gamma} - 1,$$

where $\zeta(.)$ is the Riemann zeta function, provided, $\gamma p > 1$.

Furthermore, by Corollary 5.1.4, we arrive at the following result: let a scalar-valued function $\Phi(t_1, t_2, ..., t_j)$ with an integer j be defined on the domain (1.8) and satisfy condition (1.9). Then

$$\Phi\left(\frac{1}{|z_1(f)|}, \frac{1}{|z_2(f)|}, ..., \frac{1}{|z_j(f)|}\right) < \Phi(\vartheta_1, \vartheta_2, ..., \vartheta_j).$$

In particular, let $\{d_k\}_{k=1}^{\infty}$ be a decreasing sequence of positive numbers with $d_1 = 1$. Then the previous corollary yields the inequality

$$\sum_{k=1}^{j} \frac{d_k}{|z_k(f)|} < \theta_\gamma(f) + \sum_{k=1}^{j} \frac{d_k}{(k+1)^\gamma} \quad (j = 1, 2, ...),$$

provided the relations (3.1) and (3.2) are fulfilled.

5.4   Functions represented in the Mittag-Leffler form

Consider the function defined by (1.1). Take

$$\psi_k = \frac{1}{\Gamma(1 + k/\rho)} \quad (k > 1, \rho > 1), \quad \psi_1 = 1$$

and

$$a_k = c_k \Gamma(1 + k/\rho), k \geq 1, a_0 = 1$$

where $\Gamma(.)$ is the Euler gamma function. Rewrite (1.1) as

$$f(z) = \sum_{k=0}^{\infty} \frac{a_k z^k}{\Gamma(1 + k/\rho)}. \tag{4.1}$$

We will call (4.1) *the Mittag-Leffler representation of f*.

Recall that the Mittag-Leffler function E_ρ of index $\rho > 0$ is the entire function

$$E_\rho(z) = \sum_{k=0}^{\infty} \frac{z^k}{\Gamma(1+k/\rho)}.$$

It is easy to see that $E_1(z) = e^z$.

Assume that

$$\overline{\lim} \sqrt[k]{|a_k|} = \overline{\lim} \sqrt[k]{|c_k|\Gamma(1+k/\rho)} < 1. \qquad (4.2)$$

Theorem 5.1.13 from [Berenstein and Gay, 1995, p. 310] asserts that the function f defined by (4.1) under condition (4.2) is of order ρ and finite type, provided ρ is integer.

Under consideration, we have $\theta_\psi(f) = \theta(\rho, f)$, where

$$\theta(\rho, f) := \left[|c_1|^2 + \sum_{k=2}^{\infty} |c_k\Gamma(1+k/\rho)|^2\right]^{1/2} = \left[\frac{|a_1|^2}{\Gamma^2(1+1/\rho)} + \sum_{k=2}^{\infty} |a_k|^2\right]^{1/2}.$$

Note that, we should take $\psi_1 = 1$ but $\Gamma(1+1/\rho) \neq 1$.

Denote

$$\beta_1 := \theta(\rho, f) + \frac{1}{\Gamma(1+\frac{2}{\rho})}$$

and

$$\beta_k := \frac{\Gamma(1+k/\rho)}{\Gamma(1+\frac{k+1}{\rho})}, \quad k = 2, 3, ...$$

Below in this section, we will show that

$$\beta_j \leq \kappa_\rho j^{-1/\rho} \quad (j = 2, 3, ...), \qquad (4.3)$$

where

$$\kappa_\rho := \rho^{1/\rho} exp\left[\frac{1}{\rho} + \frac{1}{2(1+\rho)}\right].$$

Theorem 5.4.1 *Under condition (4.2), the zeros of the function f defined by (4.1) satisfy the inequalities*

$$\sum_{k=1}^{j} \frac{1}{|z_k(f)|} < \sum_{k=1}^{j} \beta_k \quad (j = 1, 2, ...).$$

Proof: Taking $\psi_k = 1/\Gamma(1+k/\rho)$, $k > 1$; $\psi_1 = 1$, we have $m_{k+1} = \beta_k, k \geq 1$. Hence, thanks to Theorem 5.1.1, we arrive at the required result. Q. E. D.

Proof of inequality (4.3): By the well-known formula (6.1.13) from the book [Ram Murty, 2001, p. 341], we have

$$ln\,\Gamma(z) = (z - 1/2)\,ln(z) - z + C + \int_0^{\infty} \frac{[u] - u + 1/2}{u+z}du$$

where $[u]$ is the integer part of $u > 0$ and $C = const > 0$. Hence,

$$ln\ \Gamma(1+z) = ln\ \Gamma(z) + ln\ z = (z + 1/2)\ ln\ (z) - z$$

$$+ C + \int_0^\infty \frac{[u] - u + 1/2}{u + 1 + z} du.$$

Consequently,

$$ln\ \beta_k = ln\ \Gamma\left(1 + \frac{k}{\rho}\right) - ln\ \Gamma\left(1 + \frac{k+1}{\rho}\right) =$$

$$(k/\rho + 1/2)\ ln(k/\rho) - ((k+1)/\rho + 1/2)\ ln((1+k)/\rho) + 1/\rho + J$$

where

$$J := \int_0^\infty ([u] - u + 1/2)\left[\frac{1}{u + 1 + k/\rho} - \frac{1}{u + 1 + (k+1)/\rho}\right] du.$$

Since $[u] - u \leq 0, u \geq 0$, we obtain

$$J \leq \frac{1}{2\rho} \int_0^\infty \frac{1}{(u + k/\rho)(u + (k+1)/\rho)} du \leq \frac{1}{2\rho} \int_0^\infty \frac{1}{(u + 1 + k/\rho)^2} du$$

$$= \frac{1}{2\rho(1 + k/\rho)} = \frac{1}{2(k + \rho)}.$$

So for $k \geq 1$, we get

$$ln\ \beta_k \leq -\frac{1}{\rho}\ ln(k/\rho) + 1/\rho + \frac{1}{2(1 + \rho)}.$$

We thus arrive at the inequality

$$\beta_k \leq \rho^{1/\rho} exp\ \left[\frac{1}{\rho} + \frac{1}{2(1 + \rho)}\right].$$

As claimed. Q. E. D.

From (4.3), it follows that

$$\sum_{k=2}^\infty \beta_k^p \leq \kappa_\rho^p \sum_{k=2}^\infty k^{-\frac{p}{\rho}} \leq \kappa_\rho^p[\zeta(p/\rho) - 1] < \infty,$$

provided $p > \rho$.

Now Corollaries 5.1.2 and 5.1.3 imply

Corollary 5.4.2 *Let the function f be defined by (4.1) under condition (4.2). In addition, let $\phi(t)$ $(0 \leq t < \infty)$ be a continuous convex scalar-valued function, such that $\phi(0) = 0$. Then*

$$\sum_{k=1}^j \phi(|z_k(f)|^{-1}) < \sum_{k=1}^j \phi(\beta_k) \quad (j = 1, 3, ...).$$

In particular, for any $p \geq 1$ and $j = 2, 3, \ldots$,

$$\sum_{k=1}^{j} \frac{1}{|z_k(f)|^p} < \sum_{k=1}^{j} \beta_k^p.$$

Moreover,

$$S_p(f) < \sum_{k=1}^{\infty} \beta_k^p \leq \beta_1^p + \kappa_\rho^p[\zeta(p/\rho) - 1],$$

provided $p > \rho$.

Corollary 5.1.4 with $\psi_k = 1/\Gamma(1 + \frac{k}{\rho})$ is also valid.

Because of Theorem 5.4.1,

$$|z_j(f)| > \tilde{w}_j, \text{ where } \tilde{w}_j(f) := \frac{j}{\sum_{k=1}^{j} \beta_k} \quad (j = 1, 2, \ldots).$$

Thus, f has in $|z| \leq \tilde{w}_j$ no more than $j - 1$ zeros. In other words,

$$\nu_f(r) \leq j - 1 \quad (r \leq \tilde{w}_j).$$

5.5   An additional bound for the series of absolute values of zeros

In this section, we consider an additional approach allowing us to estimate $S_p(f)$. For the brevity, we restrict ourselves by the root-factorial representation although our reasonings are valid if a function is represented in the ψ-form or in the Mittag-Leffler form.

Again consider the entire function

$$f(\lambda) = \sum_{k=0}^{\infty} \frac{a_k \lambda^k}{(k!)^\gamma} \quad (\gamma \in (0, 1), \lambda \in \mathbf{C}, a_0 = 1) \tag{5.1}$$

with complex, in general, coefficients satisfying the condition

$$\alpha_\gamma(f) := \overline{\lim} \sqrt[k]{|a_k|} < 1. \tag{5.2}$$

Theorem 5.5.1 *Let f be defined by (5.1) and condition (5.2) hold. Then for any integer $p > 1/\gamma$, we have*

$$S_p(f) \leq \left[\sum_{k=1}^{\infty} |a_k|^{p'} \right]^{p/p'} + \zeta(p\gamma) - 1 \tag{5.3}$$

where $\zeta(.)$ is the Riemann zeta function and

$$\frac{1}{p} + \frac{1}{p'} = 1.$$

This theorem is proved in the next section. In the next section, we also prove that for any integer $p > 1/\gamma$,

$$S_p(f) \leq \sum_{k=1}^{\infty} \left[|a_k|^{p'} + \frac{1}{(k+1)^{\gamma p'}} \right]^{p/p'}. \tag{5.4}$$

Since $\gamma \in (0,1)$, for any integer $p > 1/\gamma$, we have $p \geq 2$, and therefore

$$\frac{p}{p'} = p - 1 \geq 1.$$

Take into account that for all $a > 1$ and $x > 0$, the inequality

$$(1+x)^a \leq 2^{a-1}(1+x^a)$$

is valid. Then

$$\left[|a_k|^{p'} + \frac{1}{(k+1)^{\gamma p'}} \right]^{p/p'} \leq 2^{p-2} \left(|a_k|^p + \frac{1}{(k+1)^{\gamma p}} \right).$$

Now (5.4) implies the inequality

$$S_p(f) \leq 2^{p-2} \left[\sum_{k=1}^{\infty} |a_k|^p + \zeta(p\gamma) - 1 \right]. \tag{5.5}$$

In particular, if $\gamma = \frac{1}{p-1}$, $p \geq 2$, then

$$S_p(f) \leq 2^{p-2} \left[\sum_{k=1}^{\infty} |a_k|^p + \zeta\left(\frac{p}{p-1}\right) - 1 \right].$$

Let us assume that under (5.1), the condition

$$1 < \alpha_\gamma(f) = \overline{\lim}_{k \to \infty} \sqrt[k]{|a_k|} < \infty \tag{5.6}$$

holds, and consider the function

$$h_t(\lambda) \equiv f(t\lambda) = \sum_{k=0}^{\infty} \frac{a_k(t\lambda)^k}{(k!)^\gamma} \quad (0 < t < 1/\alpha(f)).$$

Then

$$\overline{\lim}_{k \to \infty} \sqrt[k]{|t^k a_k|^p} < 1$$

and

$$\sum_{k=0}^{\infty} |a_k t^k|^p < \infty \quad (p \geq 2, \ 0 < t < 1/\alpha_\gamma(f)).$$

Clearly,

$$S_p(h_t) = \sum_{k=1}^{\infty} \frac{t^p}{|z_k(f)|^p} = t^p S_p(f).$$

Now Theorem 5.5.1 implies

Corollary 5.5.2 *Let f be defined by (5.1) and condition (5.6) hold. Then for any integer $p > 1/\gamma$,*

$$S_p(f) \leq \inf_{0 < t < 1/\alpha_\gamma(f)} \frac{1}{t^p} \Big[\Big(\sum_{k=1}^{\infty} |t^k a_k|^{p'} \Big)^{p/p'} + \zeta(p\gamma) - 1 \Big].$$

Thanks to inequality (5.5) and condition (5.6), we obtain

$$S_p(f) \leq 2^{p-2} \inf_{0 < t < 1/\alpha_\gamma(f)} \frac{1}{t^p} \Big(\sum_{k=1}^{\infty} |t^k a_k|^p + \zeta(p\gamma) - 1 \Big), \quad (5.7)$$

provided f is represented by (5.1) and the condition $p > 1/\gamma$ hold.

Furthermore, put

$$J_p(f) := \sum_{k=1}^{\infty} |Im \ \frac{1}{z_k(f)}|^p$$

and

$$\eta_p(f) := \Big[|2Im \ a_1|^{p'} + |a_2 + \frac{1}{2^\gamma}|^{p'} + \sum_{k=3}^{\infty} |a_k|^{p'} \Big]^{p/p'}.$$

In the next section, we also prove the following theorem.

Theorem 5.5.3 *Let f be defined by (5.1) and condition (5.2) hold. Then for any integer $p > 1/\gamma$, we have*

$$2^p J_p(f) \leq \eta_p(f) + \Big[|a_2 + \frac{1}{2^\gamma}|^{p'} + \frac{1}{3^{\gamma p'}} \Big]^{p/p'} + \sum_{k=3}^{\infty} \Big[|a_k|^{p'} + \frac{2}{k^{\gamma p'}} \Big]^{p/p'}$$

$$\Big(\frac{1}{p} + \frac{1}{p'} = 1 \Big).$$

5.6   Proofs of Theorems 5.5.1 and 5.5.3

For a constant $\gamma \geq 0$, consider the polynomial

$$P_\gamma(\lambda) = \sum_{k=0}^{n} \frac{a_k \lambda^{n-k}}{(k!)^\gamma} \quad (a_0 = 1) \tag{6.1}$$

and the $n \times n$-matrix

$$A_n(\gamma) = \begin{pmatrix} -a_1 & -a_2 & \cdots & -a_{n-1} & -a_n \\ \frac{1}{2^\gamma} & 0 & \cdots & 0 & 0 \\ 0 & \frac{1}{3^\gamma} & \cdots & 0 & 0 \\ \cdot & & \cdots & \cdot & \\ 0 & 0 & \cdots & \frac{1}{n^\gamma} & 0 \end{pmatrix}. \tag{6.2}$$

Recall that $\lambda_k(A_n(\gamma))$ $(k = 1, ..., n)$ are the eigenvalues of $A_n(\gamma)$ with their multiplicities. Thanks to Lemma 5.2.1, we have

$$\lambda_k(A_n(\gamma)) = z_k(P_\gamma) \quad (k = 1, ..., n). \tag{6.3}$$

Lemma 5.6.1 *Let P_γ be defined by (6.1). Then*

$$\sum_{k=1}^{n} |z_k(P_\gamma)|^p \leq \left[\sum_{k=1}^{n} |a_k|^{p'} \right]^{p/p'} + \sum_{k=1}^{n} \frac{1}{(k+1)^{p\gamma}} \quad (2 \leq p < \infty). \tag{6.4}$$

Proof: By Theorem 2.5.7, for any matrix $T = (t_{jk})_{j,k=1}^{n}$, we have

$$\sum_{k=1}^{n} |\lambda_j(T)|^p \leq \sum_{j=1}^{n} \left[\sum_{k=1}^{n} |t_{jk}|^{p'} \right]^{p/p'}. \tag{6.5}$$

If $T = A_n(\gamma)$, then

$$t_{1k} = -a_k \ (k = 1, ..., n), t_{k+1,k} = \frac{1}{(k+1)^\gamma} \ (k < n), \text{ and } t_{jk} = 0 \text{ otherwise.} \tag{6.6}$$

Hence,

$$\sum_{j=1}^{n} \left[\sum_{k=1}^{n} |t_{jk}|^{p'} \right]^{p/p'} = \left[\sum_{k=1}^{n} |a_k|^{p'} \right]^{p/p'} + \sum_{k=1}^{n-1} \frac{1}{(k+1)^{p\gamma}}.$$

This proves the required result. Q. E. D.

Proof of Theorem 5.5.1:   Consider the polynomials P_γ defined by (6.1) and

$$f_n(\lambda) := \sum_{k=0}^{n} \frac{a_k \lambda^k}{(k!)^\gamma}.$$

Clearly, $\lambda^n f_n(1/\lambda) = P_\gamma(\lambda)$. So $z_k(P_\gamma) = 1/z_k(f_n)$. Take into account that the zeros continuously depend on the coefficients. Then, we get the required result, letting $n \to \infty$ in the previous lemma. Q. E. D.

To prove inequality (5.4), we need the following result.

Lemma 5.6.2 *Let P_γ $(\gamma \geq 0)$ be the polynomial defined by (6.1). Then*

$$\sum_{k=1}^{n} |z_k(P_\gamma)|^p \leq \sum_{k=1}^{n} (|a_k|^{p'} + \tilde{\vartheta}_{kn}^{p'})^{p/p'} \quad (2 \leq p < \infty), \qquad (6.7)$$

where

$$\tilde{\vartheta}_{kn} = \frac{1}{(k+1)^\gamma} \quad (k = 1, ..., n-1); \quad \tilde{\vartheta}_{nn} = 0. \qquad (6.8)$$

Proof: Since $|\lambda_k(T)| = |\lambda_k(T^*)|$, from (6.5), we get

$$\sum_{k=1}^{n} |\lambda_k(T)|^p \leq \sum_{k=1}^{n} [\sum_{j=1}^{n} |t_{jk}|^{p'}]^{p/p'}.$$

Hence, according to (6.6),

$$\sum_{k=1}^{n} [\sum_{j=1}^{n} |t_{jk}|^{p'}]^{p/p'} = \sum_{k=1}^{n} [|a_k|^{p'} + \tilde{\vartheta}_{kn}^{p'}]^{p/p'}.$$

This proves the required result. Q. E. D.

Proof of inequality (5.4) is similar to the proof of Theorem 5.5.1 with Lemma 5.6.2 instead of Lemma 5.6.1.

To prove Theorem 5.5.3, we need the following lemma.

Lemma 5.6.3 *Let P_γ $(\gamma \geq 0)$ be the polynomial defined by (6.1) with $n \geq 3$. Then for any integer $p \geq 2$*

$$2^p \sum_{k=1}^{n} |Im\, z_k(P_\gamma)|^p \leq [|2Im\, a_1|^{p'} + |\frac{1}{2^\gamma} + a_2|^{p'} + \sum_{k=3}^{n} |a_k|^{p'}]^{p/p'} +$$

$$[|\frac{1}{2^\gamma} + a_2|^{p'} + \frac{1}{3^{p'\gamma}}]^{p/p'} + \sum_{k=3}^{n} [|a_k|^{p'} + \frac{2}{k^{p'\gamma}}]^{p/p'}.$$

Proof: By the Weyl inequalities (Lemma 1.1.4), for any matrix $T = (t_{jk})_{j,k=1}^{n}$, we have

$$\sum_{k=1}^{n} |Im\, \lambda_k(T)|^p \leq \sum_{k=1}^{n} |\lambda_k(T_I)|^p$$

where $T_I = (T - T^*)/2i$. Hence,

$$2^p \sum_{k=1}^{n} |Im\, \lambda_k(T)|^p \le \sum_{k=1}^{n} |\lambda_k(T - T^*)|^p.$$

According to (6.6),

$$\sum_{k=1}^{n} |\lambda_k(T - T^*)|^p \le \sum_{j=1}^{n} \Big[\sum_{k=1}^{n} |v_{jk}|^{p'}\Big]^{p/p'},$$

where $v_{jk} = t_{jk} - \bar{t}_{kj}$ are the entries of $T - T^*$. That is,

$$|v_{11}| = 2|Im\, a_1|, |v_{12}| = |v_{21}| = |a_2 + \frac{1}{2^\gamma}|, |v_{1j}| = |v_{j1}| = |a_j|, j \ge 3;$$

$$v_{k+1,k} = v_{k+1,k} = \frac{1}{(k+1)^\gamma} \ (2 \le k \le n - 1), \ \text{and } v_{jk} = 0 \text{ otherwise.}$$

Thus, we obtain

$$\sum_{j=1}^{n} \Big[\sum_{k=1}^{n} |v_{jk}|^{p'}\Big]^{p/p'}$$

$$= \Big[|2Im\, a_1|^{p'} + |\frac{1}{2^\gamma} + a_2|^{p'} + \sum_{k=3}^{n} |a_k|^{p'}\Big]^{p/p'}$$

$$+ \Big[|\frac{1}{2^\gamma} + a_2|^{p'} + \frac{1}{3^{p'\gamma}}\Big]^{p/p'} + \sum_{k=3}^{n-1} \Big[|a_k|^{p'} + \frac{1}{k^{p'\gamma}} + \frac{1}{(k+1)^{p'\gamma}}\Big]^{p/p'} + \Big[|a_n|^{p'} + \frac{1}{n^{p'\gamma}}\Big]^{p/p'}.$$

This proves the result. Q. E. D.

Proof of Theorem 5.5.3: Again use the equality $z_k(P_\gamma) = 1/z_k(f_n)$ established in the proof of Theorem 5.5.1. Take into account that the zeros continuously depend on the coefficients. Then we get the required result, letting $n \to \infty$ in the previous lemma. Q. E. D.

5.7 Partial sums of imaginary parts of zeros

Let

$$f(\lambda) = 1 + \sum_{k=1}^{\infty} a_k \psi_k \lambda^k \ (\lambda \in \mathbf{C}). \tag{7.1}$$

Recall that $\psi_1 = 1$ and $\psi_k, k = 2, 3, ...$ are positive numbers, such that the sequence

$$m_1 = 1, \ m_j := \frac{\psi_j}{\psi_{j-1}} \quad (j = 2, 3, ...)$$

is nonincreasing and tends to zero. Assume that

$$\sum_{k=1}^{\infty} |a_k|^2 < \infty \tag{7.2}$$

and set

$$\eta_\psi(f) := \Big[\sum_{k=3}^{\infty} |a_k|^2 \Big]^{1/2},$$

and

$$v_1 := \frac{1}{2} (|\sigma_1 \, Im \, a_1 + \sqrt{|Im \, a_1|^2 + |m_2 + a_2|^2} \, |),$$

$$v_2 := \frac{1}{2} (|\sigma_1 \, Im \, a_1 - \sqrt{|Im \, a_1|^2 + |m_2 + a_2|^2} \, |), \ v_k = 0 \ (k \geq 3),$$

where $\sigma_1 = +1$ if $Im \, a_1 \geq 0$ and $\sigma_1 = -1$ if $Im \, a_1 < 0$.

Theorem 5.7.1 *Let f be defined by (7.1) and condition (7.2) hold. Let the zeros $z_k(f), \ k = 1, 2, ...$ of f taken with their multiplicities be enumerated in the decreasing order:*

$$|Im \, \frac{1}{z_k(f)}| \geq |Im \, \frac{1}{z_{k+1}(f)}| \ (k = 1, 2, ...).$$

Then

$$\sum_{k=1}^{j} |Im \, \frac{1}{z_k(f)}| \leq (1 + j)\eta_\psi(f))/2 + \sum_{k=1}^{j} (m_{k+2} + v_k) \ (j = 1, 2, ...).$$

(If f has $l < \infty$ nonreal zeros, we put

$$Im \, \frac{1}{z_k(f)} = 0 \ for \ k = l+1, l+2, ...).$$

To prove this theorem, for an integer $n \geq 3$ again consider the polynomial

$$P(\lambda) = \lambda^n + \sum_{k=1}^{n} a_k \psi_k \lambda^{n-k}$$

with the zeros $z_k(P)$ ordered in the following way:

$$|Im \, z_k(P)| \geq |Im \, z_{k+1}(P)| \ (k = 1, ..., n - 1).$$

Put

$$\eta_\psi(P) := \Big[\sum_{k=3}^{n} |a_k|^2 \Big]^{1/2}.$$

Lemma 5.7.2 *The zeros of P satisfy the inequalities*

$$\sum_{k=1}^{j} |Im\ z_k(P)| \le (1+j)\eta_\psi(P)/2 + \sum_{k=1}^{j}(v_k + m_{k+2})\ (j = 1, ..., n).$$

Proof: Again use the $n \times n$-matrix A_n defined by (2.2) and relations (2.3). Let $Im\ A_n = (A_n - A_n^*)/2i$. Because of Lemma 1.1.4,

$$\sum_{k=1}^{j} |Im\ \lambda_k(A_n)| \le \sum_{k=1}^{j} s_k(Im\ A_n)\ (j = 1, ..., n).$$

Clearly, $Im\ A_n = B + (C - C^*)/2i + (D - D^*)/2i$, where

$$B = \begin{pmatrix} -Im\ a_1 & i(a_2 + m_2)/2 & 0 & ... & 0 & 0 \\ -i(\bar{a}_2 + m_2)/2 & 0 & 0 & ... & 0 & 0 \\ 0 & 0 & 0 & ... & 0 & 0 \\ & . & & . & . \\ 0 & 0 & 0 & ... & 0 & 0 \end{pmatrix},$$

$$C = \begin{pmatrix} 0 & 0 & -a_3 & ... & -a_n \\ 0 & 0 & 0 & ... & 0 \\ . & . & . & ... & . \\ 0 & 0 & 0 & ... & 0 \end{pmatrix} \text{ and } D = \begin{pmatrix} 0 & 0 & 0 & ... & 0 & 0 \\ 0 & 0 & 0 & ... & 0 & 0 \\ 0 & m_3 & 0 & ... & 0 & 0 \\ . & . & . & ... & . & . \\ 0 & 0 & 0 & ... & m_n & 0 \end{pmatrix}.$$

Obviously,

$$CC^* = \begin{pmatrix} \eta_\psi^2(P) & 0 & 0 & ... & 0 \\ 0 & 0 & 0 & ... & 0 \\ 0 & 0 & 0 & ... & 0 \\ . & . & . & ... & . \\ 0 & 0 & 0 & ... & 0 \end{pmatrix}$$

and

$$DD^* = \begin{pmatrix} 0 & 0 & 0 & ... & 0 & 0 \\ 0 & 0 & 0 & ... & 0 & 0 \\ 0 & 0 & m_3^2 & ... & 0 & 0 \\ . & . & . & ... & . & . \\ 0 & 0 & 0 & ... & 0 & m_n^2 \end{pmatrix}.$$

Since the diagonal entries of diagonal matrices are the eigenvalues, and $\lambda_k(CC^*) = s_k^2(C^*)$, where $s_k(.)$ are singular numbers enumerated in the decreasing way, we can write out

$$s_1(C^*) = \eta_\psi(P),\ s_k(C^*) = 0\ (k = 2, ..., n).$$

But

$$s_k(C) \le \|C\| = s_1(C^*)\ (k = 1, ..., n)$$

and $s_k(B) = v_k$. In addition,

$$s_k(D) = m_{k+2} \ (k = 1, ..., n-2), \ s_{n-1}(D) = s_n(D).$$

and $s_k(D) = s_k(D^*)$. Take into account that

$$\sum_{k=1}^{j} s_k(Im \ A_n) = \sum_{k=1}^{j} s_k(B + (C - C^*)/2i + (D - D^*)/2i)$$

$$\leq \sum_{k=1}^{j} s_k(B) + s_k(C)/2 + s_k(C^*)/2 + s_k(D)$$

(see Lemma 1.1.2). So

$$\sum_{k=1}^{j} s_k(Im \ A_n) \leq (1+j)\eta_\psi/2 + \sum_{k=1}^{j} v_k + m_{k+2}.$$

This proves the lemma. Q. E. D.

Proof of Theorem 5.7.1: Take the polynomial

$$f_n(\lambda) = 1 + a_1\lambda + ... + \psi_n a_n \lambda^n \ (3 \leq n < \infty)$$

with the zeros $z_k(f_n)$ ordered in the following way:

$$\left| Im \ \frac{1}{z_k(f_n)} \right| \geq \left| Im \ \frac{1}{z_{k+1}(f_n)} \right| \ (k = 1, ..., n-1).$$

Take into account that $z_k^{-1}(f_n) = z_k(P)$. Hence thanks to the Hurwitz theorem, we easily arrive at the required result. Q. E. D.

5.8   Representation of e^{z^r} in the root-factorial form

Consider the function

$$e^{z^r} = \sum_{k=0}^{\infty} \frac{z^{kr}}{k!} \ (r = 2, 3, ...)$$

and denote

$$\chi_k = \frac{[(rk)!]^{1/2r}}{k!}.$$

By the Stierling formula, we easily obtain

$$\sqrt[k]{\chi_k} \to 0, \; k \to \infty.$$

Besides, we can write the considered function in the root-factorial form

$$e^{z^r} = \sum_{k=0}^{\infty} \frac{\chi_k z^{kr}}{[(rk)!]^{1/2r}}.$$

5.9 The generalized Cauchy theorem for entire functions

Lemma 5.9.1 *Consider the entire function*

$$f(z) = \sum_{k=0}^{\infty} c_k z^k \;\; (c_0 = 1) \tag{9.1}$$

with complex coefficients $c_k, k \geq 1$. Suppose that

$$\sum_{k=1}^{\infty} |c_k| x^k \leq h(x) \;\; (x \geq 0),$$

where $h(x)$ is a monotonically increasing continuous function of an argument $x \geq 0$ with $h(0) < 1$ and $h(\infty) = \infty$. Let x_0 be the unique positive root of the equation $h(x) = 1$. Then any zero $z(f)$ of f satisfies the inequality

$$|z(f)| \geq x_0.$$

Proof: Since $c_0 = 1$, for any root $z(f)$ of f, we have

$$1 = -\sum_{k=1}^{\infty} c_k z^k(f) = |\sum_{k=1}^{\infty} c_k z^k(f)|.$$

But

$$h(x_0) = 1 = |\sum_{k=1}^{\infty} c_k z^k(f)| \leq \sum_{k=1}^{\infty} |c_k||z(f)|^k \leq h(|z(f)|).$$

Hence the required result follows, since $h(x)$ monotonically increases on the positive half-line. Q. E. D.

The next result can be easily derived by Lemma 5.9.1.

Corollary 5.9.2 *Let f be defined by (9.1) and*

$$|c_k| \leq b_k \;\; (k = 1, 2, ...).$$

Let the function

$$v(z) = 1 - \sum_{k=1}^{\infty} b_k z^k$$

be entire. Then

$$\inf_{k=1,2,\ldots} |z_k(f)| \geq \inf_{k=1,2,\ldots} |z_k(v)|.$$

This corollary can be considered as a generalization of the Cauchy theorem for polynomials (Theorem 4.1.1).

5.10 The Gerschgorin type domains for entire functions

Theorem 5.10.1 *All the zeros of the entire function defined by (9.1) lie in the union of the sets*

$$|z| \geq 1 \ and \ |\frac{1}{z} + c_1| \leq \sum_{k=2}^{\infty} |c_k|.$$

Proof: Consider the polynomial

$$P(\lambda) = \sum_{k=0}^{n} c_k \lambda^{n-k} \quad (c_0 = 1; \ n \geq 2).$$

Thanks to Corollary 4.8.1, any zero $z(P)$ of P, either satisfies the inequality $|z(P)| \leq 1$, or

$$|z(P) + c_1| \leq \sum_{k=2}^{n} |c_k|.$$

Now consider the polynomial

$$f_n(\lambda) = \sum_{k=0}^{n} c_k \lambda^k = \lambda^n \sum_{k=0}^{n} c_k \lambda^{k-n} = \lambda^n P(1/\lambda). \qquad (10.1)$$

We have $z(f_n) = 1/z(P)$. Thus

$$\left| \frac{1}{z(f_n)} + c_1 \right| \leq \sum_{k=2}^{n} |c_k|.$$

Letting $n \to \infty$, by the Hurwitz theorem, we get the required result. Q. E. D.

Furthermore, let us prove the following result.

Theorem 5.10.2 *Let f be the entire function defined by (9.1) and $c_k \neq 0$ for at least one $k \geq 2$. Then all the zeros of f lie in the union of the sets*

$$|z| \geq 1 \ and \ |(\frac{1}{z} - r_1)(\frac{1}{z} - r_2)| \leq p_3(f),$$

where

$$r_{1,2} = -\frac{c_1}{2} \pm \sqrt{\frac{c_1^2}{4} - c_2}$$

and

$$p_3(f) = \sum_{k=3}^{\infty} |c_k|.$$

Proof: Again consider the polynomial $P(\lambda) = \lambda^n + c_1 \lambda^{n-1} + ... + c_n$ $(n \geq 3)$. Let $c_k \neq 0$ at for least one $k = 2, ..., n$. Then thanks to Corollary 4.8.3 any zero $z(P)$ of P, either satisfies the inequality $|z(P)| \leq 1$, or

$$|z^2(P) + c_1 z(P) + c_2| = |(z(P) - r_1)(z(P) - r_2)| \leq p_3(P)$$

where

$$p_3(P) = \sum_{k=3}^{n} |c_k|.$$

Now again use the polynomial defined by (10.1). Since $z(f_n) = 1/z(P)$, letting $n \to \infty$ by the Hurwitz theorem, we get the required result. Q. E. D.

Example 5.10.3 *Let*

$$f(z) = \cos\left(\frac{2\pi z}{3}\right).$$

Then its roots are

$$z_j(f) = \frac{3}{2}(j + 1/2), \ j = 0, \pm 1, \pm 2, ...$$

In this case,

$$c_1 = 0, c_2 = -\frac{2\pi^2}{9}, r_{1,2} = \pm\sqrt{2}\frac{\pi}{3}.$$

By Theorem 5.10.2, either $z_j(f) \geq 1$, or

$$\left|\frac{1}{z_j^2(f)} - \frac{2\pi^2}{9}\right| \leq p_3(f) = \sum_{k=2}^{\infty} \frac{(2\pi)^{2k}}{3^{2k}(2k)!} = \cosh(2\pi/3) - 1 - \frac{2\pi^2}{9} \leq 0.93.$$

For the zero $z(f) = 2/3 < 1$ this gives

$$\left|\frac{9}{4} - \frac{2\pi^2}{9}\right| \leq 0.93.$$

5.11   The series of powers of zeros and traces of matrices

Again consider the entire function

$$f(\lambda) = \sum_{k=0}^{\infty} c_k \lambda^k \quad (c_0 = 1) \tag{11.1}$$

with $\rho(f) < \infty$. For an integer $m \geq 2$, we will use the $m \times m$-matrix

$$A_m = \begin{pmatrix} -c_1 & -c_2 & \cdots & -c_{m-1} & -c_m \\ 1 & 0 & \cdots & 0 & 0 \\ 0 & 1 & \cdots & 0 & 0 \\ \cdot & \cdot & \cdots & \cdot & \cdot \\ 0 & 0 & \cdots & 1 & 0 \end{pmatrix}.$$

If f is a polynomial, then it is assumed that $deg\ f > m$. Set

$$\hat{s}_m(f) := \sum_{k=1}^{\infty} \frac{1}{z_k^m(f)}.$$

Theorem 5.11.1 *Let f be defined by (11.1). Then for any integer $m > \rho(f)$, the equality*

$$\hat{s}_m(f) = Trace\ A_m^m$$

is true.

Proof: With $n > m$, again consider the polynomial

$$P(\lambda) = \sum_{k=0}^{n} c_k \lambda^{n-k} \quad (c_0 = 1).$$

Because of Theorem 4.4.1,

$$\sum_{k=1}^{n} z_k^m(P) = Trace\ A_m^m.$$

Again use the function $f_n(\lambda) = c_0 + c_1\lambda + ... + c_n\lambda^n$. Besides, $z(f_n) = 1/z(P)$. Then

$$\hat{s}_m(f_n) = Trace\ A_m^m. \tag{11.2}$$

Because of the Hadamard theorem,

$$\sum_{k=1}^{\infty} \frac{1}{|z_k(f)|^m} < \infty.$$

Take into account that the zeros continuously depend on the coefficients. Then $\hat{s}_m(f_n) \to \hat{s}_m(f)$ as $n \to \infty$. We thus get the required result, letting in (11.2) $n \to \infty$. Q. E. D.

In particular, if $\rho(f) < 1$, then by the previous theorem $\hat{s}_1(f) = -c_1$. If $\rho(f) < 2$, then

$$\hat{s}_2(f) = c_1^2 - 2c_2.$$

If $\rho(f) < 3$, then

$$\hat{s}_3(f) = -c_1^3 + 3c_2c_1 - 3c_3.$$

Moreover, if $\rho(f) < 1$, then

$$\sum_{k \neq j} \frac{1}{z_j(f)\, z_k(f)} = [\hat{s}_1^2(f) - \hat{s}_2(f)]/2 = c_2.$$

Example 5.11.2 *Consider the function* $f(\lambda) = e^\lambda$.

Then $\rho(f) = 1$. Thanks to Theorem 5.11.1,

$$\hat{s}_2(f) = c_1^2 - 2c_2 = 0, \ \hat{s}_3(f) = -1 + \frac{3}{2} - \frac{3}{3!} = 0.$$

Example 5.11.3 *Consider the function*

$$f(\lambda) = \frac{sin\,(\lambda)}{\lambda}.$$

Then $\rho(f) = 1, c_1 = 0, c_2 = -\frac{1}{3!}$. Because of Theorem 5.11.1,

$$\hat{s}_2(f) = c_1^2 - 2c_2 = \frac{1}{3}.$$

But in the considered case

$$z_k(f) = \pi k \ \ (k = \pm 1, \pm 2, ...).$$

So

$$\hat{s}_2(f) = \frac{2\zeta(2)}{\pi^2},$$

where $\zeta(.)$ is the Riemann zeta function. We thus have proved that $\zeta(2) = \frac{\pi^2}{6}$.

5.12 Zero-free sets

Again, let $\psi_1 = 1$ and $\psi_k, k = 2, ...$ be positive numbers having the following property: the sequence

$$m_1 = 1, \ m_j := \frac{\psi_j}{\psi_{j-1}} \ \ (j = 2, 3, ...)$$

is nonincreasing and tends to zero, and

$$f(\lambda) = 1 + \sum_{k=1}^{\infty} a_k \psi_k \lambda^k \quad (\lambda \in \mathbf{C}). \tag{12.1}$$

Besides,

$$\alpha_\psi(f) = \overline{\lim}_{k\to\infty} \sqrt[k]{|a_k|} < 1. \tag{12.2}$$

Denote

$$\tilde{\theta}_{\psi,1}(f) = \sum_{k=1}^{\infty} |a_k|.$$

By Lemmas 5.2.1, 1.1.3, and the Hurwitz theorem, we get our next result.

Theorem 5.12.1 *Let f be defined by (12.1) and condition (12.2) hold. Then*

$$\inf_j |z_j(f)| \geq \frac{1}{\max\{\tilde{\theta}_{\psi,1}, m_2\}}$$

and

$$\inf_j |z_j(f)| \geq \frac{1}{\max_{k=1,2,\dots}(|a_k| + m_{k+1})}.$$

In particular, one can take $\psi_j = 1/(j!)^\gamma$.

Furthermore, let f_1 and f_2 be arbitrary entire functions and $f = f_1 + f_2$. Then we can write out

$$f(z) = det\, T_2(z)$$

where $T_2(z)$ is the 2×2-matrix defined by

$$T(z) = \begin{pmatrix} f_1^\alpha(z) & -f_2^\beta(z) \\ f_2^{1-\beta}(z) & f_1^{1-\alpha}(z) \end{pmatrix}$$

with $\beta, \alpha \in [0,1]$. By the Gerschgorin theorem, *all the zeros of $f(z)$ lie in the union of the sets*

$$\{z \in \mathbf{C} : |f_1(z)|^\alpha \leq |f_2(z)|^{1-\beta}\}$$

and

$$\{z \in \mathbf{C} : |f_1(z)|^{1-\alpha} \leq |f_2(z)|^\beta\}.$$

Here we take the main branches of the fractional powers.

In particular, one can take $\alpha = \beta = 1$. Then all the zeros of $f(z)$ lie the set

$$\{z \in \mathbf{C} : |f_1(z)| \leq 1\} \cup \{z \in \mathbf{C} : |f_2(z)| \geq 1\}.$$

This relation is obvious.

Furthermore, taking $\alpha = \beta = \frac{1}{2}$, then we can assert that all the zeros of $f(z)$ lie the set

$$\{z \in \mathbf{C} : |f_1(z)| \leq |f_2(z)|\}.$$

The last relation is also obvious: for any zero $z(f)$ of f, we have: $f_1(z(f)) + f_2(z(f)) = 0$. Or $|f_1(z(f))| = |f_2(z(f))|$.

5.13   Taylor coefficients of some infinite-order entire functions

Let us consider the entire function

$$f(z) = \sum_{k=0}^{\infty} c_k z^k, \tag{13.1}$$

assuming that

$$|f(z)| \le exp\,[Aexp\,|z|] \quad (A = const > 0, z \in \mathbf{C}). \tag{13.2}$$

Clearly, to this case can be reduced the function satisfying

$$|f(z)| \le Cexp\,[Aexp\,B|z|] \quad (B, C > 0)$$

if we take $f_1(z) = f(z)/C$ and $z_1 = Bz$.

Again, let $M_f(r) = \max_{|z|=r} |f(z)|$. By the well-known inequality for the coefficients of a power series

$$|c_n| \le \frac{M_f(r)}{r^n}. \tag{13.3}$$

By (13.2), we have

$$M_f(r) < e^{Ae^r}$$

and thus

$$|c_n| \le h(r) := \frac{e^{Ae^r}}{r^n}.$$

Let us use the usual method for finding extrema. Clearly,

$$h'(r) = \left(\frac{e^{Ae^r}}{r^n}\right)' = e^{Ae^r}\left(\frac{r^n Ae^r - nr^{n-1}}{r^{2n}}\right).$$

So $h'(r) = 0$ if and only if

$$rAe^r - n = 0.$$

Denote by x the unique positive root of this equation. For an $n \ge Ae$, we have $x \ge 1$. Hence, $Ae^x \le n$, or

$$x \le ln\,\frac{n}{A}.$$

Furthermore, since $a \le e^{a-1}$, $a > 0$, we obtain

$$n = xAe^x \le Ae^{-1}e^{2x}.$$

So

$$x \ge \frac{1}{2}\,ln\,\left[\frac{ne}{A}\right].$$

Hence,

$$\max_r h(r) = \frac{e^{Ae^x}}{x^n} \le \frac{(2e)^n}{(ln\,[\frac{ne}{A}])^n}.$$

Thus, we arrive at our next result.

Lemma 5.13.1 *Let the function f defined by (13.1) satisfy the inequality (13.2). Then for all $n \ge Ae$, the inequality*

$$|c_n| \le \frac{(2e)^n}{(ln\,[\frac{ne}{A}])^n}$$

is valid.

Furthermore, note that

$$exp\,[Aexp\,z] = \sum_{k=0}^{\infty} \frac{A^k}{k!} exp\,kz = \sum_{k=0}^{\infty} \frac{A^k}{k!} \sum_{j=0}^{\infty} \frac{k^j z^j}{j!}.$$

So the Taylor coefficients of the function $exp\,[Aexp\,z]$ equal to

$$\frac{1}{j!} \sum_{k=0}^{\infty} \frac{k^j A^k}{k!}.$$

Now the previous lemma implies.

Corollary 5.13.2 *For all $j \ge Ae$, we have the inequality*

$$\sum_{k=0}^{\infty} \frac{k^j A^k}{k!} \le j! \frac{(2e)^j}{(ln\,[\frac{je}{A}])^j}.$$

Now let us briefly consider the function satisfying the inequality

$$|f(z)| \le exp\,[exp\,[exp\,|z|]] \quad (z \in \mathbf{C}). \tag{13.4}$$

Consequently,

$$M_f(r) < e^{e^{e^r}}$$

and thus by (13.3)

$$|c_n| \le v(r) := \frac{1}{r^n} e^{e^{e^r}}.$$

Clearly,

$$v'(r) = e^{e^{e^r}} \frac{1}{r^{2n}} (r^n e^{e^r} e^r - n r^{n-1}).$$

The equation $v'(r) = 0$ is equivalent to the following one:

$$r e^{e^r} e^r - n = 0.$$

Denote by y the unique positive root of this equation. For an $n \geq e^{e+1}$, we have $y \geq 1$. Hence,

$$e^{e^y} \leq n,$$

or $y \leq \ln \ln n$. Since $(\ln a + a)' \leq e^a$ $(a \geq 1)$, we get $\ln a + a \leq e^a$ $(a \geq 1)$. Or

$$ae^a \leq e^{e^a} \quad (a \geq 1).$$

So

$$n = ye^y e^{e^y} \leq e^{2e^y} \quad (n \geq e^e)$$

and therefore

$$y \geq \ln \left(\frac{\ln n}{2} \right).$$

Thus,

$$\max_{r>0} v(r) = \frac{1}{y^n} e^{e^{e^y}} \leq \frac{e^n}{[\ln \left(\frac{\ln n}{2} \right)]^n}.$$

We now arrive at the following result.

Lemma 5.13.3 *Let the function f defined by (13.1) satisfy condition (13.4). Then for all $n \geq e^{e+1}$, the inequality*

$$|c_n| \leq \frac{e^n}{[\ln \frac{\ln n}{2}]^n}$$

is valid.

Furthermore, we have

$$\exp [\exp [\exp z]] = \sum_{k=0}^{\infty} \frac{\exp (ke^z)}{k!} = \sum_{k=0}^{\infty} \frac{1}{k!} \sum_{j=0}^{\infty} \frac{k^j e^{jz}}{j!}$$

$$= \sum_{k=0}^{\infty} \frac{1}{k!} \sum_{j=0}^{\infty} \frac{k^j}{j!} \sum_{m=0}^{\infty} \frac{j^m z^m}{m!}.$$

So the Taylor coefficients of the considered function are

$$\frac{1}{m!} \sum_{k=0}^{\infty} \frac{1}{k!} \sum_{j=0}^{\infty} \frac{k^j j^m}{j!}.$$

Now the previous lemma implies.

Corollary 5.13.4 *For all $m \geq e^{e+1}$, we have the inequality*

$$\sum_{k=0}^{\infty} \sum_{j=0}^{\infty} \frac{k^j j^m}{j! k!} \leq \frac{m! e^m}{[\ln \frac{\ln m}{2}]^m}.$$

5.14   Comments to Chapter 5

Chapter 5 is based on the papers [Gil', 2000b, 2001, 2005a] and [Gil', 2007a].

The literature on the location of the zeros of analytic functions is very rich. Of course, we could not survey the whole subject here and refer the reader to the interesting papers [Abian, 1981], [Bakan and Ruscheweyh, 2002], [Barnard, Pearce, and Wheeler, 2001], [Buckholtz and Shaw, 1972], [Cardon and de Gaston, 2005], [Chanane, 2003], [Csordas and Smith, 2000], [Dewan and Govil, 1990], [Ganelius, 1953], [Hanson, 1985], [Ioakimidis, 1988], [Ioakimidis and Anastasselou, 1985], and [Zheng and Yang, 1995], and references therein.

Chapter 6

Perturbations of Finite-Order Entire Functions

Let f and h be finite-order entire functions. In this chapter we investigate the following problem: if the Taylor coefficients of f and h are close, how close are their zeros? As a particular case of our results we consider the distance between the zeros of an entire function and the zeros of its derivative. We also investigate the distance between the zeros of the Taylor series of an entire function and the zeros of its tail.

6.1 Variations of zeros

Consider two entire functions f and h in the ψ-form

$$f(\lambda) = 1 + \sum_{k=1}^{\infty} a_k \psi_k \lambda^k \text{ and } h(\lambda) = 1 + \sum_{k=1}^{\infty} b_k \psi_k \lambda^k \qquad (1.1)$$

with complex coefficients a_k, b_k. As above $\psi_1 = 1$ and $\psi_k, k = 2, 3, \ldots$ are positive numbers, such that the sequence

$$m_1 = 1, \; m_j := \frac{\psi_j}{\psi_{j-1}} \; (j = 2, 3, \ldots)$$

is nonincreasing and tends to zero. It is assumed that

$$\alpha_\psi(f) = \overline{\lim}_{k \to \infty} \sqrt[k]{|a_k|} < 1, \alpha_\psi(h) = \overline{\lim}_{k \to \infty} \sqrt[k]{|b_k|} < 1 \qquad (1.2)$$

and

$$\sum_{k=1}^{\infty} m_k^{2p} < \infty \qquad (1.3)$$

for an integer $p \geq 1$.

In particular, if the functions are represented in the root-factorial form

$$f(\lambda) = \sum_{k=0}^{\infty} \frac{a_k \lambda^k}{(k!)^\gamma}, h(\lambda) = \sum_{k=0}^{\infty} \frac{b_k \lambda^k}{(k!)^\gamma} \quad (a_0 = b_0 = 1, \ 0 < \gamma < 1), \qquad (1.4)$$

then they have the form (1.1) with

$$\psi_k = \frac{1}{(k!)^\gamma}, \ m_k = \frac{1}{k^\gamma}.$$

Besides, under the condition

$$p > \frac{1}{2\gamma} \qquad (1.5)$$

we have (1.3), since

$$\sum_{k=1}^{\infty} \frac{1}{k^{2p\gamma}} = \zeta(2\gamma p) < \infty,$$

where $\zeta(s)$ is the Riemann zeta function.

Similarly, if the functions are represented in the Mittag-Leffler form

$$f(z) := \sum_{k=0}^{\infty} \frac{a_k z^k}{\Gamma(1 + k/\rho)}, \ h(z) := \sum_{k=0}^{\infty} \frac{b_k z^k}{\Gamma(1 + k/\rho)} \qquad (1.6)$$

for a $\rho > 1$, then they have the form (1.1) with

$$\psi_k = \frac{1}{\Gamma(1 + k/\rho)} \quad (k > 1).$$

As it is shown in Section 5.4,

$$\frac{\Gamma(1 + k/\rho)}{\Gamma(1 + \frac{k+1}{\rho})} \leq \kappa_\rho k^{-1/\rho} \quad (k = 2, 3, ...),$$

where

$$\kappa_\rho = \rho^{1/\rho} exp \left[\frac{1}{\rho} + \frac{1}{2(1+\rho)} \right].$$

Thus, under consideration, we have

$$\sum_{k=1}^{\infty} m_k^{2p} = 1 + \sum_{k=2}^{\infty} \frac{\Gamma^{2p}(1 + k/\rho)}{\Gamma^{2p}(1 + \frac{k+1}{\rho})}$$

$$\leq 1 + \kappa_\rho^{2p} \sum_{k=2}^{\infty} k^{-2p/\rho} \leq \kappa_\rho^{2p} \zeta(2p/\rho) < \infty,$$

provided

$$2p > \rho. \qquad (1.7)$$

The formally more general than (1.2) case

$$\alpha_\psi(f) := \overline{\lim}_{k\to\infty} \sqrt[k]{|a_k|} < \infty \text{ and } \alpha_\psi(h) := \overline{\lim}_{k\to\infty} \sqrt[k]{|b_k|} < \infty$$

can be reduced to condition (1.2) by the substitution

$$z_1 = 2z \, max \, \{\alpha_\psi(f), \, \alpha_\psi(h)\}.$$

Recall that if f has $l < \infty$ finite zeros, then, we set

$$\frac{1}{z_k(f)} = 0 \ (k > l).$$

Definition 6.1.1 *The quantity*

$$rv_f(h) = \sup_j \inf_k \left| \frac{1}{z_k(f)} - \frac{1}{z_j(h)} \right|$$

will be called the relative variation of the zeros of h with respect to the zeros of f.

Put

$$\varpi_p(f) := 2 \left[\sum_{k=1}^{\infty} |a_k|^2 \right]^{1/2} + 2 \left[\sum_{k=2}^{\infty} m_k^{2p} \right]^{1/2p}.$$

Finally, denote

$$q := \left[\sum_{k=1}^{\infty} |a_k - b_k|^2 \right]^{1/2}$$

and

$$\xi_p(f,y) := \sum_{k=0}^{p-1} \frac{\varpi_p^k(f)}{y^{k+1}} exp \left[\frac{1}{2} + \frac{\varpi_p^{2p}(f)}{2y^{2p}} \right] \ (y > 0).$$

Now we are in a position to formulate the main result of the present chapter.

Theorem 6.1.2 *Let f and h be defined by (1.1). Let conditions (1.2) and (1.3) be fulfilled. In addition, let $y_p(q, f)$ be the unique (positive) root of the equation*

$$q\xi_p(f,y) = 1. \tag{1.8}$$

Then

$$rv_f(h) \le y_p(q, f).$$

The proof of this theorem is presented in the next section. Since $\xi_p(f, .)$ is a monotonically decreasing function, the previous theorem implies

Corollary 6.1.3 *Let f and h be defined by (1.1) and conditions (1.2) and (1.3) be fulfilled. Then all the zeros of h lie in the union of the sets $W_k(p)$ ($k = 1, 2, 3, ...$) where*

$$W_k(p) := \left\{ \lambda \in \mathbf{C} : q\xi_p \left(f, \left| \frac{1}{z_k(f)} - \frac{1}{\lambda} \right| \right) \ge 1 \right\}.$$

In particular if f has $l < \infty$ finite zeros, then all the zeros of h lie in the set

$$\cup_{k=0}^{l} W_k(p),$$

where

$$W_0(p) = \{\lambda \in \mathbf{C} : q\xi_p(f, \frac{1}{|\lambda|}) \geq 1\}.$$

Note that

$$\xi_p\left(f, \frac{1}{|\lambda|}\right) = \sum_{k=0}^{p-1} \varpi_p^k(f) |\lambda|^{k+1} \, exp\left[\frac{1}{2}(1 + \varpi_p^{2p}(f)|\lambda|^{2p})\right].$$

Substitute the equality $y = x\varpi_p(f)$ into equation (1.8) and apply Lemma 1.6.4. Then, we have the inequality

$$y_p(q, f) \leq \delta_p(q, f), \tag{1.9}$$

where

$$\delta_p(q, f) := \begin{cases} epq & \text{if } \varpi_p(f) \leq epq, \\ \varpi_p(f) \, [ln \, (\varpi_p(f)/qp)]^{-1/2p} & \text{if } \varpi_p(f) > epq \end{cases}.$$

Now Theorem 6.1.2 yields the inequality $rv_f(h) \leq \delta_p(q, f)$.

If f has an infinite set of finite zeros, then according to Theorem 6.1.2 and (1.9), for any zero $z(h)$ of h, there is a zero $z(f)$ of f, such that

$$|z(h) - z(f)| \leq y_p(q, f)|z(h)z(f)| \leq \delta_p(q, f)|z(h)z(f)|.$$

These relations imply the inequalities

$$|z(f)| - |z(h)| \leq y_p(q, f)|z(h)z(f)| \leq \delta_p(f, q)|z(h)z(f)|.$$

Hence,

$$|z(h)| \geq \frac{|z(f)|}{y_p(q, f)|z(f)| + 1}$$

$$\geq \frac{|z(f)|}{\delta_p(q, f)|z(f)| + 1}.$$

Let us assume that under (1.1), for a constant $R \in (0, 1)$, we have

$$\overline{\lim}_{k \to \infty} \sqrt[k]{|a_k|} < \frac{1}{R}$$

and

$$\overline{\lim}_{k \to \infty} \sqrt[k]{|b_k|} < \frac{1}{R},$$

and consider the functions

$$\tilde{f}(\lambda) = \sum_{k=0}^{\infty} a_k \psi_k (R\lambda)^k \quad \text{and} \quad \tilde{h}(\lambda) = \sum_{k=0}^{\infty} \psi_k b_k (R\lambda)^k.$$

That is, $\tilde{f}(\lambda) \equiv f(R\lambda)$ and $\tilde{h}(\lambda) \equiv h(R\lambda)$. So functions $\tilde{f}(\lambda)$ and $\tilde{h}(\lambda)$ can be written in the form (1.1) and satisfy conditions (1.2). Moreover, under consideration,

$$\varpi_p(\tilde{f}) = 2\Big[\sum_{k=1}^{\infty} R^{2k}|a_k|^2\Big]^{1/2} + 2\Big[\sum_{k=2}^{\infty} m_k^{2p}\Big]^{1/2p}.$$

Thus, we can apply Theorem 6.1.2 and its corollaries to functions $\tilde{f}(\lambda), \tilde{h}(\lambda)$ and to take into account that

$$z_k(\tilde{f})R = z_k(f), \ z_k(\tilde{h})R = z_k(h).$$

6.2 Proof of Theorem 6.1.2

For a finite integer $n \geq 2$, put

$$\varpi_{p,n}(f) := 2\Big[\sum_{k=1}^{n}|a_k|^2\Big]^{1/2} + 2\Big[\sum_{k=2}^{n}m_k^{2p}\Big]^{1/2p},$$

and

$$\xi_{p,n}(f,y) := \sum_{k=0}^{p-1} \frac{\varpi_{p,n}^k(f)}{y^{k+1}}exp\Big[\frac{1}{2} + \frac{\varpi_{p,n}^{2p}(f)}{2y^{2p}}\Big] \ (y > 0),$$

and

$$q_n := \Big[\sum_{k=1}^{n}|a_k - b_k|^2\Big]^{1/2}.$$

Consider the polynomials

$$P(\lambda) = \lambda^n + \sum_{k=1}^{n}a_k\psi_k\lambda^{n-k} \ \text{and} \ Q(\lambda) = \lambda^n + \sum_{k=1}^{n}b_k\psi_k\lambda^{n-k} \ (a_0 = b_0 = 1).$$

Lemma 6.2.1 *For any zero $z(Q)$ of Q, there is a zero $z(P)$ of P, such that*

$$|z(P) - z(Q)| \leq y_p(q_n, P),$$

where $y_p(q_n, P)$ be the unique (positive) root of the equation

$$q_n\xi_{p,n}(f,y) = 1. \tag{2.1}$$

Proof: Introduce the matrices

$$
A_n =
\begin{pmatrix}
-a_1 & -a_2 & \cdots & -a_{n-1} & -a_n \\
m_2 & 0 & \cdots & 0 & 0 \\
0 & m_3 & \cdots & 0 & 0 \\
\cdot & \cdot & \cdots & \cdot & \cdot \\
0 & 0 & \cdots & m_n & 0
\end{pmatrix}
$$

and

$$
B_n =
\begin{pmatrix}
-b_1 & -b_2 & \cdots & -b_{n-1} & -b_n \\
m_2 & 0 & \cdots & 0 & 0 \\
0 & m_3 & \cdots & 0 & 0 \\
\cdot & \cdot & \cdots & \cdot & \cdot \\
0 & 0 & \cdots & m_n & 0
\end{pmatrix}.
$$

Thanks to Lemma 5.2.1, we have

$$
\lambda_k(A_n) = z_k(P), \quad \lambda_k(B_n) = z_k(Q) \quad (k = 1, 2, ..., n), \tag{2.2}
$$

where $\lambda_k(.), k = 1, ..., n$ are the eigenvalues with their multiplicities. Besides,

$$
q_n = \|A_n - B_n\|,
$$

where $\|.\|$ is the operator Euclidean norm. Because of Theorem 2.12.4, for any $\lambda_j(B_n)$ there is an $\lambda_i(A_n)$, such that

$$
|\lambda_j(B_n) - \lambda_i(A_n)| \le y_p(q_n), \tag{2.3}
$$

where $y_p(q_n)$ is the unique (positive) root of the equation

$$
q_n \sum_{k=0}^{p-1} \frac{(2N_{2p}(A_n))^k}{y^{k+1}} exp \left[\frac{1}{2} + \frac{(2N_{2p}(A_n))^{2p}}{2y^{2p}}\right] = 1.
$$

Recall that

$$
N_{2p}(A) = [Trace(AA^*)^p]^{1/2p}
$$

and the asterisk means the adjointness. But $A_n = M + C$, where

$$
M =
\begin{pmatrix}
-a_1 & -a_2 & \cdots & -a_{n-1} & -a_n \\
0 & 0 & \cdots & 0 & 0 \\
\cdot & \cdot & \cdots & \cdot & \cdot \\
0 & 0 & \cdots & 0 & 0
\end{pmatrix}
$$

and

$$
C =
\begin{pmatrix}
0 & 0 & \cdots & 0 & 0 \\
m_2 & 0 & \cdots & 0 & 0 \\
0 & m_3 & \cdots & 0 & 0 \\
\cdot & \cdot & \cdots & \cdot & \cdot \\
0 & 0 & \cdots & m_n & 0
\end{pmatrix}.
$$

Therefore, with

$$\theta_P = \Big[\sum_{k=1}^{n} |a_k|^2\Big]^{1/2},$$

we have

$$MM^* = \begin{pmatrix} \theta_P^2 & 0 & \ldots & 0 & 0 \\ 0 & 0 & \ldots & 0 & 0 \\ . & . & \ldots & . & . \\ 0 & 0 & \ldots & 0 & 0 \end{pmatrix}$$

and

$$CC^* = \begin{pmatrix} 0 & 0 & \ldots & 0 & 0 \\ 0 & m_2^2 & \ldots & 0 & 0 \\ 0 & 0 & \ldots & 0 & 0 \\ . & . & \ldots & . & . \\ 0 & 0 & \ldots & 0 & m_n^2 \end{pmatrix}.$$

Hence,

$$s_k(C^*) = m_{k+1}$$

and

$$N_{2p}(A_n) \le N_{2p}(M^*) + N_{2p}(C^*) = \theta_P + \Big[\sum_{k=2}^{n} m_k^{2p}\Big]^{1/2p} = \varpi_{p,n}(f)/2.$$

Consequently, $y_p(q_n) \le y_p(q_n, f)$. Therefore, (2.3) and (2.2) imply the required result. Q. E. D.

Proof of Theorem 6.1.2: Consider the polynomials

$$f_n(\lambda) = 1 + \sum_{k=1}^{n} a_k \psi_k \lambda^k \text{ and } h_n(\lambda) = 1 + \sum_{k=1}^{n} b_k \psi_k \lambda^k.$$

Since

$$\lambda^n f_n(1/\lambda) = P(\lambda), \quad h_n(1/\lambda)\lambda^n = Q(\lambda),$$

we have

$$z_k(P) = \frac{1}{z_k(f_n)} \text{ and } z_k(Q) = \frac{1}{z_k(h_n)}. \tag{2.4}$$

Take into account that the roots continuously depend on coefficients, we have the required result, letting in the previous lemma $n \to \infty$. Q. E. D.

6.3 Approximations by partial sums

Again, consider the function

$$f(\lambda) = 1 + \sum_{k=1}^{\infty} a_k \psi_k \lambda^k. \tag{3.1}$$

It is assumed that the conditions (1.3) and

$$\alpha_\psi(f) = \overline{\lim}_{k\to\infty} \sqrt[k]{|a_k|} < 1 \tag{3.2}$$

are fulfilled. Recall that $\varpi_{p,n}(f)$ and $\xi_{p,n}(f, y)$ for an integer $p > 1$ are defined in the previous section, and

$$f_n(\lambda) = 1 + \sum_{k=1}^{n} a_k \psi_k \lambda^k \quad (2 \le n < \infty).$$

Put

$$\nu_n = \Big[\sum_{k=n+1}^{\infty} |a_k|^2 \Big]^{1/2}.$$

Theorem 6.3.1 *Under conditions (3.2) and (1.3), all the zeros of the function f defined by (3.1) are in the set*

$$\cup_{j=0}^{n} W_j(\nu_n),$$

where

$$W_0(\nu_n) = \big\{ z \in \mathbf{C} : \nu_n \xi_{p,n}\big(f, \tfrac{1}{|z|}\big) \ge 1 \big\}$$

and

$$W_j(\nu_n) = \big\{ z \in \mathbf{C} : \nu_n \xi_{p,n}\big(f, \big|\tfrac{1}{z_j(f_n)} - \tfrac{1}{z}\big|\big) \ge 1 \big\} \quad (j = 1, ..., n).$$

Proof: Put in Corollary 6.1.3 $f = f_n$ and $h = f$. Then $q = \nu_n$ and we get the required result. Q. E. D.

Since $\xi_{p,n}(f, y)$ monotonically decreases in $y > 0$, from the previous theorem it follows

Corollary 6.3.2 *Let function f be defined by (3.1) and conditions (3.2), and (1.3) be fulfilled. In addition, let $y_p(f_n)$ be the unique (positive) root of the equation*

$$\nu_n \xi_{p,n}(f, y) = 1. \tag{3.3}$$

Then for any zero $z(f)$ of f, either there is a zero $z(f_n)$ of f_n, such that

$$\Big| \frac{1}{z(f)} - \frac{1}{z(f_n)} \Big| \le y_p(f_n),$$

or

$$|z(f)| \geq \frac{1}{y_p(f_n)}.$$

Substitute the equality $y = x\varpi_p(f_n)$ into (3.3) and apply Lemma 1.6.4. Then, we have $y_p(f_n) \leq \delta_p(f_n)$, where

$$\delta_p(f_n) := \begin{cases} ep\nu_n & \text{if } \varpi_{p,n}(f) \leq ep\nu_n, \\ \varpi_p(f_n) \left[ln \left(\varpi_{p,n}(f)/q\nu_n \right) \right]^{-1/2p} & \text{if } \varpi_{p,n}(f_n) > ep\nu_n \end{cases}.$$

Now Theorem 6.3.1 yields

Corollary 6.3.3 *Let conditions (1.3) and (3.2) be fulfilled. Then, for any zero $z(f)$ of the function f defined by (3.1), either there is a zero $z(f_n)$ of f_n, such that*

$$\left| \frac{1}{z(f_n)} - \frac{1}{z(f)} \right| \leq \delta_p(f_n), \ or \ |z(f)| \geq \frac{1}{\delta_p(f_n)}.$$

6.4 Preservation of multiplicities

Again consider the functions defined by (1.1). Take a zero $z_m(f)$ of a multiplicity μ_m. Denote

$$\tilde{\beta}_m := \inf_{z_k(f) \neq z_m(f)} \left| \frac{1}{z_m(f)} - \frac{1}{z_k(f)} \right|/2.$$

Recall that ξ_p and $\xi_{p,n}$ are defined in Sections 6.1 and 6.2, respectively.

Theorem 6.4.1 *Let f and h be defined by (1.1) under conditions (1.2) and (1.3). Let a zero $z_m(f)$ of f have a multiplicity μ_m. In addition, for an $r \leq \tilde{\beta}_m$, let*

$$q\xi_p(f,r) < 1.$$

Then h has in the set

$$\left\{ \lambda \in \mathbf{C} : \left| \frac{1}{z_m(f)} - \frac{1}{\lambda} \right| \leq r \right\}$$

zeros whose total multiplicity is also equal to μ_m.

To prove this theorem, we need the matrices A_n and B_n, and the polynomials P and Q introduced in Section 6.2.

Lemma 6.4.2 *Let a zero $z_m(P)$ of P have a multiplicity μ_m. In addition, with the notation*

$$\beta_m(P) := \inf_{z_k(P) \neq z_m(P)} |z_m(P) - z_k(P)|/2,$$

6. *Perturbations of Finite-Order Entire Functions*

for a positive $r \leq \beta_m$, let

$$q_n \xi_{p,n}(f,r) < 1.$$

Then Q has in $|z_m(P) - \lambda| \leq r$ zeros, whose total multiplicity is also equal to μ_m.

Proof: The proof of this lemma is similar to the proof of Lemma 6.2.1 with Corollary 2.13.3 taken instead of Theorem 2.12.4. Q. E. D.

Proof of Theorem 6.4.1: Consider the polynomials f_n and h_n defined as in Section 6.2. Thanks to equations (2.4) and the previous lemma, we have the required result for f_n and h_n. Now taking into account that the roots continuously depend on coefficients, and letting $n \to \infty$, we prove the theorem. Q. E. D.

Since $\xi_p(f,.)$ is a monotonically decreasing function, Theorem 6.4.1 implies

Corollary 6.4.3 *Let f and h be defined by (1.1) under conditions (1.2) and (1.3). Let a zero $z_m(f)$ of function f have a multiplicity μ_m. In addition, let the unique (positive) root $y_p(q, f)$ of the equation*

$$q \xi_p(f,y) = 1$$

satisfy the inequality $y_p(q,f) \leq \tilde{\beta}_m$. Then h has in the set

$$\left\{ \lambda \in \mathbf{C} : \left| \frac{1}{z_m(f)} - \frac{1}{\lambda} \right| \leq y_p(q,f) \right\}$$

zeros whose total multiplicity is also equal to μ_m.

As it is proved in Section 6.1, $y_p(q,f) \leq \delta_p(q,f)$. Now Corollary 6.4.3 yields the following result: if a zero $z_m(f)$ of f has a multiplicity μ_m and $\delta_p(q,f) \leq \tilde{\beta}_m$, then h has in the set

$$\left\{ \lambda \in \mathbf{C} : \left| \frac{1}{z_m(f)} - \frac{1}{\lambda} \right| \leq \delta_p(q,f) \right\}$$

zeros whose total multiplicity is also equal to μ_m.

6.5 Distances between roots and critical points

For the brevity, in this section we consider, a function in the root-factorial form

$$f(\lambda) = \sum_{k=0}^{\infty} \frac{a_k \lambda^k}{(k!)^\gamma} \quad (a_0 = 1, \ \gamma \in (0,1)) \tag{5.1}$$

but similarly, functions in the ψ-form and in the Mittag-Leffler form can be considered. It is assumed that

$$\overline{\lim}_{k\to\infty} \sqrt[k]{|a_k|} < 1. \tag{5.2}$$

By (5.1), we obtain

$$f'(\lambda) = \sum_{k=1}^{\infty} \frac{a_k k \lambda^{k-1}}{(k!)^\gamma} = \sum_{k=1}^{\infty} \frac{a_k k^{1-\gamma} \lambda^{k-1}}{[(k-1)!]^\gamma}.$$

Assume that

$$a_1 \neq 0 \tag{5.3}$$

and put

$$h(\lambda) = \frac{1}{a_1} f'(\lambda) = \sum_{k=0}^{\infty} \frac{b_k \lambda^k}{(k!)^\gamma}$$

with

$$b_k = (k+1)^{1-\gamma} \frac{a_{k+1}}{a_1}.$$

Clearly, the zeros of f' and h coincide and $b_0 = 1$.

Furthermore, for an integer $p > 1/2\gamma$, under consideration, we have $\varpi_p(f) = \tilde{\varpi}_p(f)$ where

$$\tilde{\varpi}_p(f) := 2\left[\sum_{k=1}^{\infty} |a_k|^2\right]^{1/2} + 2\left[\zeta(2\gamma p) - 1\right]^{1/2p}.$$

In addition, $\xi_p(f, y) = \tilde{\xi}_p(f, y)$ where

$$\tilde{\xi}_p(f, y) := \sum_{k=0}^{p-1} \frac{\tilde{\varpi}_p^k(f)}{y^{k+1}} exp\left[\frac{1}{2} + \frac{\tilde{\varpi}_p^{2p}(f)}{2y^{2p}}\right] \quad (y > 0).$$

Finally, denote

$$q(f') := \left[\sum_{k=1}^{\infty} |a_k - (k+1)^{1-\gamma} \frac{a_{k+1}}{a_1}|^2\right]^{1/2}.$$

Now Theorem 6.1.2 yields

Corollary 6.5.1 *Let f be defined by (5.1). Let conditions (5.2) and (5.3) be fulfilled. In addition, let $y_p(f')$ be the unique (positive) root of the equation*

$$q(f')\tilde{\xi}_p(f, y) = 1. \tag{5.4}$$

Then $rv_f(f') \leq y_p(f')$.

Since $\tilde{\xi}_p(f, .)$ is a monotonically decreasing function, the previous corollary implies

Corollary 6.5.2 *Let f be defined by (5.1) and conditions (5.2), and (5.3) be fulfilled. Then all the zeros of f' lie in the union of the sets $W_k(p, f')$, where*

$$W_k(p, f) := \{\lambda \in \mathbf{C} : q(f')\tilde{\xi}_p(f, |\frac{1}{z_k(f)} - \frac{1}{\lambda}|) \geq 1\} \quad (k = 1, 2, 3, ...).$$

In particular, if f has $l < \infty$ finite zeros, then for $k > l$, we have

$$W_k(p, f') \equiv W_0(p, f') := \{\lambda \in \mathbf{C} : q(f')\tilde{\xi}_p(f, \frac{1}{|\lambda|}) \geq 1\}$$

and thus all the zeros of f' lie in the set

$$\cup_{k=0}^{l} W_k(p, f').$$

Note that

$$\tilde{\xi}_p(f, \frac{1}{|\lambda|}) = \sum_{k=0}^{p-1} \tilde{\varpi}_p^k(f)|\lambda|^{k+1} \, exp\,[\frac{1}{2}(1 + \tilde{\varpi}_p^{2p}(f)|\lambda|^{2p})].$$

Substitute in (5.4) the equality $y = x\tilde{\varpi}_p(f)$ and apply Lemma 1.6.4. Then, we have $y_p(f') \leq \delta_p(f')$, where

$$\delta_p(f') := \begin{cases} epq(f') & \text{if } \tilde{\varpi}_p(f) \leq epq(f'), \\ \tilde{\varpi}_p(f) \, [ln\,(\tilde{\varpi}_p(f)/q(f')p)]^{-1/2p} & \text{if } \tilde{\varpi}_p(f) > epq(f') \end{cases} .$$

Corollary 6.5.1 now yields the inequality

$$rv_f(f') \leq \delta_p(f').$$

6.6   Tails of Taylor series

Again consider the function $f(\lambda)$ defined by (5.1) under condition (5.2). Consider also the n-th tail

$$t_n(\lambda) := \sum_{k=n}^{\infty} \frac{a_k \lambda^k}{(k!)^\gamma}.$$

Assuming that

$$a_n \neq 0, \tag{6.1}$$

put

$$h(\lambda) = \frac{(n!)^\gamma}{\lambda^n a_n} t_n(\lambda) = \sum_{k=0}^{\infty} \frac{b_k \lambda^k}{(k!)^\gamma}$$

with

$$b_k = \frac{(k!n!)^\gamma a_{k+n}}{[(k+n)!]^\gamma a_n}.$$

Because of (5.2),

$$\sum_{k=0}^{\infty} |b_k|^2 \leq \frac{(n!)^{2\gamma}}{|a_n|^2} \sum_{k=0}^{\infty} |a_k|^2 < \infty.$$

Clearly, the nonzero zeros of t_n and h coincide.

Furthermore, for an integer $p > 1/2\gamma$ again use $\tilde{\varpi}_p(f)$ and $\tilde{\xi}_p(f, y)$ (see Section 6.5). Finally, denote

$$q(t_n) := \left[\sum_{k=1}^{\infty} |a_k - b_k|^2\right]^{1/2} =$$

$$\left[\sum_{k=1}^{\infty} |a_k - \frac{(k!n!)^\gamma a_{k+n}}{[(k+n)!]^\gamma a_n}|^2\right]^{1/2}.$$

Now Theorem 6.1.2 yields the following result.

Corollary 6.6.1 *Let f be defined by (5.1). Let conditions (5.2) and (6.1) be fulfilled. In addition, let $y_p(t_n)$ be the unique (positive) root of the equation*

$$q(t_n)\tilde{\xi}_p(f, y) = 1. \tag{6.2}$$

Then $rv_f(t_n) \leq y_p(t_n)$.

Since $\tilde{\xi}_p(f, .)$ is a monotonically decreasing function, the previous corollary implies

Corollary 6.6.2 *Let f be defined by (5.1). Let conditions (5.2) and (6.1) be fulfilled. Then all the zeros of t_n lie in the set*

$$\cup_{k=1}^{\infty} W_k(p, t_n),$$

where

$$W_k(p, t_n) := \left\{\lambda \in \mathbf{C} : q(t_n)\tilde{\xi}_p\left(f, |\frac{1}{z_k(f)} - \frac{1}{\lambda}|\right) \geq 1\right\} \quad (k = 1, 2, ...).$$

In particular, if f has $l < \infty$ finite zeros, then for $k > l$, we have

$$W_k(p, t_n) \equiv W_0(p, t_n) = \left\{\lambda \in \mathbf{C} : q(t_n)\tilde{\xi}_p\left(f, \frac{1}{|\lambda|}\right) \geq 1\right\}$$

and thus all the zeros of t_n lie in the set

$$\cup_{k=0}^{l} W_k(p, t_n).$$

Substitute in (6.2) the equality $y = x\tilde{\varpi}_p(f)$ and apply Lemma 1.6.4. Then, we have $y_p(f) \leq \delta_p(t_n)$, where

$$\delta_p(t_n) := \begin{cases} epq(t_n) & \text{if } \tilde{\varpi}_p(f) \leq epq(t_n), \\ \tilde{\varpi}_p(f) \left[ln \left(\tilde{\varpi}_p(f)/q(t_n)p \right) \right]^{-1/2p} & \text{if } \tilde{\varpi}_p(f) > epq(t_n) \end{cases}.$$

Corollary 6.6.1 now yields the inequality $rv_f(t_n) \leq \delta_p(t_n)$.

Similarly, one can consider the relative variation $rv_{t_n}(f)$. Moreover, one can investigate $rv_f(t_n)$ for functions represented in the Mittag-Leffler form instead of (5.1).

6.7   Comments to Chapter 6

The variation of the zeros of analytic functions under perturbations has been investigated rather intensively. In particular, P. Rosenbloom [1969] established the perturbation result that provides the existence of a zero of a perturbed function in a given domain. Location of the zeros of derivatives was explored in the interesting papers [Clunie and Edrei, 1991], [Craven, Csordas, and Smith, 1987], [Edwards and Hellerstein, 2002], [Genthner, 1985], [Harel, Namn and Sturm, 1999], [Schmieder and Szynal, 2002], and [Sheil-Small, 1989]. The zeros of the tails of Taylor series, were investigated in the well-known papers [Ostrovskii, I.V., 2000] and [Yildirim, 1994].

In the present chapter, a new approach to the mentioned problems is proposed. Chapter 6 is based on the papers [Gil', 2000a and 2000c].

Chapter 7

Functions of Order Less than Two

In this chapter for entire functions whose order is less than two we make sharper the perturbation results proved in Chapter 6. In addition, we establish an identity between the sums

$$\sum_{k=1}^{\infty} \left(Im\ \frac{1}{z_k(f)} \right)^2 \text{ and } \sum_{k=1}^{\infty} \left(Re\ \frac{1}{z_k(f)} \right)^2.$$

7.1 Relations between real and imaginary parts of zeros

Let

$$f(\lambda) = 1 + \sum_{k=1}^{\infty} c_k \lambda^k \ \ (\lambda \in \mathbf{C}), \tag{1.1}$$

be an entire function with complex coefficients of order less than two. Let $\psi_1 = 1$ and $\psi_k, k = 2, 3, \dots$ be positive numbers having the following property: the sequence

$$m_1 = 1,\ m_j := \frac{\psi_j}{\psi_{j-1}} \ \ (j = 2, 3, \dots)$$

is nonincreasing and

$$\sum_{k=2}^{\infty} m_k^2 = \sum_{j=2}^{\infty} \frac{\psi_j^2}{\psi_{j-1}^2} < \infty. \tag{1.2}$$

Put

$$a_k = \frac{c_k}{\psi_k},\ k = 1, 2, \dots.$$

Then we have the ψ-representation

$$f(\lambda) = 1 + \sum_{k=1}^{\infty} a_k \psi_k \lambda^k \quad (\lambda \in \mathbf{C}). \tag{1.3}$$

It is assumed that

$$\theta_\psi(f) := \sqrt{\sum_{k=1}^{\infty} |a_k|^2} = \sqrt{\sum_{k=1}^{\infty} \frac{|c_k|^2}{\psi_k^2}} < \infty. \tag{1.4}$$

Recall that if f has $l < \infty$ finite zeros, we set

$$\frac{1}{z_k(f)} = 0 \quad (k = l+1, l+2, ...).$$

Denote

$$S_2(f) := \sum_{k=1}^{\infty} \frac{1}{|z_k(f)|^2}$$

and

$$\tau_\psi(f) := \theta_\psi^2(f) + \sum_{k=2}^{\infty} m_k^2,$$

and

$$\chi_\psi(f, t) := Re\,(a_1 e^{it})^2 - 2m_2 Re\,a_2 e^{2it} + \tau_\psi(f)$$

for a $t \in [0, 2\pi)$.

Theorem 7.1.1 *Let the function f defined by (1.3) satisfy conditions (1.2) and (1.4) . Then for any $t \in [0, 2\pi)$, the relations*

$$\tau_\psi(f) - S_2(f) = \chi_\psi(f, t) - 2 \sum_{k=1}^{\infty} \left(Re\, \frac{e^{it}}{z_k} \right)^2 \geq 0$$

are valid.

This theorem is proved in the next section.
     Note that

$$\chi_\psi(f, 0) = Re\, a_1^2 - 2m_2 Re\, a_2 + \tau_\psi(f).$$

In addition,

$$\chi_\psi(f, \pi/2) = -Re\, a_1^2 + 2m_2 Re\, a_2 + \tau_\psi(f).$$

Denote

$$J_2(f) := \sum_{k=1}^{\infty} (Im\, \frac{1}{z_k(f)})^2 \text{ and } R_2(f) := \sum_{k=1}^{\infty} \left(Re\, \frac{1}{z_k(f)} \right)^2.$$

Now Theorem 7.1.1 implies

Corollary 7.1.2 *Let f be defined by (1.1), and conditions (1.2) and (1.4) hold. Then*

$$\tau_\psi(f) - S_2(f) = \chi_\psi(f, 0) - 2R_2(f) = \chi_\psi(f, \pi/2) - 2J_2(f) \geq 0,$$

and therefore, the inequalities

$$S_2(f) \leq \tau_\psi(f),$$

$$2R_2(f) \leq \chi_\psi(f, 0), \quad and \quad 2J_2(f) \leq \chi_\psi(f, \pi/2)$$

are valid.

In particular, let

$$\psi_k = \frac{1}{(k!)^\gamma} \quad (\gamma \in (1/2, 1]). \tag{1.5}$$

Then f takes the root-factorial form

$$f(\lambda) = \sum_{k=0}^{\infty} \frac{a_k \lambda^k}{(k!)^\gamma} \quad (\gamma \in (1/2, 1], \ \lambda \in \mathbb{C}, \ a_0 = 1). \tag{1.6}$$

Note that the case $\gamma = 1$ is also considered in the next chapter. In the case (1.5) under condition (1.4), we have

$$\rho(f) \leq \frac{1}{\gamma} < 2.$$

Moreover, for any function f with $f(0) = 1$, whose order is $\rho(f) < 2$, we can take $\gamma > 1/\rho(f)$, such that representation (1.6) holds with condition (1.4) and ψ_k defined by (1.5). Then

$$\tau_\psi(f) = \tau_\gamma(f) \text{ and } \chi_\psi(f, t) = \chi_\gamma(f, t),$$

where

$$\tau_\gamma(f) := \sum_{k=1}^{\infty} |a_k|^2 + \zeta(2\gamma) - 1$$

and

$$\chi_\gamma(f, t) := Re \, (a_1 e^{it})^2 - 2^{1-\gamma} Re \, a_2 e^{2it} + \tau_\gamma(f)$$

for a $t \in [0, 2\pi)$. Note that

$$\chi_\gamma(f, 0) = Re \, a_1^2 - 2^{1-\gamma} Re \, a_2 + \tau_\gamma(f)$$

and

$$\chi_\gamma(f, \pi/2) = -Re \, a_1^2 + 2^{1-\gamma} Re \, a_2 + \tau_\gamma(f).$$

Corollary 7.1.2 implies that for a function f be defined by (1.6) under condition (1.4), for any $t \in [0, 2\pi)$, the relations

$$\tau_\gamma(f) - S_2(f) = \chi_\gamma(f, t) - 2 \sum_{k=1}^{\infty} (Re \, \frac{e^{it}}{z_k})^2 \geq 0$$

are valid. Moreover,

$$\tau_\gamma(f) - S_2(f) = \chi_\gamma(f,0) - 2R_2(f) = \chi_\gamma(f,\pi/2) - 2J_2(f) \geq 0,$$

and therefore,

$$S_2(f) \leq \tau_\gamma(f),$$

and

$$2R_2(f) \leq \chi_\gamma(f,0), \text{ and } 2J_2(f) \leq \chi_\gamma(f,\pi/2).$$

Similarly, one can take

$$\psi_k = \frac{1}{\Gamma(1 + k/\rho)} \quad (0 < \rho < 2)$$

and use the Mittag-Leffler form of f.

7.2   Proof of Theorem 7.1.1

Again consider the polynomial

$$P(\lambda) = \lambda^n + \sum_{k=1}^{n} a_k \psi_k \lambda^{n-k}. \tag{2.1}$$

Denote

$$\tau(P) = \sum_{k=1}^{n} |a_k|^2 + \sum_{k=2}^{n} m_k.$$

For a $t \in [0, 2\pi)$ put

$$\chi(P,t) := Re\,(a_1^2 e^{2it}) - 2m_2 Re\,(a_2 e^{2it}) + \tau(P).$$

Lemma 7.2.1 Let P be defined by (2.1). Then, for any $t \in [0, 2\pi)$, we have

$$\tau(P) - \sum_{k=1}^{n} |z_k(P)|^2 = \chi(P,t) - \frac{1}{2} \sum_{k=1}^{n} |e^{it} z_k(P) + e^{-it} \overline{z}_k(P)|^2.$$

Proof: Again use the matrix

$$A_n = \begin{pmatrix} -a_1 & -a_2 & \cdots & -a_{n-1} & -a_n \\ m_2 & 0 & \cdots & 0 & 0 \\ 0 & m_3 & \cdots & 0 & 0 \\ \cdot & \cdot & \cdots & \cdot & \cdot \\ 0 & 0 & \cdots & m_n & 0 \end{pmatrix}. \tag{2.2}$$

By Lemma 5.2.1,
$$\lambda_k(A_n) = z_k(P) \quad (k = 1, ..., n). \tag{2.3}$$

Because of Theorem 1.9.1,

$$N_2^2(A_n) - \sum_{k=1}^{n} |\lambda_k(A_n)|^2$$

$$= N_2^2(e^{it}A + e^{-it}A^*)/2 - \frac{1}{2}\sum_{k=1}^{n} |e^{it}\lambda_k(A_n) + e^{-it}\overline{\lambda}_k(A_n)|^2, \tag{2.4}$$

where $N_2(.)$ is the Hilbert-Schmidt norm. But

$$e^{it}A_n + A_n^*(\gamma)e^{-it} =$$

$$\begin{pmatrix} -a_1e^{it} - e^{-it}\overline{a}_1 & e^{-it}m_2 - a_2e^{it} & ... & -a_{n-1}e^{it} & -a_ne^{it} \\ e^{it}m_2 - \overline{a}_2e^{-it} & 0 & ... & 0 & 0 \\ -\overline{a}_3e^{-it} & e^{it}m_3 & ... & 0 & 0 \\ . & . & ... & . & . \\ -\overline{a}_ne^{-it} & 0 & ... & e^{it}m_n & 0 \end{pmatrix}.$$

Simple calculations show that

$$N_2^2(A_n) = \tau(P) \tag{2.5}$$

and

$$N_2^2(e^{it}A + e^{-it}A^*)/2 = |a_1e^{it} + \overline{a}_1e^{-it}|^2/2$$

$$+ |m_2 - a_2e^{2it}|^2 + \sum_{k=3}^{n} (|a_k|^2 + m_k^2).$$

So

$$N_2^2(e^{it}A + e^{-it}A^*)/2 = |a_1|^2 + (a_1^2e^{2it} + \overline{a}_1^2e^{-2it})/2$$

$$+ m_2^2 + |a_2|^2 - m_2(a_2e^{2it} + \overline{a}_2e^{-2it})$$

$$+ \sum_{k=3}^{n} (|a_k|^2 + m_k^2) = \chi(P, t).$$

Hence (2.3) and (2.4) imply the required result. Q. E. D.

Proof of Theorem 7.1.1: Consider the polynomial

$$f_n(\lambda) = 1 + \sum_{k=1}^{n} a_k\psi_k\lambda^k.$$

Clearly, $\lambda^n f_n(1/\lambda) = P(\lambda)$. So $z_k(P) = 1/z_k(f_n)$. Taking into account that the roots continuously depend on coefficients, we have the required result, letting in the previous lemma $n \to \infty$. Q. E. D.

7.3   Perturbations of functions of order less than two

Consider the functions

$$f(\lambda) = 1 + \sum_{k=1}^{\infty} a_k \lambda^k \psi_k \text{ and } h(\lambda) = 1 + \sum_{k=1}^{\infty} b_k \psi_k \lambda^k, \qquad (3.1)$$

where ψ_k are the same as in Section 7.1; a_k, b_k are complex numbers satisfying the conditions

$$\sum_{k=1}^{\infty} |a_k|^2 < \infty, \ \sum_{k=1}^{\infty} |b_k|^2 < \infty \text{ and } \sum_{k=2}^{\infty} m_k^2 < \infty \qquad (3.2)$$

with $m_k = \psi_k / \psi_{k-1}$ $(k = 2, 3, ...)$. Again put

$$\tau_\psi(f) = \sum_{k=1}^{\infty} |a_k|^2 + \sum_{k=2}^{\infty} m_k^2,$$

and denote

$$g_\psi(f) := \Big[\tau_\psi(f) - \sum_{k=1}^{\infty} \frac{1}{|z_k(f)|^2} \Big]^{1/2}$$

and

$$q := \Big[\sum_{k=1}^{\infty} |a_k - b_k|^2 \Big].$$

Recall that

$$rv_f(h) = \sup_j \inf_k \Big| \frac{1}{z_k(f)} - \frac{1}{z_j(h)} \Big|.$$

Thanks to Theorem 5.11.1 and the equalities $a_2 m_2 = a_2 \psi_2 = c_2$, we obtain

$$\sum_{k=1}^{\infty} \frac{1}{z_k^2(f)} = a_1^2 - 2a_2 m_2. \qquad (3.3)$$

Since

$$\sum_{k=1}^{\infty} \frac{1}{|z_k(f)|^2} \geq \Big| \sum_{k=1}^{\infty} \frac{1}{z_k^2(f)} \Big|,$$

due to (3.3), we have

$$g_\psi(f) \leq [\tau_\psi(f) - |a_1^2 - a_2 2m_2|]^{1/2} \leq \sqrt{\tau_\psi(f)}.$$

Everywhere in this section one can replace $g_\psi(f)$ by the simple calculated quantity

$$[\tau_\psi(f) - |a_1^2 - a_2 2m_2|]^{1/2}.$$

Theorem 7.3.1 *Let the functions f and h defined by (3.1) satisfy conditions (3.2). In addition, let $y(q, f)$ be the unique positive (simple) root of the equation*

$$\frac{q\sqrt{e}}{y} exp\left[\frac{g_\psi^2(f)}{2y^2}\right] = 1. \qquad (3.4)$$

Then

$$rv_f(h) \le y(q, f).$$

This theorem is proved in the next section. Substitute into (3.4) the equality $y = xg_\psi(f)$ and apply Lemma 1.6.4. Then, we have

$$y_\psi(q, f) \le \tilde{\delta}_\psi(q, f), \qquad (3.5)$$

where

$$\tilde{\delta}_\psi(q, f) := \begin{cases} qe & \text{if } g_\psi(f) \le qe \\ g_\psi(f) \left[ln \left(g_\psi(f)/q\right)\right]^{-1/2} & \text{if } g_\psi(f) > qe \end{cases}.$$

Theorem 7.3.1 and (3.5) yield the inequality

$$rv_f(h) \le \tilde{\delta}_\psi(q, f).$$

Furthermore, put

$$\xi_\psi(f, y) = \frac{q\sqrt{e}}{y} exp\left[\frac{g_\psi^2(f)}{2y^2}\right] \quad (y > 0).$$

Taking into account that this function monotonically decreases in y, we get

Corollary 7.3.2 *Let functions f and h be defined by (3.1), and conditions (3.2) be fulfilled. Then all the zeros of h lie in the set*

$$\cup_{k=1}^\infty \omega_k,$$

where

$$\omega_k := \{\lambda \in \mathbf{C} : q\xi_\psi\left(f, \left|\frac{1}{z_k(f)} - \frac{1}{\lambda}\right|\right) \ge 1\} \quad (k = 1, 2, ...).$$

In particular if f has $l < \infty$ finite zeros, then $\omega_k \equiv \omega_0$ for $k > l$, where

$$\omega_0 := \{\lambda \in \mathbf{C} : q\xi_\psi\left(f, \frac{1}{|\lambda|}\right) \ge 1\}$$

and thus all the zeros of h lie in the set

$$\cup_{k=0}^l \omega_k.$$

Furthermore, let us consider functions f and h in the root-factorial form

$$f(\lambda) = \sum_{k=0}^{\infty} \frac{a_k \lambda^k}{(k!)^{\gamma}} \text{ and } h(\lambda) = \sum_{k=0}^{\infty} \frac{b_k \lambda^k}{(k!)^{\gamma}} \quad (\gamma \in (1/2, 1]; \ a_0 = b_0 = 1). \quad (3.6)$$

Again assume that conditions (3.2) hold. In the case (3.6) $\psi_k = 1/(k!)^{\gamma}$ and $m_k = 1/k^{\gamma}$. We thus have $g_{\psi}(f) = g_{\gamma}(f)$ where

$$g_{\gamma}(f) = \left[\tau_{\gamma}(f) - \sum_{k=1}^{\infty} \frac{1}{|z_k(f)|^2} \right]^{1/2}$$

with

$$\tau_{\gamma}(f) = \sum_{k=1}^{\infty} |a_k|^2 + \zeta(2\gamma) - 1.$$

Besides, q is again defined by the equality

$$q := \left[\sum_{k=1}^{\infty} |a_k - b_k|^2 \right]^{1/2}.$$

According to (3.3),

$$g_{\gamma}(f) \leq [\tau_{\gamma}(f) - |a_1^2 - a_2 2^{1-\gamma}|]^{1/2}.$$

Similarly, one can take

$$\psi_k = \frac{1}{\Gamma(1 + k/\rho)} \quad (0 < \rho < 2)$$

and use the Mittag-Leffler form of f and h.

7.4   Proof of Theorem 7.3.1

To prove Theorem 7.3.1, again consider the polynomials

$$P(\lambda) = \lambda^n + \sum_{k=1}^{n} \psi_k a_k \lambda^{n-k} \text{ and } Q(\lambda) = \lambda^n + \sum_{k=1}^{n} b_k \psi_k \lambda^{n-k} \quad (2 \leq n < \infty).$$

$$(4.1)$$

Let

$$q_n = \left[\sum_{k=1}^{n} |a_k - b_k|^2 \right]^{1/2}$$

and

$$g_1(P) = \left[\sum_{k=1}^{n} |a_k|^2 - |z_k(P)|^2 + \sum_{k=2}^{n} m_k^2 \right]^{1/2}.$$

Lemma 7.4.1 *For any zero $z(Q)$ of Q, there is a zero $z(P)$ of P, such that*

$$|z(P) - z(Q)| \leq y(P, q_n),$$

where $y(P, q_n)$ is the unique positive (simple) root of the equation

$$y^n = q_n \left[1 + \frac{1}{n-1}(y^2 + g_1^2(P)) \right]^{(n-1)/2}. \tag{4.2}$$

Proof: Let A_n be the matrix defined by (2.2) and

$$B_n = \begin{pmatrix} -b_1 & -b_2 & \cdots & -b_{n-1} & -b_n \\ m_2 & 0 & \cdots & 0 & 0 \\ 0 & m_3 & \cdots & 0 & 0 \\ \cdot & \cdot & \cdots & \cdot & \cdot \\ 0 & 0 & \cdots & m_n & 0 \end{pmatrix}.$$

By (2.3),

$$\lambda_k(A_n) = z_k(P), \lambda_k(B_n) = z_k(Q), \tag{4.3}$$

where $\lambda_k(.)$ $(k = 1, ..., n)$ are the eigenvalues with their multiplicities. Moreover,

$$q_n = \|A_n - B_n\|,$$

where the norm is Euclidean. Recall that for an $n \times n$-matrix A,

$$g(A) = \left(N_2^2(A) - \sum_{k=1}^{n} |\lambda_k(A)|^2 \right)^{1/2}$$

(see Section 1.7). Because of Lemma 1.7.3, for any $\lambda_k(B_n)$, there is a $\lambda_k(A_n)$, such that

$$|\lambda_k(A_n) - \lambda_k(B_n)| \leq \tilde{y}(q_n), \tag{4.4}$$

where $\tilde{y}(q_n)$ is is the unique positive (simple) root of the equation

$$\frac{q_n}{y} \left[1 + \frac{1}{n-1} (1 + \frac{g^2(A_n)}{y^2}) \right]^{(n-1)/2} = 1.$$

Take into account that

$$g(A_n) = g_1(P).$$

This, (4.3) and (4.4) prove the result. Q. E. D.

Proof of Theorem 7.3.1:   Again consider the polynomials

$$f_n(\lambda) = 1 + \sum_{k=1}^{n} a_k \psi_k \lambda^k \text{ and } h_n(\lambda) = 1 + \sum_{k=1}^{n} b_k \psi_k. \tag{4.5}$$

Since $\lambda^n f_n(1/\lambda) = P(\lambda)$ and $h_n(1/\lambda)\lambda^n = Q(\lambda)$, we have

$$z_k(P) = 1/z_k(f_n) \text{ and } z_k(Q) = 1/z_k(h_n). \tag{4.6}$$

Take into account that the roots continuously depend on coefficients, we have the required result, letting in the previous lemma $n \to \infty$, since

$$\frac{q_n}{y}\Big[1 + \frac{1}{n-1}\big(1 + \frac{g_1^2(P))}{y^2}\big)\Big]^{(n-1)/2} \to \frac{q}{y}exp\,\Big[\frac{1}{2}(1 + \frac{g_\psi^2(f)}{y^2})\Big]$$

for any $y > 0$. Q. E. D.

7.5   Approximations by polynomials

In this section and in the next one, for brevity, we restrict ourselves by the root-factorial form although our reasonings are valid if a function is represented in the ψ-form or in the Mittag-Leffler form.

Let

$$f(\lambda) = \sum_{k=0}^{\infty} \frac{a_k \lambda^k}{(k!)^\gamma} \quad (\gamma \in (1/2, 1], \ \lambda \in \mathbf{C}, \ a_0 = 1) \tag{5.1}$$

under the condition

$$\sum_{k=1}^{n} |a_k|^2 < \infty. \tag{5.2}$$

Introduce the polynomial

$$f_n(\lambda) = \sum_{k=0}^{n} \frac{a_k \lambda^k}{(k!)^\gamma},$$

and denote

$$\nu_n = \Big[\sum_{k=n+1}^{\infty} |a_k|^2\Big]^{1/2}.$$

We have

$$g_\gamma(f_n) = \Big[\sum_{k=1}^{n} |a_k|^2 - \sum_{k=1}^{n} \frac{1}{|z_k(f_n)|^2} + \zeta(2\gamma) - 1\Big]^{1/2}.$$

Thanks to Theorem 5.11.1,

$$\sum_{k=1}^{n} \frac{1}{|z_k(f_n)|^2} \geq \Big|\sum_{k=1}^{n} \frac{1}{z_k^2(f_n)}\Big| = |a_1^2 - a_2 2^{1-\gamma}|,$$

and thus,

$$g_\gamma(f_n) \leq \Big[\sum_{k=1}^{n} |a_k|^2 + \zeta(2\gamma) - 1 - |a_1^2 - a_2 2^{1-\gamma}|\Big]^{1/2}$$

$$\leq \left[\sum_{k=1}^{n} |a_k|^2 + \zeta(2\gamma) - 1 \right]^{1/2}.$$

So everywhere in this section one can replace $g_\gamma(f_n)$ by

$$\left[\sum_{k=1}^{n} |a_k|^2 + \zeta(2\gamma) - 1 - |a_1^2 - a_2 2^{1-\gamma}| \right]^{1/2}.$$

Furthermore, put

$$\xi_\gamma(f_n, y) := \frac{1}{y} exp \left[\frac{1}{2} + \frac{g_\gamma^2(f_n)}{2y^2} \right] \ (y > 0).$$

Theorem 7.5.1 *Under condition (5.2), all the zeros of the function f defined by (5.1) are in the set*

$$\cup_{j=0}^{n} \Omega_{j,n},$$

where

$$\Omega_{0,n} = \left\{ z \in \mathbf{C} : \nu_n \xi_\gamma \left(f_n, \frac{1}{|z|} \right) \geq 1 \right\}$$

and

$$\Omega_{j,n} = \left\{ z \in \mathbf{C} : \nu_n \xi_\gamma \left(f_n, \left| \frac{1}{z_j(f_n)} - \frac{1}{z} \right| \right) \geq 1 \right\} \ (j = 1, ..., n).$$

Indeed, this result is due to Corollary 7.3.2 with $f = f_n$ and $h = f$. Note that

$$\xi_\gamma \left(f_n, \frac{1}{|z|} \right) = |z| exp \left[\frac{1}{2} \left(1 + g_\gamma^2(f_n)|z|^2 \right) \right].$$

Since $\xi_\gamma(f_n, y)$ monotonically decreases in y, from the previous theorem it follows

Corollary 7.5.2 *Let f be defined by (5.1) and condition (5.2) be fulfilled. In addition, let $\tilde{y}(f_n)$ be the unique (positive) root of the equation*

$$\nu_n \xi_\gamma(f_n, y) = 1. \tag{5.3}$$

Then for any zero $z(f)$ of f, either there is a zero $z(f_n)$ of f_n, such that

$$\left| \frac{1}{z(f)} - \frac{1}{z(f_n)} \right| \leq \tilde{y}(f_n),$$

or

$$|z(f)| \geq \frac{1}{\tilde{y}(f_n)}.$$

Substitute in (5.3) the equality $y = x g_\gamma(f_n)$ and apply Lemma 1.6.4. Then, we have

$$\tilde{y}(f_n) \leq \delta_\gamma(f_n),$$

where

$$\delta_\gamma(f_n) := \begin{cases} e\nu_n & \text{if } g_\gamma(f_n) \le e\nu_n \\ g_\gamma(f_n) \left[ln \left(g_\gamma(f_n)/\nu_n \right) \right]^{-1/2} & \text{if } g_\gamma(f_n) > e\nu_n \end{cases}.$$

Now Corollary 7.5.2 yields that for any zero $z(f)$ of f, either there is a zero $z(f_n)$ of f_n, such that

$$\left| \frac{1}{z(f_n)} - \frac{1}{z(f)} \right| \le \delta_\gamma(f_n), \text{ or } |z(f)| \ge \frac{1}{\delta_\gamma(f_n)}.$$

7.6   Preservation of multiplicities in the case $\rho(f) < 2$

Let us consider entire functions f and h in the root-factorial form

$$f(\lambda) = \sum_{k=0}^{\infty} \frac{a_k \lambda^k}{(k!)^\gamma} \text{ and } h(\lambda) = \sum_{k=0}^{\infty} \frac{b_k \lambda^k}{(k!)^\gamma} \quad (\gamma \in (1/2, 1]; \ a_0 = b_0 = 1) \quad (6.1)$$

under the conditions

$$\sum_{k=1}^{\infty} |a_k|^2 < \infty \text{ and } \sum_{k=1}^{\infty} |b_k|^2 < \infty. \quad (6.2)$$

Recall that

$$q = \left[\sum_{k=1}^{\infty} |a_k - b_k|^2 \right]^{1/2}, \tau_\gamma(f) = \sum_{k=1}^{\infty} |a_k|^2 + \zeta(2\gamma) - 1,$$

$$g_\gamma(f) = \left[\tau_\gamma(f) - \sum_{k=1}^{\infty} \frac{1}{|z_k(f)|^2} \right]^{1/2} \le [\tau_\gamma(f) - |a_1^2 - a_2 2^{1-\gamma}|]^{1/2}$$

and

$$\xi_\gamma(f, y) := \frac{1}{y} exp \left[\frac{1}{2} + \frac{g_\gamma^2(f)}{2y^2} \right] \quad (y > 0).$$

Theorem 7.6.1 *Let functions f and h be defined by (6.1) and conditions (6.2) hold. Let a zero $z_m(f)$ of f have a multiplicity μ_m. Put*

$$\tilde{\beta}_m := \frac{1}{2} \inf_{k \ne m} \left| \frac{1}{z_k(f)} - \frac{1}{z_m(f)} \right|.$$

If, for a positive $r \le \tilde{\beta}_m$, we have the inequality

$$q\xi_\gamma(f, r) < 1,$$

then function h has in the set

$$\left\{ \lambda \in \mathbf{C} : \left| \frac{1}{z_m(f)} - \frac{1}{\lambda} \right| \leq r \right\}$$

zeros whose total multiplicity is also equal to μ_m.

Proof: The proof of this theorem is similar to the proof of Theorem 7.3.1 with Lemma 1.8.1 taken instead of Lemma 1.7.3. Q. E. D.

Since $\xi_\gamma(f, .)$ is a monotonically decreasing function, Theorem 7.6.1 implies

Corollary 7.6.2 *Let f and h be defined by (6.1) under conditions (6.2). Let a zero $z_m(f)$ of function f have a multiplicity μ_m. In addition, let the unique (positive) root $y_\gamma(q, f)$ of the equation*

$$q\xi_\gamma(f, y) = 1$$

satisfy the inequality

$$y_\gamma(q, f) \leq \tilde{\beta}_m.$$

Then function h has in the set

$$\left\{ \lambda \in \mathbf{C} : \left| \frac{1}{z_m(f)} - \frac{1}{\lambda} \right| \leq y_\gamma(q, f) \right\}$$

zeros whose total multiplicity is also equal to μ_m.

According to (3.5) $y(q, f) \leq \tilde{\delta}_\gamma(q, f)$, where

$$\tilde{\delta}_\gamma(q, f) := \begin{cases} qe & \text{if } g_\gamma(f) \leq qe \\ g_\gamma(f) \left[ln \left(g_\gamma(f)/q \right) \right]^{-1/2} & \text{if } g_\gamma(f) > qe \end{cases}.$$

So by Corollary 7.6.2, if f and h are defined by (6.1) under (6.2), and a zero $z_m(f)$ of function f has a multiplicity μ_m, then the inequality $\tilde{\delta}_\gamma(q, f) \leq \tilde{\beta}_m$ implies that function h has in

$$\left\{ \lambda \in \mathbf{C} : \left| \frac{1}{z_m(f)} - \frac{1}{\lambda} \right| \leq \tilde{\delta}_\gamma(q, f) \right\}$$

zeros whose total multiplicity is also equal to μ_m.

Example 7.6.3 *Consider the function*

$$h(z) = c_0 + c_1 z + c_2 z^2 + l_1 e^{-z\gamma_1} + l_2 e^{-z\gamma_2} \quad (0 \leq \gamma_1, \gamma_2 = const < 1) \quad (6.3)$$

with positive coefficients c_0, c_1, c_2, l_1, l_2. Without any loss of generality, assume that $h(0) = 1$. That is,

$$c_0 + l_1 + l_2 = 1.$$

Then

$$h(z) = 1 + (c_1 - l_1\gamma_1 - l_2\gamma_2)z + z^2(2c_2 + l_1\gamma_1^2 + l_2\gamma_2^2)/2$$

$$+ \sum_{k=3}^{\infty} [l_1(-\gamma_1)^k + l_2(-\gamma_2)^k]\frac{z^k}{k!}.$$

Rewrite the considered function h in the form (6.1) with $\gamma = 1$,

$$b_1 = c_1 - l_1\gamma_1 - l_2\gamma_2, \quad b_2 = 2c_2 + l_1\gamma_1^2 + l_2\gamma_2^2$$

and

$$b_k = (-1)^k[l_1\gamma_1^k + l_2\gamma_2^k] \ (k \geq 3).$$

Take

$$f(z) = 1 + b_1 z + \frac{b_2 z^2}{2} \tag{6.4}$$

assuming that

$$2b_2 > b_1^2. \tag{6.5}$$

We have

$$q^2 = \sum_{k=3}^{\infty} b_k^2 \leq 2 \sum_{k=3}^{\infty} l_1^2 \gamma_1^{2k} + l_2^2 \gamma_2^{2k}$$

$$= \frac{2l_1^2\gamma_1^6}{1 - \gamma_1^2} + \frac{2l_2^2\gamma_2^6}{1 - \gamma_2^2}.$$

and

$$g_1^2(f) \leq \tau_1(f) = b_1^2 + b_2^2 + \zeta(2) - 1.$$

Under condition (6.5), the zeros of the polynomial f defined by (6.4) are complex and adjoint:

$$z_{1,2}(f) = -\frac{b_1}{b_2} \pm i\Delta$$

with

$$\Delta := \sqrt{\frac{2}{b_2} - \frac{b_1^2}{b_2^2}}.$$

Put

$$r_0 := \frac{1}{2}\left|\frac{1}{z_1(f)} - \frac{1}{z_2(f)}\right| = \frac{\Delta}{\Delta^2 + (b_1^2/b_2)^2}.$$

So because of Theorem 7.6.1, we can assert that in the set

$$\left\{z \in \mathbf{C} : \left|\frac{1}{z_j(f)} - \frac{1}{z}\right| \leq r_0\right\} \ (j = 1, 2),$$

h has exactly one simple zero, provided

$$\frac{q}{r_0} exp\left[\frac{1}{2}\left(1 + \frac{\tau_1(f)}{r_0^2}\right)\right] < 1.$$

7.7 Comments to Chapter 7

Sections 7.1 and 7.2 are based on the paper [Gil, 2006]. The material of Sections 7.3 - 7.5 is adopted from [Gil', 2000c]. Theorem 7.6.1 is probably new.

Chapter 8

Exponential Type Functions

Recall that an entire function f is said to be of the exponential type, if it satisfies the inequality

$$|f(z)| \le Me^{\alpha|z|} \quad (M, \alpha = const;\ z \in \mathbf{C}).$$

In this chapter, for the exponential type entire functions some results from Chapters 5 and 7 are detailed. Besides, an essential role is played by the Borel transform.

8.1   Application of the Borel transform

Consider the entire function

$$f(\lambda) = \sum_{k=0}^{\infty} \frac{a_k \lambda^k}{k!} \quad (\lambda \in \mathbf{C},\ a_0 = 1) \tag{1.1}$$

with complex coefficients, satisfying the condition

$$\alpha(f) := \overline{\lim} \sqrt[k]{|a_k|} < 1. \tag{1.2}$$

So, $a_k = f^{(k)}(0)$, $k = 1, 2, \dots$ and f is of exponential type. Below in Section 8.3, we consider the formally more general case $\alpha(f) < \infty$. Put

$$F_f(z) := \int_0^{\infty} e^{-zt} f(t)dt \quad (Re\ z > \alpha(f)).$$

That is, F_f is the Borel transform to $f(z)$. According to (1.1), (1.2), we easily obtain

$$F_f(z) = \sum_{k=0}^{\infty} \frac{a_k}{z^{k+1}} \quad (|z| \geq 1).$$

Let

$$\theta(f) = \left[\sum_{k=1}^{\infty} |a_k|^2 \right]^{1/2}.$$

Then, by the Parseval equality,

$$|F_f|_{L^2}^2 := \frac{1}{2\pi} \int_0^{2\pi} |F_f(e^{is})|^2 ds = \sum_{k=0}^{\infty} |a_k|^2 = \theta^2(f) + 1 \leq m_f^2 + 1 \qquad (1.3)$$

where

$$m_f := \sqrt{\sup_{|z|=1} |F_f(z)|^2 - 1}.$$

Denote

$$\tau(f) := \sum_{k=1}^{\infty} |a_k|^2 + \zeta(2) - 1 = \theta^2(f) + \zeta(2) - 1,$$

where $\zeta(.)$ is the Riemann zeta function. Clearly,

$$\tau(f) = |F_f|_{L^2}^2 + \zeta(2) - 2 \leq m_f^2 + \zeta(2) - 1.$$

For a $t \in [0, 2\pi)$ put

$$\chi(f, t) := Re \, (a_1 e^{it})^2 - Re \, (a_2 e^{2it}) + \tau(f).$$

By Theorem 7.1.1 with $\tau_1(f) = \tau(f)$, we obtain the following result.

Corollary 8.1.1 *Let function f be defined by (1.1) and condition (1.2) hold. Then for any $t \in [0, 2\pi)$, we have*

$$\tau(f) - S_2(f) = \chi(f, t) - 2 \sum_{k=1}^{\infty} \left(Re \, \frac{e^{it}}{z_k} \right)^2 \geq 0. \qquad (1.4)$$

Note that $\chi(f, 0) = Re \, a_1^2 - Re \, a_2 + \tau(f)$. In addition,

$$\chi(f, \pi/2) = -Re \, a_1^2 + Re \, a_2 + \tau(f).$$

Recall that

$$J_2(f) = \sum_{k=1}^{\infty} \left(Im \, \frac{1}{z_k(f)} \right)^2 \text{ and } R_2(f) = \sum_{k=1}^{\infty} \left(Re \, \frac{1}{z_k(f)} \right)^2.$$

Now Corollary 8.1.1 implies

Corollary 8.1.2 *Let f be defined by (1.1) and condition (1.2) hold. Then*

$$\tau(f) - S_2(f) = \chi(f,0) - 2R_2(f) = \chi(f,\pi/2) - 2J_2(f) \geq 0$$

and thus $S_2(f) \leq \tau(f)$,

$$2R_2(f) \leq \chi(f,0) \text{ and } 2J_2(f) \leq \chi(f,\pi/2).$$

By (1.3), we have

$$\tau(f) = |F_f|_{L^2}^2 + \zeta(2) - 2 \leq m_f^2 + \zeta(2) - 1.$$

Hence,

$$S_2(f) \leq m_f^2 + \zeta(2) - 1,$$

$$2J_2(f) \leq -\text{Re } a_1^2 + \text{Re } a_2 + \zeta(2) - 1 + m_f^2$$

and

$$2R_2(f) \leq \text{Re } a_1^2 - \text{Re } a_2 + m_f^2 + \zeta(2) - 1.$$

8.2   The counting function

Enumerate the zeros of f with their multiplicities in the increasing order and rearrange $a_1, a_2, \ldots$ in the descending order: $|a_{[1]}| \geq |a_{[2]}| \geq \ldots$.

Theorem 8.2.1 *Let the function f defined by (1.1) satisfy condition (1.2). Then for any* $j = 1, 2, \ldots$, *we have*

$$\sum_{k=1}^{j} \frac{1}{|z_k(f)|} < \theta(f) + \sum_{k=1}^{j} \frac{1}{k+1} \quad \text{and} \quad \sum_{k=1}^{j} \frac{1}{|z_k(f)|} < \sum_{k=1}^{j} (|a_{[k]}| + \sum_{k=1}^{j} \frac{1}{k+1}).$$

The proof of this theorem is similar to the proof of Theorem 5.1.1.

Furthermore, since $|z_j(f)| \leq |z_{j+1}(f)|$, because of the previous theorem,

$$\frac{j}{|z_j(f)|} < \theta(f) + \sum_{k=1}^{j} \frac{1}{k+1} \quad (j = 1, 2, \ldots).$$

So

$$|z_j(f)| > u_j(f) \geq \tilde{u}_j(f),$$

where

$$u_j(f) := \frac{j}{\sum_{k=1}^{j}(k+1)^{-1} + \theta(f)}$$

and

$$\tilde{u}_j(f) := \frac{j}{\sum_{k=1}^{j}(k+1)^{-1} + m_f} \qquad (j = 1, 2, ...).$$

Thus, f has in the circle $|\lambda| \leq u_j(f)$ no more than $j - 1$ zeros. Hence, we get $\nu_f(r) \leq j - 1$, provided $r \leq u_j$ or $r \leq \tilde{u}_j$. Take into account that

$$\sum_{k=1}^{j} \frac{1}{k+1} \leq \int_{1}^{j+1} \frac{dx}{x} = \ln (j+1).$$

Now Theorem 8.2.1 implies

Corollary 8.2.2 *Let f be defined by (1.1) and condition (1.2) hold. Then*

$$\sum_{k=1}^{j} \frac{1}{|z_k(f)|} < \theta(f) + \ln (1+j) \quad (j = 1, 2, ...)$$

and $\nu_f(r) \leq j - 1$, provided

$$r \leq \frac{j}{\theta(f) + \ln (1+j)}.$$

8.3   The case $\alpha(f) < \infty$

Consider the function f defined by (1.1) and assume that

$$\alpha(f) = \overline{\lim}_{k\to\infty} \sqrt[k]{|a_k|} < \infty. \qquad (3.1)$$

Introduce the function

$$h_s(\lambda) \equiv f(s\lambda) = \sum_{k=0}^{\infty} \frac{a_k(s\lambda)^k}{k!}$$

with

$$0 < s < \frac{1}{\alpha(f)}.$$

Clearly,

$$S_2(h_s) = \sum_{k=1}^{\infty} \frac{s^2}{|z_k(f)|^2} = s^2 S_2(f).$$

Similarly, $J_2(h_s) = s^2 J_2(f)$ and $R_2(h_s) = s^2 R_2(f)$. Now Corollary 8.1.2 implies

Corollary 8.3.1 *Let* f *be defined by (1.1) and condition (3.1) hold. Then* $S_2(f) \leq \tilde{\tau}(f)$, *where*

$$\tilde{\tau}(f) := \inf_{s \in (0, 1/\alpha(f))} \frac{1}{s^2} \Big[\sum_{k=1}^{\infty} |s^k a_k|^2 + \zeta(2) - 1 \Big].$$

Thanks to Corollary 8.2.2, we obtain

$$\sum_{k=1}^{j} \frac{1}{|z_k(f)|} < \inf_{0 < s < 1/\alpha(f)} s^{-1} [\theta(h_s) + ln\,(1+j)], \tag{3.2}$$

where

$$\theta(h_s) = \Big[\sum_{k=1}^{\infty} |s^k a_k|^2 \Big]^{1/2}.$$

Let us apply the Borel transformation to $f(st)$. For a positive $s < 1/\alpha(f)$, we have

$$\int_0^{\infty} exp\,[-zt] f(st) dt = \frac{1}{s} \int_0^{\infty} exp\,[-zt_1 s^{-1}] f(t_1) dt_1$$

$$= \frac{1}{s} F_f\Big(\frac{z}{s}\Big) \quad (s \in (0, 1/\alpha(f))).$$

According to (1.1),

$$F_f\Big(\frac{z}{s}\Big) = \sum_{k=0}^{\infty} a_k \frac{s^{k+1}}{z^{k+1}}.$$

So because of the Parseval equality,

$$|F_f(\frac{e^{it}}{s})|_{L^2}^2 = \frac{1}{2\pi} \int_0^{2\pi} |F_f(\frac{e^{it}}{s})|^2 dt$$

$$= \sum_{k=0}^{\infty} |a_k s^{k+1}|^2 = s^2 \sum_{k=0}^{\infty} |a_k s^k|^2.$$

Hence,

$$s^2 [\theta^2(h_s) + 1] = |F_f\Big(\frac{e^{it}}{s}\Big)|_{L^2}^2.$$

Since

$$|F_f(\frac{e^{it}}{s})|_{L^2} \leq \max_{|z|=1/s} |F_f(z)|$$

we get

$$\theta^2(h_s) \leq \frac{1}{s^2} \max_{|z|=1/s} |F_f(z)|^2 - 1 \quad \Big(0 < s < \frac{1}{\alpha(f)}\Big). \tag{3.3}$$

So from (3.2) it follows

Corollary 8.3.2 *Let f be defined by (1.1) and condition (3.1) hold. Then*

$$\sum_{k=1}^{j} \frac{1}{|z_k(f)|} \leq \inf_{0<s<1/\alpha(f)} \frac{1}{s} \Big[\, ln\,(1+j) + \frac{1}{s^2} \max_{|z|=1/s} (|F_f(z)|^2 - 1)^{1/2}\Big].$$

Example 8.3.3 *Let us consider the function*

$$f(\lambda) := \frac{sin\,\pi\,\lambda}{\pi\lambda} = \sum_{k=0}^{\infty} \frac{(\pi\lambda)^{2k}(-1)^k}{(2k+1)!}. \tag{3.4}$$

The zeros of f are $\pm 1, \pm 2, ...$ We thus have

$$\sum_{k=1}^{2j} \frac{1}{|z_k(f)|} = 2\sum_{k=1}^{j} \frac{1}{k} \quad (j=1,2,...). \tag{3.5}$$

Writing the considered function in the form (1.1), with

$$a_{2k+1} = 0, \ a_{2k} = (-1)^k \frac{\pi^{2k}}{2k+1},$$

we have $\alpha(f) = \pi$. Thanks to (3.2), with $0 < s < 1/\pi$, we get

$$s\sum_{k=1}^{j} \frac{1}{|z_k(f)|} < \sum_{k=1}^{\infty} (\pi s)^{4k} + ln\,(1+j)$$

$$= \frac{(\pi s)^4}{1-(\pi s)^4} + ln\,(1+j).$$

Take

$$s = \frac{1}{\pi\sqrt[4]{3}},$$

then

$$\sum_{k=1}^{2j} \frac{1}{|z_k(f)|} \leq \pi\sqrt[4]{3}\Big(\frac{3}{2} + ln\,(1+2j)\Big).$$

This is rather close to (3.5).

Example 8.3.4 *Let us consider the function*

$$\tilde{f}(\lambda) := cos\,(\pi\lambda) = \sum_{k=0}^{\infty} \frac{(\pi\lambda)^{2k}(-1)^k}{(2k)!}.$$

The zeros of $\tilde{f}$ are $k+1/2, k=0, \pm 1, \pm 2, ...$ We thus have

$$S_2(\tilde{f}) = \sum_{k=-\infty}^{\infty} \frac{1}{(k+\frac{1}{2})^2}. \tag{3.6}$$

Writing the considered function in the form (1.1), with

$$a_{2k+1} = 0, \ a_{2k} = (-1)^k \pi^{2k}$$

we get $\alpha(\tilde{f}) = \pi$. Thanks to Corollary 8.3.1 with $0 < s < 1/\pi$, we obtain

$$s^2 S_2(\tilde{f}) \leq \sum_{k=1}^{\infty} (s\pi)^{4k} + \zeta(2) - 1 =$$

$$\frac{(s\pi)^4}{1 - (s\pi)^4} + \zeta(2) - 1.$$

Take

$$s = \frac{1}{\pi \sqrt[4]{3}},$$

then

$$S_2(f) \leq \pi^2 \sqrt{3} [\zeta(2) + \frac{1}{2}].$$

This is rather close to equality (3.6).

8.4 Variations of roots

In this section, we show that some our above derived perturbation results can be reformulated in terms of the Borel transform. Consider the entire functions

$$f(\lambda) = \sum_{k=0}^{\infty} \frac{a_k \lambda^k}{k!} \ \text{ and } \ h(\lambda) = \sum_{k=0}^{\infty} \frac{b_k \lambda^k}{k!} \quad (\lambda \in \mathbf{C}, \ a_0 = b_0 = 1) \qquad (4.1)$$

with complex, in general, coefficients. It is assumed that the conditions

$$\alpha(f) < 1 \ \text{ and } \ \alpha(h) < 1 \qquad (4.2)$$

hold. Let

$$g_1(f) = [\tau(f) - \sum_{k=1}^{\infty} \frac{1}{|z_k(f)|^2}]^{1/2}.$$

Recall that

$$\tau(f) = |F_f|_{L^2}^2 + \zeta(2) - 2.$$

Thanks to Theorem 5.11.1,

$$\sum_{k=1}^{\infty} \frac{1}{|z_k(f)|^2} \geq |a_1^2 - a_2|.$$

Thus

$$g_1(f) \leq [\tau(f) - |a_1^2 - a_2|]^{1/2} \leq [\tau(f)]^{1/2}.$$

So, in our reasonings, we can replace $g_1(f)$ by

$$[\tau(f) - |a_1^2 - a_2|]^{1/2} \text{ or by } [\tau(f)]^{1/2}.$$

Again put

$$q := \Big[\sum_{k=1}^{\infty} |a_k - b_k|^2\Big]^{1/2}.$$

Furthermore, let $F_f(z)$ and $F_h(z)$ be the Borel transforms of f and h, respectively. According to (4.1) and (4.2)

$$F_f(z) = \sum_{k=0}^{\infty} \frac{a_k}{z^{k+1}} \text{ and } F_h(z) = \sum_{k=0}^{\infty} \frac{b_k}{z^{k+1}} \ (|z| \geq 1).$$

Thanks to the Parseval equality, we obtain

$$|F_f(e^{is}) - F_h(e^{is})|_{L^2} \equiv \Big[\frac{1}{2\pi} \int_0^{2\pi} |F_f(e^{is}) - F_h(e^{is})|^2 ds\Big]^{1/2}$$

$$= \Big[\sum_{k=1}^{\infty} |a_k - b_k|^2\Big]^{1/2} = q.$$

Thus

$$q \leq \max_{|z|=1} |F_f(z) - F_h(z)|.$$

Denote

$$\xi(f, y) := \frac{1}{y} exp \Big[\frac{1}{2} + \frac{g_1^2(f)}{2y^2}\Big] \ (y > 0).$$

Then, from Theorem 7.3.1, our next result follows.

Theorem 8.4.1 *Let functions f and h be defined by (4.1) and condition (4.2) be fulfilled. In addition, let $y(q, f)$ be the unique (positive) root of the equation*

$$q\xi(f, y) = 1. \tag{4.3}$$

Then $rv_f(h) \leq y(q, f)$.

Since $\xi(f, .)$ is a monotonically decreasing function, Theorem 8.4.1 implies

Corollary 8.4.2 *Under the hypothesis of Theorem 8.4.1, all the zeros of h lie in the set*

$$\cup_{k=1}^{\infty} W_{k1}$$

where

$$W_{k1} := \{\lambda \in \mathbf{C} : q\xi(f, |\frac{1}{z_k(f)} - \frac{1}{\lambda}|) \geq 1\}.$$

In particular, if f has $l < \infty$ zeros, then for $k > l$, $W_{k1} = W_{01}$, were

$$W_{01} := \{\lambda \in \mathbf{C} : q\xi(f, \frac{1}{|\lambda|}) \geq 1\} = \{\lambda \in \mathbf{C} : q|\lambda| \, exp\left[\frac{1}{2}(1+\tilde{g}^2(f)|\lambda|^2)\right] \geq 1\}$$

and all the zeros of h lie in the set

$$\cup_{k=0}^{l} W_{k1}.$$

Furthermore, substitute in (4.3) the equality $y = xg_1(f)$ and apply Lemma 1.6.4. Then, we have

$$y(q, f) \leq \delta(q, f), \tag{4.4}$$

where

$$\delta(q, f) := \begin{cases} qe & \text{if } g_(f) \leq qe \\ g_1(f) \, [ln \, (g_1(f)/q)]^{-1/2} & \text{if } g_1(f) > qe \end{cases}.$$

Now Theorem 8.4.1 yields the inequality

$$rv_f(h) \leq \delta(q, f).$$

8.5   Functions close to *cos z* and e^z

Let us consider the function

$$h(\lambda) = \sum_{k=0}^{\infty} \frac{c_{2k}\lambda^{2k}(-1)^k}{(2k)!}, \quad c_0 = 1 \tag{5.1}$$

with the condition

$$q_c(d) := \left[\sum_{k=1}^{\infty} |d^{2k} - c_{2k}|^2\right]^{1/2} < \infty \tag{5.2}$$

for a positive $d < 1$.

Take $f(\lambda) = cos\,(\lambda d)$. Then with the notations of the previous section, we have $a_{2k} = (-1)^k d^{2k}$, $a_{2k+1} = 0$, $q = q_c(d)$. Since the zeros of $cos\,(dz)$ are

$$z_k(f) = \frac{\pi}{2d}(2k+1), \quad (k = 0, \pm 1, \pm 2, ...)$$

we obtain the equality $g_1(f) = \eta_c(d)$ where

$$\eta_c(d) := \left[\sum_{k=1}^{\infty} d^{4k} + \zeta(2) - 1 - \frac{4d^2}{\pi^2} \sum_{k=-\infty}^{\infty} \frac{1}{(2k+1)^2}\right]^{1/2} =$$

$$\left[\frac{d^4}{1-d^4} + \zeta(2) - 1 - \frac{4d^2}{\pi^2} \sum_{k=-\infty}^{\infty} \frac{1}{(2k+1)^2}\right]^{1/2}.$$

Now Theorem 8.4.1 yields the inequality

$$rv_{cos\ (zd)}(h) \leq y_c(d),$$

where $y_c(d)$ is the unique positive root of the equation

$$\frac{q_c(d)}{y} exp\ \left[\frac{1}{2} + \frac{\eta_c^2(d)}{2y^2}\right] = 1.$$

We thus obtain the next result.

Corollary 8.5.1 *Let h be defined by (5.1) under condition (5.2). Then all the zeros of h lie in the union of the sets*

$$\{z \in \mathbf{C} : \left|\frac{1}{z} - \frac{d}{\pi(k+\frac{1}{2})}\right| \leq y_c(d)\}, k = 0, \pm 1, \pm 2, ...$$

Set

$$\delta_c(d) := \begin{cases} q_c(d)e & \text{if } \eta_c(d) \leq q_c(d)e, \\ \eta_c(d)\ [ln\ (\eta_c(d)/q_c(d))]^{-1/2} & \text{if } \eta_c(d) > q_c(d)e \end{cases} \quad (5.3)$$

Then by (4.4) $y_c(d) \leq \delta_c(d)$.

Similarly, we can consider the zeros of the function

$$\sum_{k=0}^{\infty} \frac{c_k \lambda^{2k+1}(-1)^k}{(2k+1)!}, \quad c_0 = 1$$

comparing it with the function

$$\frac{sin\ (\lambda d)}{\lambda d}.$$

Now let us consider the function

$$h_e(\lambda) = \sum_{k=0}^{\infty} \frac{b_k \lambda^k}{k!}, \quad b_0 = 1 \quad (5.4)$$

under the condition

$$q_e(d) := \left[\sum_{k=1}^{\infty} |d^k - b_k|^2\right]^{1/2} < \infty \quad (0 < d < 1). \quad (5.5)$$

Compare it with the function $e^{\lambda d}$. Then, we have $a_k = d^k$, $q = q_e(d)$, and $g_1(f) = \eta_e(d)$, where

$$\eta_e^2(d) := \sum_{k=1}^{\infty} d^{2k} + \zeta(2) - 1 = \frac{d^2}{1-d^2} + \zeta(2) - 1.$$

Theorem 8.4.1 yields the inequality

$$rv_{exp\,(zd)}(h_e) \le y_e(d),$$

where $y_e(d)$ is the unique positive root of the equation

$$\frac{q_e(d)}{y} exp\left[\frac{1}{2} + \frac{\eta_e^2(d)}{2y^2}\right] = 1. \tag{5.6}$$

Since e^z does not have finite zeros, the following result is valid.

Corollary 8.5.2 *Let h_e be defined by (5.4) under condition (5.5). Then all its zeros lie in the set*

$$\{z \in \mathbf{C} : |z| \ge \frac{1}{y_e(d)}\}.$$

Denote

$$\delta_e(d) := \begin{cases} q_e(d)e & \text{if } \eta_e(d) \le q_e(d)e, \\ \eta_e(d)\,[ln\,(\eta_e(d)/q_e(d))]^{-1/2} & \text{if } \eta_e(d) > q_e(d)e \end{cases}.$$

Then by (4.4), $y_e(d) \le \delta_e(d)$.

8.6 Estimates for functions on the positive half-line

We need the following two lemmas.

Lemma 8.6.1 *Let D be the closed convex hull of points $x_0, x_1, ..., x_n \in \mathbf{C}$ (part of the points or all the points may coincide) and let a scalar-valued function u be regular on a neighborhood D_1 of D. In addition, let $\Gamma \subset D_1$ be a Jordan closed contour surrounding the points $x_0, x_1, ..., x_n$. Then the inequality*

$$\left|\frac{1}{2\pi i}\int_\Gamma \frac{u(\lambda)d\lambda}{(\lambda - x_0)...(\lambda - x_n)}\right| \le \frac{1}{n!} \sup_{\lambda \in D} |u^{(n)}(\lambda)|$$

is valid.

Proof: First, let all the points be distinct: $x_j \ne x_k$ for $j \ne k$ $(j, k = 0, ..., n)$, and let $D_u(x_0, x_1, ..., x_n)$ be a divided difference of u at points $x_0, x_1, ..., x_n$ of the complex plane. Since u is regular on a neighborhood of the closed convex hull D of the points $x_0, x_1, ..., x_n$, then the divided difference admits the representation

$$D_u(x_0, x_1, ..., x_n) = \frac{1}{2\pi i}\int_\Gamma \frac{u(\lambda)d\lambda}{(\lambda - x_0)...(\lambda - x_n)} \tag{6.1}$$

(see [Gel'fond, 1967, formula (54)]). But, on the other hand, the following estimate is well known:

$$| D_u(x_0, x_1, ..., x_n) | \leq \frac{1}{n!} \sup_{\lambda \in D} |u^{(n)}(\lambda)|$$

[Gel'fond, 1967, formula (49)]. Combining this inequality with relation (6.1), we arrive at the required result. If $x_j = x_k$ for some $j \neq k$, then the claimed inequality can be obtained by small perturbations and the previous reasonings. Q. E. D.

Lemma 8.6.2 *Let $x_0 \leq x_1 \leq ... \leq x_n$ be real points (part of the points or all the points may coincide), and let a function u be regular on a neighborhood D_1 of $[x_0, x_n]$ and real on that segment. In addition, let $\Gamma \subset D_1$ be a Jordan closed contour surrounding $[x_0, x_n]$. Then there is a point $\eta \in [x_0, x_n]$, such that the equality*

$$\frac{1}{2\pi i} \int_\Gamma \frac{u(\lambda)d\lambda}{(\lambda - x_0)...(\lambda - x_n)} = \frac{1}{n!} u^{(n)}(\eta)$$

is true.

Proof: First suppose that all the points are distinct: $x_0 < x_1 < ... < x_n$. Then the divided difference $D_u(x_0, x_1, ..., x_n)$ of u in the points $x_0, x_1, ..., x_n$ admits the representation

$$D_u(x_0, x_1, ..., x_n) = \frac{1}{n!} u^{(n)}(\eta)$$

with some point $\eta \in [x_0, x_n]$ [Gel'fond, 1967, formula (43)], [Ostrowski, 1973, p. 5]. Combining this equality with representation (6.1), we arrive at the required result. If $x_j = x_k$ for some $j \neq k$, then the claimed inequality can be obtained by small perturbations and the previous reasonings. Q. E. D.

8.7 Difference equations

This section is devoted to the scalar difference equation

$$\sum_{k=0}^{n} c_{n-k} x(j + k) = f_j \quad (j = 0, 1, ...) \tag{7.1}$$

with the constant real coefficients c_k $(k = 1, ..., n)$, $c_0 = 1$, a given bounded sequence f_j and the zero initial condition

$$x(j) = 0 \quad (j = 0, 1, ..., n - 1). \tag{7.2}$$

Recall that the Z-transform $\tilde{f}(z)$ of a number sequence $f = \{f_k\}_{k=0}^{\infty}$ under the condition

$$\overline{\lim}_{k\to\infty} \sqrt[k]{|f_k|} \leq R < \infty$$

is defined by

$$\tilde{f}(z) = \sum_{k=0}^{\infty} \frac{f_k}{z^k} \quad (|z| > R)$$

cf. [Doetsch, 1961, Chapter 4, Section 26]. We will write

$$Z\{f_k\} = \tilde{f}(z).$$

The inverse Z-transform is given by

$$f_k = \frac{1}{2\pi i} \int_{C_0} \lambda^{k-1} \tilde{f}(\lambda)\, d\lambda.$$

Here C_0 is a closed Jordan contour containing all the singularities of $\tilde{f}(z)$.
Since

$$\sum_{k=0}^{\infty} \frac{f_{k+1}}{z^k} = z \sum_{k=0}^{\infty} \frac{f_{k+1}}{z^{k+1}} = z(\tilde{f}(z) - f_0),$$

we have

$$Z\{f_{k+1}\} = z(\tilde{f}(z) - f_0) = z(Z\{f_{k+1}\} - f_0).$$

Hence

$$Z\{f_{k+2}\} = z(Z\{f_{k+1}\} - f_1) = z(z(Z\{f_k\} - f_0) - f_1) =$$
$$z^2(Z\{f_k\} - f_0) - zf_1.$$

Similarly, for an arbitrary positive integer m, we have

$$Z\{f_{k+m}\} = z^m Z\{f_k\} - z^m f_0 - z^{m-1} f_1 - \ldots - z f_{m-1}.$$

Applying the Z-transform to problem (7.1), (7.2), we obtain

$$P(z)\tilde{x}(z) = \tilde{f}(z)$$

where $\tilde{x}(z)$ is the Z-transform of the solution $x(k)$, and

$$P(\lambda) = \lambda^n + c_1\lambda^{n-1} + \ldots + c_n$$

is the characteristic polynomial. By the inverse Z-transform, a solution of problem (7.1), (7.2) can be represented by

$$x(j) = \frac{1}{2\pi i} \int_C \lambda^{j-1} \frac{\tilde{f}(\lambda)}{P(\lambda)}\, d\lambda.$$

Here C is a closed Jordan contour containing all the singularities of the integrand.

Let the polynomial $P(\lambda)$ have the zeros $z_1, ..., z_n$ and

$$r_0 := \max_k |z_k|.$$

Then with $\tilde{c} = const > r_0$, the function

$$G(j) := \frac{1}{2\pi i} \int_{|z|=\tilde{c}} \frac{z^{j-1}}{P(z)} \, dz$$

is said to be the Green function to equation (7.1).

Since for two sequences $\{f_k\}, \{h_k\}$ the equality

$$Z\{\sum_{j=0}^{k} f_{k-j} h_j\} = Z\{f_k\} \, Z\{h_k\}$$

holds [Doetsch, 1961, Chapter 4, Section 26], we can assert that the solution of problem (7.1), (7.2) can be represented as

$$x(j) = \sum_{k=1}^{j} G(j-k) f_k.$$

By virtue of Lemma 8.6.1, we easily obtain

$$|G(j)| \leq \frac{1}{(n-1)!} \left| \frac{d^{n-1} z^{j-1}}{dz^{n-1}} \right|_{z=r_0} =$$

$$\frac{1}{(n-1)!} (j-1)...(j+1-n) r_0^{j-n} \quad (j \geq n).$$

Now let all the zeros of $P(\lambda)$ be positive:

$$0 < z_1 \leq z_2 \leq ... \leq z_n.$$

From Lemma 8.6.2, it follows that for some $s = s(j) \in [z_1, z_n]$,

$$G(j) = \frac{1}{(n-1)!} \frac{d^{n-1} z^{j-1}}{dz^{n-1}} \Big|_{z=s} = \frac{1}{(n-1)!} (j-1)...(j+1-n) s^{j-n} \quad (j \geq n).$$

So if the sequence f_j is non-negative and all the zeros of $P(\lambda)$ are nonnegative, then a solution of problem (7.1), (7.2) is nonnegative.

8.8   Comments to Chapter 8

The material of Sections 8.1 - 8.3 is adopted from the paper [Gil', 2000c]. Sections 8.4 and 8.5 are based on the paper [Gil, 2007a]. Theorem 8.6.3 is probably new. For more information about exponential type functions, see for instance the excellent books [Berenstein and Gay, 1995] and [Levin, 1980 and 1996].

Chapter 9

Quasipolynomials

A quasipolynomial is an entire function of the form

$$f(z) = \sum_{j=0}^{n} \sum_{k=1}^{m} c_{jk} z^j e^{\lambda_k z} \quad (\lambda_k, c_{jk} = const).$$

The Borel transform of a quasipolynomial is easily calculated. This fact enables us to obtain explicit bounds for the zeros of quasipolynomials. In addition, in this chapter, we derive two-sided estimates and positivity conditions for some quasipolynomials on the positive half-line.

9.1 Sums of absolute values of zeros

Consider the quasipolynomial

$$f(\lambda) = \sum_{j=0}^{n} \sum_{k=1}^{m} c_{jk} \lambda^j e^{\lambda_k \lambda} \tag{1.1}$$

with complex coefficients c_{jk} and "frequencies" λ_k. Without loss of generality suppose that

$$\max_k |\lambda_k| < 1 \tag{1.2}$$

and

$$f(0) = 1. \tag{1.3}$$

If conditions (1.2) and (1.3) do not hold and $f(0) \neq 0$, we can consider the function

$$f_1(w) = C_1 f(C_2 w)$$

with fitting constants C_1 and C_2. Simple calculations show that the Borel transform of the function defined by (1.1) is

$$F_f(z) = \sum_{j=0}^{n} \sum_{k=1}^{m} \frac{c_{jk} j!}{(z - \lambda_k)^{j+1}}. \tag{1.4}$$

Take into account that

$$|F_f(e^{it})|_{L^2} := \left[\frac{1}{2\pi} \int_0^{2\pi} |F_f(e^{it})|^2 dt\right]^{1/2} \leq \sup_{|z|=1} |F_f(z)|.$$

Clearly,

$$\sup_{|z|=1} |F_f(z)| \leq w(f), \tag{1.5}$$

where

$$w(f) := \sum_{k=1}^{m} \sum_{j=0}^{n} \frac{j! |c_{jk}|}{(1 - |\lambda_k|)^{j+1}}.$$

Recall that

$$S_2(f) = \sum_{k=1}^{\infty} \frac{1}{|z_k(f)|^2},$$

$$J_2(f) = \sum_{k=1}^{\infty} \left(Im \frac{1}{z_k(f)}\right)^2 \text{ and } R_2(f) = \sum_{k=1}^{\infty} \left(Re \frac{1}{z_k(f)}\right)^2.$$

Now we can directly apply our results from Sections 8.1 and 8.2. In particular, Corollary 8.1.2 implies

Corollary 9.1.1 *Let f be defined by (1.1) and conditions (1.2), (1.3) hold. Then*

$$S_2(f) \leq w^2(f) + \zeta(2) - 2,$$

$$2R_2(f) \leq Re\, a_1^2 - Re\, a_2 + w^2(f) + \zeta(2) - 2$$

and

$$2J(f) \leq -Re\, a_1^2 + Re\, a_2 + w^2(f) + \zeta(2) - 2.$$

Here $a_k = f^{(k)}(0)$ $(k = 0, 1, ...)$.

Let us derive a bound for the counting function $\nu_f(r)$ of a quasipolynomial f. Recall that

$$\theta(f) = \sqrt{|F_f|_{L^2}^2 - 1}$$

(see Section 8.1). Thus,

$$\theta(f) \leq \sqrt{w^2(f) - 1}.$$

Enumerate the zeros of f in the increasing order. Now Corollary 8.2.2 implies

Corollary 9.1.2 *Let f be defined by (1.1) and conditions (1.2) and (1.3) hold. Then*

$$\sum_{k=1}^{j} \frac{1}{|z_j(f)|} < ln\,(1+j) + \sqrt{w^2(f) - 1}$$

and $\nu_f(r) \leq j - 1$ for any

$$r \leq \frac{j}{ln\,(1+j) + \sqrt{w^2(f) - 1}} \quad (j = 1, 2, \ldots).$$

9.2 Variations of roots

Let us consider the quasipolynomials

$$f(\lambda) = \sum_{j=0}^{n}\sum_{k=1}^{m} c_{jk}\lambda^j e^{\lambda_k \lambda} \text{ and } h(\lambda) = \sum_{j=0}^{n}\sum_{k=1}^{m} d_{jk}\lambda^j e^{\lambda_k \lambda}, \quad (2.1)$$

with complex coefficients c_{jk}, d_{jk}. In addition,

$$f(0) = h(0) = 1 \text{ and } \max_k |\lambda_k| < 1. \quad (2.2)$$

We have

$$F_h(z) = \sum_{j=0}^{n}\sum_{k=1}^{m} \frac{d_{jk}j!}{(z - \lambda_k)^{j+1}}.$$

Again put

$$q = |F_f(e^{is}) - F_h(e^{is})|_{L^2}$$

(see Section 8.4). Then

$$q = \Big|\sum_{j=0}^{n}\sum_{k=1}^{m} \frac{(c_{jk} - d_{jk})j!}{e^{is} - \lambda_k)^{j+1}}\Big|_{L^2} \leq \sup_{0 \leq s \leq 2\pi} |F_f(e^{is}) - F_h(e^{is})| \leq \tilde{q},$$

where

$$\tilde{q} := \sum_{j=0}^{n}\sum_{k=1}^{m} \frac{j!|c_{jk} - d_{jk}|}{(1 - |\lambda_k|)^{j+1}}.$$

Recall that

$$g_1(f) = \Big[\theta^2(f) + \zeta(2) - 1 - \sum_{k=1}^{\infty} \frac{1}{|z_k(f)|^2}\Big]^{1/2}.$$

As it is shown in Section 8.4,

$$g_1(f) \leq [\theta^2(f) + \zeta(2) - 1 - |a_1^2 - a_2|]^{1/2}.$$

Thus $g_1(f) \leq \tilde{g}(f)$, where

$$\tilde{g}(f) := [w^2(f) + \zeta(2) - 2 - |a_1^2 - a_2|]^{1/2}.$$

Everywhere in this section one can replace $g_1(f)$ by the simple calculated quantity $\tilde{g}(f)$, and q one can replace by $\tilde{q}$. Put

$$\Psi(f, y) := \frac{1}{y} exp \left[\frac{1}{2} + \frac{\tilde{g}^2(f)}{2y^2} \right] \quad (y > 0).$$

Since $\Psi(f, y)$ decreases in $y > 0$, Theorem 8.4.1 yields our next result.

Corollary 9.2.1 *Let f and h be defined by (2.1), and conditions (2.2) hold. In addition, let $y(q, f)$ be the unique (positive) root of the equation*

$$q\Psi(f, y) = 1. \tag{2.3}$$

Then the relative variation of zeros of h with respect to zeros of f satisfies the inequality $rv_f(h) \leq y(q, f)$. Thus, all the zeros of h lie in the set

$$\cup_{j=1}^{\infty} W_j,$$

where

$$W_j := \left\{ \lambda \in \mathbf{C} : q\Psi(f, |\frac{1}{z_j(f)} - \frac{1}{\lambda}|) \geq 1 \right\} \quad (j = 1, 2, ...).$$

In particular, if f has $l < \infty$ zeros, then all the zeros of h lie in the set

$$\cup_{j=0}^{l} W_j$$

where

$$W_0 := \left\{ \lambda \in \mathbf{C} : q|\lambda| exp \left[\frac{1}{2}(1 + |\lambda|^2 \tilde{g}^2(f)) \right] \geq 1 \right\}.$$

Put

$$\tilde{\delta}(q, f) := \begin{cases} qe & \text{if } \tilde{g}(f) \leq qe, \\ \tilde{g}(f) \left[ln \left(\tilde{g}(f)/q \right) \right]^{-1/2} & \text{if } \tilde{g}(f) > qe \end{cases}.$$

Then thanks to inequality (4.4) from Chapter 8 we get the inequality $rv_f(h) \leq \tilde{\delta}(q, f)$.

For instance, let us consider the trigonometric sum

$$f(\lambda) = \sum_{k=1}^{m} c_{1k} cos \left(\omega_k \lambda \right) + c_{2k} sin \left(\omega_k \lambda \right)$$

with the conditions (1.3) and

$$\omega_k \in (0, 1) \quad (k = 1, ..., m). \tag{2.4}$$

So under consideration $\lambda_k = i\omega_k$. Clearly,

$$F_f(z) = \sum_{k=1}^{m} \frac{c_{1k} z}{z^2 + \omega_k^2} + \frac{\omega_k c_{2k}}{z^2 + \omega_k^2}.$$

Thus,

$$|F_f(e^{it})|_{L^2} \le \sup_{|z|=1} |F_f(z)| \le \tilde{w}(f)$$

with

$$\tilde{w}(f) = \sum_{k=1}^{m} \frac{|c_{1k}| + \omega_k|c_{2k}|}{1 - \omega_k^2}.$$

Furthermore, let

$$h(\lambda) = \sum_{k=1}^{m} d_{k1}\cos(\omega_k\lambda) + d_{k2}\sin(\omega_k\lambda) \tag{2.5}$$

be another trigonometric sum with the conditions (2.4) and $h(0) = 1$. Then

$$F_h(z) = \sum_{k=1}^{m} \frac{d_{1k}z + \omega_k d_{2k}}{z^2 + \omega_k^2}.$$

So

$$q = |F_g - F_h|_{L^2} \le \sum_{k=1}^{m} \frac{|c_{1k} - d_{1k}| + \omega_k|c_{2k} - d_{2k}|}{1 - \omega_k^2}.$$

Now we can apply the results from Sections 8.1 and 8.2.

Example 9.2.2 *Consider the function*

$$h(\lambda) = 1 + \lambda(d_{10} + d_{11}e^{b\lambda})$$

$$+\lambda^2(d_{20} + d_{21}e^{b\lambda} + d_{22}e^{2b\lambda}) \ (b \in (0, 1/3))$$

with real coefficients.

Take $f(\lambda) = 1 + \lambda d_{11}e^{b\lambda} + \lambda^2 d_{22}e^{2b\lambda}$. Then

$$F_f(z) = \frac{1}{z} + \frac{d_{11}}{(z-b)^2} + \frac{2d_{22}}{(z-2b)^3}$$

and

$$(F_h - F_f)(z) = \frac{d_{10}}{z^2} + \frac{2d_{20}}{z^3} + \frac{2d_{21}}{(z-b)^3}.$$

Thus,

$$w(f) = 1 + \frac{|d_{11}|}{(1-b)^2} + \frac{2|d_{22}|}{(1-2b)^3}$$

and

$$q \le |d_{10}| + 2|d_{20}| + \frac{2|d_{21}|}{(1-b)^3}.$$

Let $r_{1,2}$ be the zeros of the polynomial $1 + zd_{11} + z^2 d_{22}$. Then we have

$$f(\lambda) = d_{22}(\lambda e^{b\lambda} - r_1)(\lambda e^{b\lambda} - r_2).$$

So if we know the zeros of the simple quasipolynomials $ze^{bz} - r_{1,2}$, we can define the domains containing the zeros of h by Corollary 9.2.1.

9.3   Trigonometric polynomials

In this section and in the next one we do not assume that condition (1.2) holds.

For an integer $n \geq 2$, let us consider the trigonometric polynomial

$$T(\lambda) = \sum_{k=0}^{n} c_k e^{i(n-k)\lambda} \quad (c_0 = 1). \tag{3.1}$$

Put

$$P(z) = \sum_{k=0}^{n} c_k z^{n-k}.$$

Clearly, for any zero $z(T)$ of T, there is a zero $z(P)$ of P and an integer j, such that

$$iz(T) = Ln\, z(P) = ln\, z(P) + 2\pi i j \; (j = \pm 1, \pm 2, ...), \tag{3.2}$$

where

$$ln\, z(P) = ln\, |z(P)| + arg\, z(P) \quad (arg\, z(P) \in [0, 2\pi)).$$

In the sequel, we take the main values $z_k(T) = -i\, ln\, z_k(P), k = 1, 2, ..., n$.

Again put

$$A_m = \begin{pmatrix} -c_1 & -c_2 & ... & -c_{m-1} & -c_m \\ 1 & 0 & ... & 0 & 0 \\ 0 & 1 & ... & 0 & 0 \\ . & . & ... & . & . \\ 0 & 0 & ... & 1 & 0 \end{pmatrix}. \tag{3.3}$$

Thanks to Theorem 4.4.1 we get

Corollary 9.3.1 *For any $m \leq n$ the equality*

$$\sum_{k=1}^{n} e^{iz_k(T)m} = Trace\, A_m^m$$

is valid.

In particular, from the previous corollary, it follows that

$$\sum_{k=1}^{n} e^{2iz_k(T)} = c_1^2 - 2c_2 \quad (n \geq 2)$$

and

$$\sum_{k=1}^{n} e^{3iz_k(T)} = -c_1^3 + 3c_1 c_2 - 3c_3 \quad (n \geq 3).$$

Set

$$\tau_T = \sum_{k=1}^{n} |c_k|^2 + n - 1$$

and

$$\chi_T(t) = Re\,(c_1 e^{it})^2 - 2Re\,(c_2 e^{2it}) + \tau_T \quad (t \in [0, 2\pi)).$$

Let

$$u_k = Re\,z_k(T) \text{ and } w_k = Im\,z_k(T).$$

Since

$$e^{iz_k(T)} = e^{-w_k}[\cos\,(u_k) + i\,\sin\,(u_k)],$$

Theorem 4.2.1 implies

Corollary 9.3.2 *For any $t \in [0, 2\pi)$ we have*

$$\tau_T - \sum_{k=1}^{n} e^{-2w_k} = \chi_T(t) - 2\sum_{k=1}^{n} (Re\,e^{i(t+z_k(T))})^2 \geq 0.$$

Hence, the inequality

$$\sum_{k=1}^{n} e^{-2w_k}\,(u_k) \leq \tau_T$$

follows.
  Furthermore, let

$$|T(.)|_{L^2} = [\frac{1}{2\pi} \int_0^{2\pi} |T(t)|^2 dt]^{1/2}.$$

Then because of the Parseval equality,

$$\sum_{k=0}^{n} |c_k|^2 = |T(.)|_{L^2}^2 \leq \max_{t \in [0, 2\pi]} |T(t)|^2.$$

Therefore,

$$\tau_T = |T(.)|_{L^2}^2 + n - 2 \leq \max_{t \in [0, 2\pi]} |T(t)|^2 + n - 2.$$

Enumerate the zeros of $T(\lambda)$ with their multiplicities in the following way:

$$w_k \leq w_{k+1} \quad (k = 1, ..., n-1).$$

Now by Theorem 4.3.1 we arrive at the following result.

Corollary 9.3.3 *The zeros of the trigonometric polynomial T defined by (3.1) satisfy the inequalities*

$$\sum_{k=1}^{j} e^{-w_k} \leq \sqrt{|T(.)|_{L^2}^2 - 1 + j} \quad (j = 1, ..., n).$$

9.4 Estimates for quasipolynomials on the positive half-line

Again consider the quasipolynomial $f(z)$ defined by (1.1) under condition (1.3). Rewrite (1.4) as

$$F_f(z) = \frac{\mu(z)}{P(z)}, \qquad (4.1)$$

where

$$P(z) = \prod_{k=1}^{m} (z - \lambda_k)^{n_k}.$$

Here n_k is the multiplicity of λ_k and $\mu(z)$ is a polynomial of the degree less than the degree of P. Put

$$\nu := \deg P(z) - 1.$$

Theorem 9.4.1 *Let the Borel transform F_f of the quasipolynomial defined by (1.1) under condition (1.3) be represented by (4.1). Then*

$$|f(t)| \leq \frac{1}{\nu!} \sup_{z \in co(P)} |\frac{d^\nu}{dz^\nu}(\mu(z)e^{zt})| \ \ (t \geq 0),$$

where $co(P)$ is the closed convex hull of the numbers $\lambda_1, \lambda_2, ..., \lambda_m$.

If, in addition, all the numbers $\lambda_1, \lambda_2, ..., \lambda_m$ are real and belong to a segment $[a, b]$ and μ is real on $[a, b]$, then there is an $\eta(t) \in [a, b]$, such that

$$f(t) = \frac{1}{\nu!} \frac{d^\nu}{dx^\nu}(\mu(x)e^{xt})|_{x=\eta(t)} \ \ (t \geq 0).$$

Proof: The required resul is due to Lemmas 8.6.1 and the Jordan lemma. Q. E. D.

9.5 Differential equations

Consider the linear differential equation

$$P(D)y(t) = 0 \ \ (D = d/dt, t > 0), \qquad (5.1)$$

where $P(\lambda) = \lambda^n + c_1\lambda^{n-1} + ... + c_n$ is a real polynomial, whose zeros are z_k $(k = 1, ..., n)$.

Take the initial conditions

$$y^{(k)}(0) = 0 \ (k = 0, ..., n - 2), \ y^{(n-1)}(0) = 1. \tag{5.2}$$

A solution of problem (5.1), (5.2) is called *the Green function* of equation (5.1). We denote it by $G(t)$. Any solution $x(t)$ of the equation

$$P(D)x(t) = f(t) \tag{5.3}$$

with a given continuous function $f(t)$ and the zero initial conditions

$$x^{(k)}(0) = 0 \ (k = 0, ..., n - 1) \tag{5.4}$$

can be represented by the Variation of Constants Formula

$$x(t) = \int_0^t G(t - t_1)f(t_1)dt_1 \ (t > 0). \tag{5.5}$$

Let

$$\alpha := \max_k \operatorname{Re} z_k.$$

Then, by the Laplace transform, we have

$$G(t) := \frac{1}{2\pi i} \int_{d_0 - i\infty}^{d_0 + i\infty} \frac{e^{zt}dz}{P(z)} \ (d_0 = const > \alpha; \ t \geq 0).$$

According to Theorem 9.4.1, we can write

$$|G(t)| \leq \frac{e^{\alpha t}t^{n-1}}{(n-1)!} \ (t \geq 0).$$

So a solution $x(t)$ of problem (5.3), (5.4) satisfies the inequality

$$|x(t)| \leq \frac{1}{(n-1)!} \int_0^t e^{\alpha(t-s)}(t - s)^{n-1}|f(s)|ds \ (t > 0).$$

Now let all the zeros z_k of $P(\lambda)$ be real:

$$z_1 \leq z_2 \leq ... \leq z_n.$$

In this case by Theorem 9.4.1, we can assert that

$$0 \leq \frac{e^{z_1 t}t^{n-1}}{(n-1)!} \leq G(t) \leq \frac{e^{z_n t}t^{n-1}}{(n-1)!} \ (t \geq 0).$$

Let $f(t)$ be non-negative on the positive half-line. Then a solution of problem (5.3), (5.4) is also non-negative. Moreover, it satisfies the two-sided estimate

$$\frac{1}{(n-1)!} \int_0^t e^{z_1(t-s)}(t - s)^{n-1}f(s)ds \leq x(t)$$

$$\leq \frac{1}{(n-1)!} \int_0^t e^{z_n(t-s)}(t-s)^{n-1} f(s)ds \quad (t>0).$$

Now, we are going to establish some positivity conditions for the Green function in the case when the characteristic polynomial has complex zeros. To this end, assume that the polynomial

$$P_3(z) = \sum_{k=0}^{3} c_k z^{3-k} \quad (c_0 = 1, \ z \in \mathbf{C})$$

has a pair of complex conjugate zeros: $\gamma \pm i\omega$ $(\gamma, \omega > 0)$, and a real zero z_0. In addition, let

$$z_0 > \gamma. \tag{5.6}$$

Introduce the Green function

$$G_3(t) := \frac{1}{2\pi i} \int_{\tilde{c}-i\infty}^{\tilde{c}+i\infty} \frac{e^{zt}dz}{P_3(z)} \quad (\tilde{c} > \gamma, t \geq 0)$$

of the equation

$$P_3(D)x(t) = 0. \tag{5.7}$$

Besides,

$$G_3(0) = G_3'(0) = 0 \ , \ G_3''(0) = 1.$$

Lemma 9.5.1 *Let* $P_3(\lambda)$ *have a pair of complex conjugate zeros* $\gamma \pm i\omega$ $(\gamma \in \mathbf{R}, \omega > 0)$*, and a real zero* z_0*. In addition, let condition (5.6) hold. Then* $G_3(t) \geq 0$ $(t \geq 0)$*.*

Proof: By the residue theorem, we arrive at the equality

$$G_3(t) = m_0[e^{z_0 t} - e^{\gamma t}(\cos(\omega t) + b_0 \sin(\omega t))],$$

where

$$b_0 := \frac{z_0 - \gamma}{\omega} \quad \text{and} \quad m_0 := \frac{1}{(z_0 - \gamma)^2 + \omega^2}.$$

Put $b_1 = z_0 - \gamma$. So $b_1 = \omega b_0$. Since

$$G_3(t) = m_0 e^{\gamma t} f_1(t),$$

where

$$f_1(t) = e^{b_1 t} - \cos(\omega t) - b_0 \sin(\omega t),$$

it is enough to check that f_1 is positive. By virtue of the Taylor series,

$$e^{b_1 t} = 1 + b_1 t + g(t) \quad (g(t) \geq 0).$$

Since

$$|\cos s| \leq 1, |\sin s| \leq s \ (s \geq 0)$$

we have

$$f_1(t) = 1 + b_1 t + g(t) - \cos \omega t - b_0 \sin \omega t \geq 0$$

for all $t \geq 0$. As claimed. Q. E. D.

Now let us consider the differential equation in the complex plane:

$$\sum_{k=0}^{n} c_k \frac{d^{n-k} x(z)}{dz^{n-k}} = 0, \; c_0 = 1. \tag{5.8}$$

The Green function to this equation is an analytic continuation of $G(t)$ $(t \geq 0)$ to all $z \in \mathbf{C}$. Without any loss of the generality, assume that all the zeros of P are inside the circle $|z| < 1$. Since the Borel transform of $G(.)$ is

$$F_G(z) = \frac{1}{P(z)},$$

thanks to Corollary 8.2.2, the zeros $z_k(G)$ of the Green function $G(z)$ satisfy the inequality

$$\sum_{k=1}^{j} \frac{1}{|z_k(G)|} < \sqrt{w^2(G) - 1} + \ln(1 + j), \; j = 1, 2, \dots ,$$

where

$$w(G) = \frac{1}{\inf_{|z|=1} |P(z)|}.$$

9.6   Positive Green functions of functional differential equations

9.6.1   The first-order equations

For non-negative constants h, a and b, let us introduce the quasipolynomial

$$K_0(\lambda) := \lambda + a e^{-h\lambda} + b \; (\lambda \in \mathbf{C}).$$

Consider the following function generated by that quasipolynomial:

$$G_0(t) = \frac{1}{2\pi i} \int_{c_0 - i\infty}^{c_0 + i\infty} \frac{e^{zt}}{K_0(z)} dz. \tag{6.1}$$

Here and below in the present section c_0 is a real constant such that all the singularities of the integrand are in the half-plane $Re \, z < c_0$.

It is well known that $G_0(t)$ is the Green function of the linear first-order differential delay equation

$$\dot{u}(t) + au(t) + bu(t - h) = 0 \quad (t > 0, \; \dot{u}(t) = du(t)/dt), \qquad (6.2)$$

cf. [Hale, 1977]. That is, $G_0(t)$ is a solution of equation (6.2) with the initial conditions

$$G_0(0) = 1, \; G_0(t) = 0 \, (-h \leq t < 0). \qquad (6.3)$$

Besides K_0 is the characteristic quasipolynomial of (6.2).

Lemma 9.6.1 *Let the condition*

$$hbe^{ah} < e^{-1} \qquad (6.4)$$

hold. Then $G_0(t) \geq 0, t \geq 0$.

Proof: First, we consider the Green function $G(t)$ of the equation

$$\dot{u} + bu(t - h) = 0 \quad (t \geq 0). \qquad (6.5)$$

Let

$$bh < e^{-1}. \qquad (6.6)$$

Since

$$\max_{\tau \geq 0} \tau e^{-\tau} = e^{-1},$$

there is a positive solution w_0 of the equation $we^{-w} = bh$. Taking $c = h^{-1}w_0$, we get a solution c of the equation

$$c = be^{hc}.$$

Put in (6.5) $u(t) = e^{-ct}z(t)$. Then

$$\dot{z}(t) - cz(t) + be^{hc}z(t - h) = \dot{z} + c(z(t - h) - z(t)) = 0.$$

But because of (6.3) $z(0) = 1, z(t) = 0 \; (t < 0)$. So the latter equation is equivalent to the following one:

$$z(t) = 1 + \int_0^t c[z(s) - z(s - h)]ds = 1 + c\int_0^t z(s)ds - c\int_0^{t-h} z(s)ds.$$

Consequently,

$$z(t) = 1 + c\int_{t-h}^t z(s)ds.$$

Because of the Neumann series it follows that $z(t)$ and therefore, the Green function $G(t)$ of (6.5) are positive. Substituting $u(t) = e^{-at}v(t)$ in (6.2), we have

$$\dot{v}(t) + be^{ah}v(t - h) = 0.$$

According to (6.6), condition (6.4) provides the positivity of the Green function of the latter equation. Hence, the required result follows. Q. E. D.

Now introduce the exponential type function

$$K_\mu(\lambda) := \lambda + \int_0^h e^{-s\lambda} d\mu(s),$$

where μ is a nondecreasing function having a bounded variation $Var\ (\mu)$. Consider the following function:

$$G_\mu(t) = \frac{1}{2\pi i} \int_{c_0-i\infty}^{c_0+i\infty} \frac{e^{zt}}{K_\mu(z)} dz.$$

It is not hard to check that G_μ is the Green function $G_\mu(t)$ of the linear first-order differential delay equation

$$\dot{x}(t) + \int_0^h x(t-s)d\mu(s) = 0 \quad (t > 0). \tag{6.7}$$

That is, G_μ is a solution of (6.7) with the initial conditions

$$G_\mu(0) = 1; \ G_\mu(t) = 0 \ (-h \le t < 0).$$

Besides K_μ is the characteristic quasipolynomial of (6.7).
   Let us derive the positivity conditions for that Green function.

Theorem 9.6.2 *Let the condition*

$$eh\, Var\ (\mu) < 1$$

hold. Then $G_\mu(t)$ *is non-negative for all* $t \ge 0$. *Moreover,*

$$G_\mu(t) \ge G_+(t) \ge 0 \ (t \ge 0),$$

where $G_+(t)$ *is the Green functions of equation (6.5) with* $b = Var(\mu)$.

Proof: According to the initial conditions, for a sufficiently small $t_0 > h$,

$$G_\mu(t) \ge 0, \ \dot{G}_\mu(t) \le 0 \ (0 \le t \le t_0).$$

Thus,

$$G_\mu(t-h) \ge G_\mu(t-s) \ (0 \le s \le h \le t \le t_0).$$

Hence,

$$Var\ (\mu)G_\mu(t-h) \ge \int_0^h G_\mu(t-s)d\mu(s) \ (t \le t_0).$$

We thus have

$$\dot{G}_\mu(t) + Var\ (\mu)G_\mu(t-h) = f(t)$$

with

$$f(t) = Var\,(\mu)G_\mu(t-h) - \int_0^h G_\mu(t-s)d\mu(s) \geq 0 \ (0 \leq t \leq t_0).$$

Hence, by virtue of the Variation of Constants formula, we arrive at the relation

$$G_\mu(t) = G_+(t) + \int_0^t G_+(t-s)f(s)ds \geq G_+(t) \ (0 \leq t \leq t_0).$$

Extending this inequality to the whole half-line, by Lemma 9.6.1, we get the required inequality. Q. E. D.

9.6.2 The second-order equations

Let

$$K_2(z) := z^2 + Az + Bze^{-z} + C + De^{-z} + Ee^{-2z}$$

with non-negative constants A, B, C, D, E. Put

$$G_2(t) := \frac{1}{2\pi i} \int_{c_0-i\infty}^{c_0+i\infty} \frac{e^{tz}dz}{K_2(z)}.$$

G_2 is the Green function of the equation

$$\ddot{u}(t) + A\dot{u}(t) + B\dot{u}(t-1) + Cu(t) + Du(t-1) + Eu(t-2) = 0. \qquad (6.8)$$

Namely, G_2 satisfies this equation and the initial conditions

$$G_2'(0) = 1, \ G_2'(t) = 0 \ (-2h \leq t < 0), \ G_2(t) = 0 \ (-2h \leq t \leq 0).$$

Assume that

$$B^2/4 > E, \ A^2/4 > C \qquad (6.9)$$

and put

$$r_\pm(A,C) = -A/2 \pm \sqrt{A^2/4 - C},$$

and

$$r_\pm(B,E) = -B/2 \pm \sqrt{B^2/4 - E}.$$

Lemma 9.6.3 *Let the conditions (6.9) and*

$$D \leq r_+(B,E)r_+(A,C) + r_-(B,E)r_-(A,C), \qquad (6.10)$$

and

$$r_-(B,E)e^{1-r_+(A,C)} < 1 \ and \ r_+(B,E)e^{1-r_-(A,C)} < 1 \qquad (6.11)$$

hold. Then the Green function of equation (6.8) is non-negative.

Proof: Let a_k, b_k be numbers, such that

$$a_1 + a_2 = -A, \quad a_1 a_2 = C; \tag{6.12}$$

$$b_1 + b_2 = -B, \quad b_1 b_2 = E. \tag{6.13}$$

The roots of (6.12) are

$$a_{1,2} = r_{\pm}(A, C),$$

and the roots of (6.13) are

$$b_{1,2} = r_{\pm}(B, E).$$

Put

$$\tilde{K}(z) := (z - r_+(A, C) - r_-(B, E)e^{-z})(z - r_-(A, C) - r_+(B, E)e^{-z}).$$

Then

$$\tilde{K}(z) = z^2 + Az + Bze^{-z} + C + D_0 e^{-z} + Ee^{-2z},$$

where

$$D_0 = r_+(B, E)r_+(A, C) + r_-(B, E)r_-(A, C).$$

Taking into account that a product of the Laplace transforms of several functions corresponds to the convolution of these functions, we have due to Lemma 9.6.1 that under conditions (6.11), the Laplace original $\tilde{G}(t), t \geq 0$ of the function

$$\frac{1}{\tilde{K}(z)}$$

is non-negative. Besides, $\tilde{G}$ is the Green function to the equation

$$\ddot{u}(t) + A\dot{u}(t-1) + B\dot{u}(t-1) + Cu(t) + D_0 u(t-1) + Eu(t-2) = 0. \tag{6.14}$$

Put $v(t) = G_2(t) - \tilde{G}(t)$. Then subtracting equation (6.14) from (6.8), we arrive at the equation

$$\ddot{v}(t) + A\dot{v}(t) + B\dot{v}(t-1) + Cv(t) + D_0 v(t-1) + Ev(t-2) = (D_0 - D)G_2(t-1).$$

So

$$G_2(t) = \tilde{G}(t) + \int_0^t \tilde{G}(t-s)(D_0 - D)G_2(s-1)ds.$$

Thus,

$$G_2(t) = \tilde{G}(t) + \int_0^{t-1} \tilde{G}(t-t_1-1)(D_0 - D)G_2(t_1)dt_1.$$

Applying the Neumann series and taking into account that $\tilde{G}(t)$ is non-negative, we arrive at the inequalities

$$G_2(t) \geq \tilde{G}(t) \geq 0. \tag{6.15}$$

As claimed. Q. E. D.

9.6.3   Higher-order equations

Consider the function

$$K_n(\lambda) := \prod_{j=1}^{n} (\lambda + a_j + b_j e^{-h_j \lambda}) \quad (a_j, b_j = const; \ j = 1, ..., n)$$

with given non-negative numbers h_j, a_j and b_j. Put

$$h = h_1 + ... + h_n$$

and introduce the operator S_k by

$$(S_k v)(t) = v(t - h_k) \quad (t \geq 0)$$

at a continuous function v defined on $[-h, \infty)$. Then the Laplace original $G_n(t)$ of $1/K_n(z)$ is the Green function to the equation

$$\prod_{k=1}^{n} \left(\frac{d}{dt} + a_k + b_k S_k\right) x(t) = 0 \quad (t > 0). \tag{6.16}$$

Taking into account that a product of the Laplace transforms of several functions corresponds to the convolution of these functions, we have due to Lemma 9.6.1 the following result:

Lemma 9.6.4 *Let the inequalities*

$$b_j h_j e^{1 + a_j h_j} < 1 \quad (j = 1, ..., n)$$

hold. Then the Green function of equation (6.16) is non-negative.

9.7   Stability conditions and lower bounds for some quasipolynomials

Recall that a quasipolynomial *is stable* if all its zeros are in the open left half-plane.

Consider the function

$$K(z) = z + \sum_{k=1}^{m} b_k e^{-h_k z} \quad (h_k = const \geq 0)$$

with positive constants b_k.

Lemma 9.7.1 *With the notation*

$$c := 2 \sum_{k=1}^{m} b_k,$$

let

$$h_l c < \pi/2 \ (l = 1, ..., m). \tag{7.1}$$

Then $K(z)$ is stable and

$$\inf_{\omega \in \mathbf{R}} |K(i\omega)| \geq \sum_{k=1}^{m} b_k \cos (ch_k).$$

Moreover, the infimum is attained on $[-c, c]$.

Proof: We restrict ourselves by the case $m = 2$. In the general case, the proof is similar. Put $h_1 = v, h_2 = h$. Introduce the function

$$f(y) = |iy + b_2 e^{-ihy} + b_1 e^{-iyv}|^2.$$

Clearly,

$$f(y) = |iy + b_2 cos(hy) + b_1 cos(yv) - i(b_2 sin(hy) + b_1 sin(yv))|^2$$

$$= (b_2 cos(hy) + b_1 cos(yv))^2 + (y - b_2 sin(hy) - b_1 sin(yv))^2$$

$$= y^2 + b_2^2 + b_1^2 - 2b_2 y \ sin(hy) - 2b_1 y sin(yv)$$

$$+ 2b_2 b_1 \ sin(hy) \ sin(yv) + 2b_2 b_1 cos(yv) cos(hy).$$

So

$$f(y) = y^2 + b_2^2 + b_1^2 - 2b_2 y \ sin(hy) - 2b_1 y sin(yv) + 2b_2 b_1 cos \ y(v - h). \tag{7.2}$$

But $f(0) = (b_2 + b_1)^2$ and

$$f(y) \geq (|y| - b_2 - b_1)^2 \geq (b_2 + b_1)^2 \ (|y| > 2(b_2 + b_1)).$$

Thus, the minimum of f is attained on $[-c, c]$ with $c = 2(b_2 + b_1)$. Then thanks to (7.2)

$$f(y) \geq w(y) \ (0 \leq y \leq c),$$

where

$$w(y) = y^2 + b_2^2 + b_1^2 - 2y[b_2 \ sin(hc)$$

$$+ b_1 sin(vc)] + 2b_2 b_1 cos \ c(h - v).$$

and

$$dw(y)/dy = 2y - 2(b_2 \ sin(hc) + b_1 sin(vc)).$$

The zero of $dw(y)/dy = 0$ is

$$s = b_2 \ sin(hc) + b_1 sin(vc).$$

Thus,

$$\min_{y} w(y) = s^2 + b_2^2 + b_1^2 - 2s^2 + 2b_2 b_1 \cos \ c(h - v)$$

$$= b_2^2 + b_1^2 - (b_2 \ \sin \ (hc) + b_1 \sin(vc))^2 + 2b_2 b_1 \cos \ c(h - v).$$

Hence,

$$min_y w(y) = b_2^2 + b_1^2 - b_2^2 \sin^2 \ (ch) - b_1^2 \sin^2 \ (cv) + 2b_2 b_1 \cos \ (ch) \cos \ (cv)$$

$$= b_2^2 \cos^2 \ (hc) + b_1^2 \cos^2 \ (vc) + 2b_2 b_1 \cos(ch)\cos(cv) = (b_2 \cos(ch) + b_1 \cos(cv))^2.$$

This proves the required inequality. To prove the stability, consider the function

$$K(z, s) = z + b_1 e^{-szv} + b_2 e^{-szh} \quad (s \in [0, 1]).$$

Clearly, $K(z, 0)$ is stable and due to the just proved inequality,

$$\inf_{\omega \in \mathbf{R}} |K(i\omega, s)| \geq b_1 \cos(csv) + b_2 \cos(csh) \geq b_1 \cos(cv) + b_2 \cos(ch) > 0 \quad (s \in [0, 1]).$$

So $K(z, s)$ does not have zeros on the imaginary axis. This proves the lemma.
Q. E. D.

9.8   Comments to Chapter 9

The material of Section 9.1 is adopted from the paper [Gil', 2007a]. The results of Sections 2 and 3 are probably new. The material of Sections 9.4 and 9.5 is taken from the paper [Gil', 2007a] and the book [Gil', 2005c]. The material of Section 9.6 is taken from the papers [Gil', 2002a] and [Gil', 2007c]. About the classical results on positive solutions of differential and differential delay equations see for instance the well-known books [Agarwal, Grace, and O'Regan, 2000], [Agarwal, O'Regan and Wong, 1999], and references therein.

Lemma 9.8.7 is probably new. The theory of quasipolynomials has a long history, the literature on quasipolynomials and their applications is very rich. Many important results on the zeros of quasipolynomials and, in particular, stability criteria for quasipolynomials the reader can find in the books [Hale and Lunel, 1996], [Kolmanovskii and Myshkis, 1998], and [Wang, Lee and Tan, 1999].

Let us remind some well-known stability criteria for quasipolynomials.

In the case of discrete delays, the stability analysis can be reduced to an analysis of a quasipolynomial of the form

$$D(z) = \sum_{i=0}^{m} \sum_{j=0}^{n} a_{ij} z^i exp(b_j z)$$

with positive b_j. In particular, a quasipolynomial with commensurable delays may be represented as

$$D(z) = \sum_{i=0}^{m} \sum_{j=0}^{n} a_{ij} z^i exp(jz).$$

The most general results on the roots of that quasipolynomial are due to [Pontryagin, 1955]. We state the results of Pontryagin (Theorems 9.8.1 and 9.8.2) without proofs. Put

$$P(z, w) = \sum_{i=0}^{m} \sum_{j=0}^{n} a_{ij} z^i w^j.$$

Then $D(z) = P(z, e^z)$. A term $a_{i_0 j_0} z^{i_0} w^{j_0}$ is called the principal term of the polynomial $P(z, w)$, if $a_{i_0 j_0} \neq 0$, and if, for each other term $a_{ij} z^i w^j$ with $a_{ij} \neq 0$, we have either $j < j_0, i < i_0$, or $j = j_0, i < i_0$, or $j < j_0, i = i_0$.

Theorem 9.8.1 *If the polynomial $P(z, w)$ has no a principal term, then the equation $P(z, e^z) = 0$ has an infinite number of zeros with arbitrarily large positive real parts.*

One of the basic results for applications is the next theorem.

Theorem 9.8.2 *Let $D(z) = P(z, e^z)$, where $P(z, w)$ is a polynomial with a principal term. Suppose $D(i\omega)$ ($\omega \in \mathbf{R}$) is separated into its real and imaginary parts, $D(i\omega) = F(\omega) + iG(\omega)$. If all the zeros of $D(z)$ have negative real parts, then all the zeros of $F(\omega)$ and $G(\omega)$ are real, simple, alternate and*

$$G'(\omega)F(\omega) - F'(\omega)G(\omega) > 0 \qquad (8.1)$$

for every $\omega \in \mathbf{R}$. Conversely, all the zeros of $P(z, e^z)$ have negative real parts provided that either of the following conditions are satisfied:
(i). All the zeros of $F(\omega)$ and $G(\omega)$ are real, simple, alternate and inequality (8.1) holds for at last one ω.
(ii). All the zeros of $F(\omega)$ are real and, for each zero inequality (8.1) is satisfied.
(iii). All the zeros of $G(\omega)$ are real and, for each zero inequality (8.1) is satisfied.

The following result is due to Hayes (see the book [Hale, 1977, Appendix]).

Theorem 9.8.3 *Let a, b be real. Then all the roots of the equation*

$$(z + a)e^z + b = 0$$

have negative real parts if and only if

$$a > -1;$$

$$a + b > 0;$$

$$b < z_0 \sin z_0 - a \cos z_0,$$

where z_0 is the root of the equation $z = -a \tan z$ $(0 < z < \pi)$ if $a \neq 0$, and $z_0 = \pi/2$ if $a = 0$.

The following theorem is also proved in [Hale, 1977, Appendix].

Theorem 9.8.4 *All the roots of the equation*

$$(z^2 + az)e^z + 1 = 0$$

have negative real parts if and only if $a > (\sin z)/z$, where z is a unique root of the equation $z^2 = \cos z$, $0 < z < \pi/2$.

A calculation shows that $a > 0.8905835$.

Chapter 10

Transforms of Entire Functions and Canonical Products

This chapter contains additional results on the zeros of entire functions. In particular, we introduce transforms that generalize the Borel transform. In addition, we establish upper and lower bounds for the canonical products in terms of the sums of their zeros.

10.1 Comparison functions

Let $\psi_0 = \psi_1 = 1$ and $\psi_k, k = 2, 3, \ldots$ be positive numbers having the following property: the sequence

$$\frac{\psi_j}{\psi_{j-1}} \quad (j = 1, 2, \ldots)$$

is nonincreasing and tends to zero. Consider the entire function

$$f(z) = \sum_{k=0}^{\infty} c_k z^k \quad (c_0 = 1). \tag{1.1}$$

The function

$$\psi(t) = \sum_{k=0}^{\infty} \psi_k t^k \quad (t \geq 0) \tag{1.2}$$

is called *a comparison function of f* if for some positive number τ (depending on f),

$$|f(z)| \leq const\ \psi(\tau r),\ r \to \infty\ (r = |z|). \tag{1.3}$$

A comparison function is necessarily entire, as the ratio test for convergence shows. We denote by V_ψ the class of entire functions such that (1.3) holds. The infimum of numbers τ for which (1.3) holds *is the (exact) ψ-type of f*; we denote by $V_\psi(\tau)$ the class of functions whose ψ-type is τ or less. For example, when $\psi(t) = e^t$, $V_\psi(\tau)$ is the class of functions of exponential type 1, that is, entire functions of order 1 and type not exceeding τ, or of order less than 1.

The ψ-type of a function can be computed from the coefficients in its power series by applying the following theorem, called Nachibin's theorem, see [Boas and Buck, 1964, p. 6].

Theorem 10.1.1 *The function $f(z)$ defined by (1.1) is of ψ-type τ if and only if*

$$\overline{\lim}_{n\to\infty}|c_n/\psi_n|^{1/n} = \tau.$$

Proof: First let

$$\limsup_n |c_n/\psi_n|^{1/n} = \tau < \infty.$$

Then, for any $\tau_1 > \tau$, we may choose a constant B, so that

$$|c_n/\psi_n| \leq B\tau_1^n$$

for $n = 1, 2, ...$ Thus, on the circle $|z| = r$,

$$|f(z)| \leq \sum_{k=0}^{\infty} |c_k| r^k \leq B \sum_{k=0}^{\infty} \tau_1^k \psi_k r^k = B\psi(\tau_1 r).$$

Since τ_1 may be arbitrarily close to τ, this shows that f is of ψ-type at most τ.

In the other direction, we need a simple lemma connecting the rate of growth of ψ with that of its coefficients.

Lemma 10.1.2 *Let*

$$\gamma_n = \min_{x>0} \psi(x)x^{-n}.$$

Then, for all nonnegative integers n,

$$1 \leq \frac{\gamma_n}{\psi_n} \leq (n+1)e \tag{1.4}$$

and consequently,

$$\lim_{n\to\infty} [\frac{\gamma_n}{\psi_n}]^{1/n} = 1.$$

Proof: Since $\psi(x) \geq \psi_n x^n$, it is evident that $\gamma_n \geq \psi_n$. To obtain the right-hand side of (1.4), we estimate $\psi(x)$ for a choice of x near that which minimizes $\psi(x)x^{-n}$. Let $d_n = \psi_{n-1}/\psi_n$ and let $0 < \omega < 1$, ω is to be near 1, and will be specified later. Recalling that a restriction on ψ was that $\{d_n\}$ increases, we observe that $\psi_k \leq \psi_n d_n^{n-k}$, both for $k < n$ and $k \geq n$. Setting $x = \omega d_n$, we have

$$\psi(x) = \sum_{k=0}^{\infty} \psi_k x^k \leq \psi_n \sum_{k=0}^{\infty} d_n^{n-k}(\omega d_n)^k$$

$$\leq \frac{\psi_n d_n^n}{1 - \omega}.$$

For this choice of x, we have

$$\frac{\psi(x)}{x^n} \leq \frac{\psi_n}{\omega^n(1 - \omega)}$$

and so

$$\frac{\gamma_n}{\psi_n} \leq \frac{1}{\omega^n(1 - \omega)}.$$

Choosing ω as $n/(n+1)$ to minimize the right-hand side, we obtain

$$\frac{\gamma_n}{\psi_n} \leq (n+1)(1 + n^{-1})^n \leq (n+1)e.$$

Q. E. D.

To apply the lemma, suppose that f is of ψ-type τ. If $\tau_1 > \tau$, then for some constant M, we have $|f(z)| \leq M\psi(\tau_1 r)$. By Cauchy's inequality,

$$|c_n| \leq \frac{M\psi(\tau_1 r)}{r^n} = \frac{M\tau_1^n \psi(\tau_1 r)}{(\tau_1 r)^n}.$$

Choosing r to minimize the right-hand side, we find $|c_n| \leq \gamma_n M\tau_1^n$, and

$$\left|\frac{\gamma_n}{\psi_n}\right|^{1/n} \leq M^{1/n}\tau_1(\gamma_n/\psi_n)^{1/n}.$$

Invoking the lemma, we then have

$$\limsup_{n \to \infty} \left[\frac{\gamma_n}{\psi_n}\right]^{1/n} \leq \tau_1.$$

The conclusion of Nachibin's theorem follows on letting τ_1 approach τ.

Q. E. D.

10.2   Transforms of entire functions

10.2.1   The ψ-transform

Let $f \in V_\psi$ be defined by (1.1) and

$$\overline{\lim}_{k \to \infty} \sqrt[k]{\frac{c_k}{\psi_k}} < \frac{1}{R} \quad (0 < R < \infty). \tag{2.1}$$

Then, we will call the function

$$F(f, z) = \sum_{k=0}^{\infty} \frac{c_k}{\psi_k z^{k+1}} \quad (|z| \geq R) \tag{2.2}$$

the ψ-transform of f.

Theorem 10.2.1 *Let*

$$\psi(t) = \sum_{k=0}^{\infty} \psi_k t^k \quad (t \geq 0)$$

be a comparison function. Let

$$f(z) = \sum_{k=0}^{\infty} c_k z^k$$

belong to the class V_ψ as in (1.3), and condition (2.1) hold. Then

$$f(z) = \frac{1}{2\pi i} \int_{|w|=R} \psi(zw) F(f, w) dw \quad (z \in \mathbf{C}).$$

Proof: With $C = \{z \in \mathbf{C} : |z| = R\}$, we have

$$\frac{1}{2\pi i} \int_C \frac{\psi(zw) dw}{w^{n+1}} = \frac{1}{2\pi i} \int_C \sum_{k=0}^{\infty} z^k \psi_k w^{k-n-1} dw = \psi_n z^n.$$

Hence,

$$\frac{1}{2\pi i} \int_C \psi(zw) F(f, w) dw = \frac{1}{2\pi i} \int_C \sum_{n=0}^{\infty} \psi_n (zw)^n \sum_{k=0}^{\infty} \frac{c_k}{\psi_k w^{k+1}} dw = f(z).$$

Q. E. D.

When $\psi(t) = e^t$, (2.2) describes the correspondence between an entire function $f(z)$ of exponential type and its Laplace (or Borel) transform. The general case can be applied, by suitable choice of ψ_n, either to entire functions of arbitrary order or to functions that are regular in a prescribed region.

10.2.2 The Mittag-Leffler transform

Again consider the entire function defined by (1.1) assuming that

$$\overline{\lim}_{k \to \infty} \sqrt[k]{|c_k|\Gamma(1 + k/\rho)} < \frac{1}{R} \ (0 < R < \infty), \tag{2.3}$$

where Γ is the Euler gamma function. Then the function $M_\rho(f, z)$ defined by

$$M_\rho(f, z) := \sum_{k=0}^{\infty} \frac{\Gamma(1 + k/\rho)c_k}{z^{k+1}} \ (|z| \geq R)$$

will be called *the Mittag-Leffler transform of f*.

So under consideration $\psi_k = 1/\Gamma(1 + k/\rho), k > 1$, hence, if a function f is represented in the Mittag-Leffler form

$$f(z) := \sum_{k=0}^{\infty} \frac{a_k z^k}{\Gamma(1 + k/\rho)},$$

then

$$M_\rho(f, z) = \sum_{k=0}^{\infty} \frac{a_k}{z^{k+1}}.$$

Lemma 10.2.2 *Let f be given by (1.1) under condition (2.3). Then*

$$M_\rho(f, z) = \rho z^{\rho-1} \int_0^\infty e^{-(zt)^\rho} t^{\rho-1} f(t) dt$$

for all $z \in \mathbf{C}$ with Re $z^\rho > 0$ and $|z| \geq R$.

Proof: For an integer $k \geq 0$ denote

$$H_k(z) = \rho z^{\rho-1} \int_0^\infty e^{-(zt)^\rho} t^{\rho-1} t^k dt \ (\text{Re } z^\rho > 0).$$

Putting $s = zt$, we have

$$H_k(z) = z^{-k-1} \rho \int_0^\infty e^{-s^\rho} s^{\rho-1} s^k ds.$$

Hence, with $v = s^\rho$, we obtain

$$H_k(z) = z^{-k-1} \int_0^\infty e^{-v} v^{k/\rho} dv = z^{-k-1} \Gamma(1 + k/\rho).$$

Integrating term by term the series in (1.1) we prove the lemma. Q. E. D.

Let (2.3) hold with $R = 1$. Then

$$\theta_M^2(\rho, f) := \sum_{k=1}^{\infty} |c_k \Gamma(1 + k/\rho)|^2 < \infty.$$

Introduce the norm

$$\|M_\rho(f, e^{it})\|_{L^2} = \left[\frac{1}{2\pi} \int_0^{2\pi} |M_\rho(f, e^{it})|^2 dt\right]^{1/2}.$$

Under (2.3) with $R = 1$, by the Parseval equality and (1.1), we can write out,

$$\theta_M^2(\rho, f) = \|M_\rho(f, e^{it})\|_{L^2}^2 - 1.$$

So $\theta_M^2(\rho, f) \le \max_{|z|=1} |M_\rho(f, z)|^2 - 1$. Remind that by the notion

$$\theta^2(\rho, f) = \theta_M^2(\rho, f) + |c_1|^2(1 - 1/\Gamma^2(1 + 1/\rho)),$$

in Section 5.4, estimates for zeros are established.

Lemma 10.2.3 *Let condition (2.3) hold with $R = 1$ and $f = f_1 + f_2 + ... + f_m$, where f_k are entire functions with finite $\theta_M(\rho, f_k)$. Then*

$$\sqrt{\theta_M^2(\rho, f) + 1} \le \sum_{k=1}^m \sqrt{\theta_M^2(\rho, f_k) + 1}.$$

Proof: Indeed,

$$\sqrt{\theta_M^2(\rho, f) + 1} = \|M_\rho(f, e^{it})\|_{L^2} = \|\sum_{k=1}^m M_\rho(f_k, e^{it})\|_{L^2}$$

$$\le \sum_{k=1}^m \|M_\rho(f_k, e^{it})\|_{L^2} = \sum_{k=1}^m \sqrt{\theta_M^2(\rho, f_k) + 1}.$$

As claimed. Q. E. D.

The integral representations of the Mittag-Leffler transform and inverse one can be found in [Berenstein and Gay, 1995, Chapter 5]. Note that in that book the Mittag-Leffler transform is defined as $M_\rho(f, 1/z) \to f$, that is, conversely to our definition.

10.2.3   The root-factorial transform

Again consider the function f represented in the form (1.1) assuming that

$$\overline{\lim} \sqrt[k]{|c_k|(k!)^\gamma} < 1/R \quad (0 < R < \infty) \tag{2.4}$$

for some $\gamma > 0$. Let us introduce the transform $f \to T_\gamma(f, z)$ by

$$T_\gamma(f, z) = \sum_{k=0}^\infty \frac{c_k(k!)^\gamma}{z^{k+1}} \quad (|z| \ge R).$$

We will call it the *root-factorial transform of f*. So,

$$\psi_k = 1/(k!)^\gamma$$

and if a function f is represented in the root-factorial form

$$f(\lambda) = \sum_{k=0}^{\infty} \frac{\tilde{a}_k \lambda^k}{(k!)^\gamma} \quad (\lambda \in \mathbf{C}, \ \tilde{a}_k = c_k (k!)^\gamma)$$

then

$$T_\gamma(f, z) = \sum_{k=0}^{\infty} \frac{\tilde{a}_k}{z^{k+1}} \quad (z \in \mathbf{C}).$$

Let condition (2.4) hold with $R = 1$. Then

$$\theta_\gamma^2(f) := \sum_{k=1}^{\infty} |\tilde{a}_k|^2 < \infty.$$

Put

$$\|T_\gamma(f, e^{it})\|_{L^2} = \left[\frac{1}{2\pi} \int_0^{2\pi} |T_\gamma(f, e^{it})|^2 dt\right]^{1/2}.$$

By the Parseval equality we can write out,

$$\theta_\gamma^2(f) = \|T_\gamma(f, e^{it})\|_{L^2}^2 - 1.$$

So

$$\theta_\gamma^2(f) \le \max_{|z|=1} |T_\gamma(f, z)|^2 - 1.$$

Lemma 10.2.4 *Let condition (2.4) hold with $R = 1$ and let $f = f_1 + f_2 + \dots + f_m$, where f_k are entire functions with finite $\theta_\gamma(f_k)$. Then*

$$\sqrt{\theta_\gamma^2(f) + 1} \le \sum_{k=1}^{m} \sqrt{\theta_\gamma^2(f_k) + 1}.$$

The proof of this lemma is very similar to the proof of the Lemma 10.2.3.

10.2.4 A relation between the Mittag-Leffler and root-factorial transforms

For an integer $p \ge 2$, put

$$d_p = (pe)^{1/p}.$$

We need the following technical result.

Lemma 10.2.5 *For any integer $p \geq 1$, the inequality*

$$\frac{\Gamma(1 + k/p)}{\Gamma^{1/p}(1 + k)} \geq \frac{\sqrt{2\pi}}{d_p^k \sqrt{p}} \quad (k = 1, 2, ...)$$

is true.

Proof: By the Stierling formula [Markushevich, 1968, p. 323],

$$\Gamma(z) = \sqrt{2\pi} z^{z - \frac{1}{2}} exp \left[- z - \int_0^\infty \frac{\phi(t) dt}{(z + t)^2} \right],$$

where the function ϕ satisfies the condition

$$-\frac{1}{8} \leq \phi(t) \leq 0, \ t > 0.$$

Hence,

$$\Gamma(1 + z) = \Gamma(z)z = \sqrt{2\pi} z^{z + \frac{1}{2}} exp \left[- z - \int_0^\infty \frac{\phi(t) dt}{(z + t)^2} \right].$$

Therefore, for a positive z,

$$\Gamma(1 + z) \geq \sqrt{2\pi} z^{z + \frac{1}{2}} exp[-z].$$

However, for an integer n,

$$n! \leq \frac{1}{2^n} (n + 1)^n$$

cf. [Mitrinovich, 1970, p. 192]. For $k \geq 1$, this gives $k! \leq k^k$. Thus,

$$\Gamma(1 + k/p) \geq \sqrt{2\pi}(k/p)^{k/p + \frac{1}{2}} exp[-k/p] \geq k^{k/p} \frac{\sqrt{2\pi}}{d_p^k \sqrt{p}}$$

and

$$\Gamma(1 + k/p) \geq \Gamma^{1/p}(1 + k) \frac{\sqrt{2\pi}}{d_p^k \sqrt{p}}.$$

As claimed. Q. E. D.

The following theorem establishes a relation between the Mittag-Leffler and root-factorial transforms.

Theorem 10.2.6 *Let f be defined by (1.1) and for an integer $p > 1$, the condition*

$$\overline{lim}_{k \to \infty} \sqrt[k]{|c_k| \Gamma(1 + k/p)} < \frac{1}{d_p}$$

hold. Then

$$\|T_{1/p}(f, e^{it})\|_{L^2} \leq \frac{d_p \sqrt{p}}{\sqrt{2\pi}} \|M_p(f, d_p e^{it})\|_{L^2}.$$

Proof: By the previous lemma and the Parseval equality

$$\|M_p(f, d_p e^{it})\|_{L^2}^2 = \frac{1}{2\pi} \int_0^{2\pi} |M_p(f, d_p e^{is})|^2 ds$$

$$= \sum_{k=0}^{\infty} \frac{|c_k \Gamma(1 + k/p)|^2}{d_p^{2(1+k)}} \geq \frac{2\pi}{pd_p^2} \sum_{k=0}^{\infty} |c_k \Gamma^{1/p}(1 + k)|^2$$

$$= \frac{2\pi}{pd_p^2} \|T_{1/p}(f, e^{it})\|_{L^2}^2.$$

As claimed. Q. E. D.

10.3   Relations between canonical products and S_p

Let $\Lambda := \{z_j\}_{j=1}^{\infty}$ be a sequence of complex numbers with their multi-plicities ordered in the nondecreasing way: $|z_k| \leq |z_{k+1}|$, and satisfying the condition

$$S_p := \sum_{k=1}^{\infty} \frac{1}{|z_k|^p} < \infty \text{ for an integer } p \geq 1. \tag{3.1}$$

Consider the function

$$\Pi_p(z) = \prod_{j=1}^{\infty} G_j(z, p - 1)$$

with

$$G_j(z, p - 1) = \left(1 - \frac{z}{z_j}\right) exp\left[\sum_{m=1}^{p-1} \frac{z^m}{mz_j^m}\right] \ (p \geq 2); G_j(z, 0) = \left(1 - \frac{z}{z_j}\right).$$

For a positive r, again put

$$\Omega(r) = \{s \in \mathbf{C} : |s| \leq r\}.$$

Let $L(z) \subset \Omega(|z|)$ $(z \in \mathbf{C})$ be a *simple Jordan contour connecting 0 and z* (that is, $L(z)$ is ended by 0 and z). It is assumed that $L(z)$ does not contain the numbers z_j, such that

$$\phi_L(z) := \sup_{s \in L(z); \ k=1,2,\ldots} \frac{|z_k|}{|s - z_k|} < \infty. \tag{3.2}$$

Lemma 10.3.1 *Let conditions (3.1) and (3.2) hold. Then*

$$|ln \ \Pi_p(z)| \le S_p \psi_p(z) \ \ (z \notin \Lambda), \tag{3.3}$$

where

$$\psi_p(z) := \sup_{z_k \in \Lambda} \Big| \int_{L(z)} s^{p-1} \frac{z_k}{s - z_k} ds \Big|.$$

Proof: Let $p \ge 2$. For the brevity, put $G_j(z, p-1) = G_j(z)$. Clearly,

$$\Pi'_p(z) = \sum_{k=1}^{\infty} G'_k(z) \prod_{j=1, j \ne k}^{\infty} G_j(z)$$

and

$$G'_j(z) = \Big[-\frac{1}{z_j} + \Big(1 - \frac{z}{z_j}\Big) \sum_{m=0}^{p-2} \frac{z^m}{z_j^{m+1}} \Big] exp \Big[\sum_{m=1}^{p-1} \frac{z^m}{m z_j^m} \Big].$$

Take into account that

$$-\frac{1}{z_j} + \Big(1 - \frac{z}{z_j}\Big) \sum_{m=0}^{p-2} \frac{z^m}{z_j^{m+1}} = -\frac{z^{p-1}}{z_j^p}.$$

So

$$G'_j(z) = -\frac{z^{p-1}}{z_j^p(1 - \frac{z}{z_j})} G_j(z).$$

Hence,

$$\Pi'_p(z) = h(z)\Pi_p(z),$$

where

$$h(z) = -z^{p-1} \sum_{k=1}^{\infty} \frac{1}{z_k^p(1 - \frac{z}{z_k})}.$$

But

$$\int_{L(z)} h(s)ds = ln \ \Pi_p(z) = -\sum_{k=1}^{\infty} \frac{1}{z_k^p} \int_{L(z)} \frac{s^{p-1} \ ds}{1 - \frac{s}{z_k}}.$$

Consequently, (3.3) holds. The case $p = 1$ is similarly considered. As claimed.
Q. E. D.

Corollary 10.3.2 *Let conditions (3.1) and (3.2) hold. Then*

$$|ln \ \Pi_p(z)| \le S_p l_p(z) \phi_L(z) \ \ (z \in \mathbf{C}, \ p = 1, 2, ...), \tag{3.4}$$

where

$$l_p(z) := \int_{L(z)} |s|^{p-1} |ds|.$$

In particular, if $L(z)$ is a segment of a ray $\{w \in \mathbf{C} : \arg w = t\}$ with a fixed $t \in [0, 2\pi)$:

$$L(z) = \{s : \arg s = t,\ 0 \le |s| \le |z|\} \tag{3.5}$$

then

$$\left| \ln \Pi_p(z) \right| \le \frac{|z|^p}{p} \phi_L(z) S_p. \tag{3.6}$$

Indeed, this result is due to the previous lemma and the inequality

$$\left| \int_{L(z)} \frac{s^{p-1}\, ds}{1 - s/z_k} \right| \le \phi_L(z) \int_{L(z)} |s|^{p-1} |ds| = l_p(z) \phi_L(z).$$

If $L(z)$ is defined as in (3.5), then $l_p(z) = |z|^p/p$.

Furthermore, from (3.4) it follows that

$$|\Pi_p(z)| = e^{Re \ln \Pi_p(z)} \ge e^{-|\ln \Pi_p(z)|} \ge e^{-l_p(z)\phi_L(z)S_p}. \tag{3.7}$$

Moreover, under condition (3.5), we have

$$|\Pi_p(z)| \ge e^{-|z|^p \phi_L(z)/p}. \tag{3.8}$$

The latter corollary *is sharp*. Indeed, under (3.5) the right-hand part of inequality (3.8) tends to zero as $z \to z_k \in \Lambda$, since in this case $\phi_L(z) \to +\infty$.

10.4   Lower bounds for canonical products in terms of S_p

Again, let $\Lambda = \{z_k\}$ be the set of the zeros of $\Pi_p(z)$ ($|z_1| \le |z_2| \le ...$). For an $a > 0$, let $z_1, z_2, ..., z_m \in \Lambda$ be all the zeros satisfying $|z_j| \le a$. That is, $\nu(a) = m$, where $\nu(r)$ is the counting function of $\{z_j\}$ in $\Omega(r)$. Put

$$w_a := \prod_{j=1}^{m} \frac{1}{z_j} \quad \text{and} \quad \gamma_p := \sum_{l=1}^{p} \frac{1}{l}.$$

Theorem 10.4.1 *Let condition (3.1) hold and $m = \nu(a) \ge 1$. Then for any positive $b < a/2$, there is a system $\Psi(b)$ of at most m circles whose sum of radii is $2b$, such that*

$$|\Pi_p(z)| \ge |w_a| (be^{-\gamma_p})^m exp\left[-\frac{|z|^p S_p}{(1 - \frac{|z|}{a})} \right]$$

for all $z \in \Omega(a)$, $z \notin \Psi(b)$. If $\nu(a) = 0$, that is, if

$$\inf_{j=1,2,...} |z_j| > a,$$

then

$$|\Pi_p(z)| \ge exp\left[-\frac{S_p |z|^p}{p(1 - \frac{|z|}{a})} \right] \quad (|z| \le a). \tag{4.1}$$

This theorem is proved in the next section.

Assume that

$$\nu := \sup_{j=1,2,\ldots} \frac{j^{1/\rho_1}}{|z_j|} < \infty \tag{4.2}$$

for some $\rho_1 > 0$. Take an integer $p \geq 2$, such that

$$p > \rho_1. \tag{4.3}$$

We have

$$S_p \leq \nu^p \sum_{k=1}^{\infty} \frac{1}{k^{p/\rho_1}} = \nu^p \zeta(p/\rho_1) < \infty,$$

where ζ is the Riemann Zeta function. From Theorem 10.4.1 it follows

Corollary 10.4.2 *Under conditions (4.2), (4.3), let $\nu(a) \geq 1$. Then for any positive $b < a/2$, the inequality*

$$|\Pi_p(z)| \geq |w_a| b^m exp \left[-\frac{|z|^p \nu^p \zeta(\frac{p}{\rho_1})}{1 - \frac{|z|}{a}} - m\gamma_p \right]$$

is valid for all $z \in \Omega(a)$ except $\Psi(b)$. If $\nu(a) = 0$, then

$$|\Pi_p(z)| \geq exp \left[-\frac{\nu^p \zeta(\frac{p}{\rho_1}) |z|^p}{p(1 - \frac{|z|}{a})} \right] \quad (|z| < a).$$

10.5   Proof of Theorem 10.4.1

Let $m = \nu(a) \geq 1$. Set

$$P_a(z) = \prod_{j=1}^{m} (z - z_j),$$

$$H_a(z) = \prod_{j=1}^{m} G_j(z) = (-1)^m P_a(z) w_a exp \left[\sum_{j=1}^{m} \sum_{l=1}^{p-1} \frac{z^l}{l z_j^l} \right]$$

and

$$F_a(z) := \prod_{j=m+1}^{\infty} G_j(z).$$

Then $|\Pi_p(z)| = |F_a(z) H_a(z)|$. Function F_a does not have zeros in $|z| \leq a$. So

$$\sup_{k>m} \frac{|z_k|}{|z - z_k|} \leq \frac{1}{1 - |z|/a} \quad (|z| < a).$$

Now Corollary 10.3.2 implies

$$|F_a(z)| \geq exp\left[-\frac{|z|^p W_{p,m}}{p(1-|z|/a)}\right] \quad (|z| < a),$$

where

$$W_{p,m} := \sum_{k=m+1}^{\infty} \frac{1}{|z_k|^p}.$$

Furthermore, by the Cartan theorem (see Section 3.4), for any positive b, there is a system $\Psi(b)$ of circles whose sum of radii is $2b$, such that outside that system

$$|P_a(z)| \geq (be^{-1})^m.$$

Clearly,

$$\sum_{l=1}^{p-1} |z|^l \sum_{j=1}^{m} \frac{1}{l z_j^l} | \leq \sum_{l=1}^{p-1} |z|^l \frac{1}{l} S_{l,m},$$

where

$$S_{l,m} := \sum_{k=1}^{m} \frac{1}{|z_k|^l}.$$

Thus,

$$|H_a(z)| \geq |w_a|(be^{-1})^m exp\left[-\sum_{l=1}^{p-1} |z|^l \frac{1}{l} S_{l,m}\right] \quad (z \in \Omega(a), z \notin \Psi(b)).$$

Since $|\Pi_p(z)| = |F_a(z)H_a(z)|$, we have established the following result.

Lemma 10.5.1 *Let condition (3.1) hold and $m = \nu(a) \geq 1$. Then for any positive $b < a/2$, the inequality*

$$|\Pi_p(z)| \geq |w_a|(e^{-1}b)^m exp\left[-\frac{|z|^p W_{p,m}}{p(1-\frac{|z|}{a})}\right] exp\left[-\sum_{l=1}^{p-1} \frac{|z|^l}{l} S_{l,m}\right]$$

is valid for all $z \in \Omega(a), z \notin \Psi(b)$.

Proof of Theorem 10.4.1: If $\nu(a) = 0$, then the required result is due to (3.8). Now let $m = \nu(a) \geq 1$. By the Young inequality, we can write out

$$x \leq \frac{x^s}{s} + \frac{s-1}{s} \quad (x > 0, s > 1).$$

Hence, taking $s = p/l$, we get,

$$\sum_{l=1}^{p-1} \frac{|z|^l}{|z_j|^l l} \leq \sum_{l=1}^{p-1} \left(\frac{|z|^p}{|z_j|^p p} + \frac{p-l}{pl}\right).$$

But

$$\sum_{l=1}^{p-1} \frac{p-l}{pl} = \frac{1}{p}\sum_{l=1}^{p-1}(\frac{p}{l}-1) = \sum_{l=1}^{p-1}\frac{1}{l} - \frac{p-1}{p} = \gamma_p - 1.$$

So

$$\sum_{j=1}^{m}\sum_{l=1}^{p-1}\frac{|z|^l}{l|z_j|^l} \le \sum_{j=1}^{m}\frac{(p-1)|z|^p}{|z_j|^p p} + m(\gamma_p - 1).$$

Hence,

$$|z|^p W_{p,m} + \sum_{l=1}^{p-1}\frac{1}{l}|z|^l S_{l,m} = |z|^p W_{p,m} + \sum_{j=1}^{m}\sum_{l=1}^{p-1}\frac{|z|^l}{l|z_j|^l}$$

$$\le |z|^p(W_{p,m} + S_{p,m}) + m(\gamma_p - 1) = |z|^p S_p + m(\gamma_p - 1) \ (m \ge 1).$$

Since $m = \nu(a)$, by Lemma 10.5.1, we arrive at the required result. Q. E. D.

10.6   Canonical products and determinants

Again put

$$\Pi_p(z) = \prod_{j=1}^{\infty}\left(1 - \frac{z}{z_j}\right)exp\left[\sum_{m=1}^{p-1}\frac{z^m}{m z_j^m}\right] \ (p > 1)$$

and

$$\Pi_1(z) = \prod_{j=1}^{\infty}\left(1 - \frac{z}{z_j}\right).$$

Furthermore, in Section 2.14 it is proved that there is a constant β_p, such that

$$\left|(1 - z) \, exp\left[\sum_{m=1}^{p-1}\frac{z^m}{m}\right]\right| \le e^{\beta_p|z|^p} \ (z \in \mathbf{C}) \tag{6.1}$$

for any integer $p \ge 2$. Besides, $\beta_2 = 1/2$ and

$$\beta_p \le 3e(2 + ln \, (p-1)) \ (p \ge 3).$$

Lemma 10.6.1 *Let*

$$S_p < \infty \text{ for an integer } p \ge 1. \tag{6.2}$$

Then the inequality

$$|\Pi_p(z)| \le exp \, [\beta_p S_p |z|^p] \ (z \in \mathbf{C}, \beta_1 = 1) \tag{6.3}$$

is valid.

Proof: Obviously,

$$|\Pi_p(z)| \le \prod_{j=1}^{\infty} exp\left[\frac{\beta_p|z|^p}{|z_j|^p}\right] = exp\left[\sum_{k=1}^{\infty} \frac{\beta_p|z|^p}{|z_j|^p}\right] = exp\left[\beta_p|z|^p S_p\right].$$

Q. E. D.

Let

$$\Pi_p(z) = \sum_{k=0}^{\infty} \frac{a_k \lambda^k}{(k!)^\gamma} \quad (0 < \gamma \le 1/p) \tag{6.4}$$

be the Taylor expansion under the condition

$$\overline{\lim} \sqrt[k]{|a_k|} < \infty. \tag{6.5}$$

Recall that $det_p(I - A)$ is the regularized determinant.

Lemma 10.6.2 *Let the expansion (6.4) hold under condition (6.5). Then the equality*

$$\Pi_p(z) = det_p(I - Az) \tag{6.6}$$

is valid, where A is the compact operator defined in l^2 by the infinite matrix

$$\begin{pmatrix} -a_1 & -a_2 & -a_3 & \dots \\ 1/2^\gamma & 0 & 0 & \dots \\ 0 & 1/3^\gamma & 0 & \dots \\ 0 & 0 & 1/4^\gamma & \dots \\ \cdot & \cdot & \cdot & \dots \end{pmatrix}.$$

Proof: Indeed, let

$$A_n(\gamma) = \begin{pmatrix} -a_1 & -a_2 & \dots & -a_{n-1} & -a_n \\ 1/2^\gamma & 0 & \dots & 0 & 0 \\ 0 & 1/3^\gamma & \dots & 0 & 0 \\ 0 & 0 & \dots & 0 & 0 \\ \cdot & \cdot & \dots & \cdot & \cdot \\ 0 & 0 & \dots & 1/n^\gamma & 0 \end{pmatrix}$$

and

$$f_n(z) = 1 + a_1 z + \frac{a_2 z^2}{2^\gamma} + \dots + \frac{a_n z^n}{n^\gamma}.$$

As it was shown in Section 5.7, $z_k(f_n) = \frac{1}{\lambda_k(A_n(\gamma))}$. Hence, letting $n \to \infty$, we have

$$z_k(\Pi_p) = \frac{1}{\lambda_k(A)}, \quad k = 1, 2, \dots$$

This proves the lemma. Q. E. D.

In the previous lemma, we have restricted ourselves by the root-factorial representation, but the same arguments are valid for other representations, for example, for the Mittag-Leffler representation.

Now let us consider infinite order canonical products. Let $\{z_k\}_{k=1}^{\infty}$ be a sequence of complex numbers enumerated in the nondecreasing way. Then there is a nondecreasing sequence $\pi = \{p_j\}_{j=1}^{\infty}$ ($p_1 \geq 1$) of positive integers, such that

$$\gamma_\pi(\lambda) := \sum_{j=1}^{\infty} \beta_{p_j} \left(\frac{|\lambda|}{|z_j|}\right)^{p_j} < \infty \qquad (6.7)$$

for all $\lambda \in \mathbf{C}$. The canonical product corresponding to the sequence π define by

$$\Pi_\pi(\lambda) := \prod_{j=1}^{\infty} (1 - \frac{\lambda}{z_j}) \exp \left[\sum_{m=1}^{p_j-1} \frac{\lambda^m}{m z_j^m}\right].$$

We put

$$\sum_{m=1}^{0} = 0.$$

Theorem 10.6.3 *Under condition (6.7), the ineaquality*

$$|\Pi_\pi(\lambda)| \leq e^{\gamma_\pi(\lambda)}$$

is valid.

Proof: Thanks to (6.1), we have

$$|\Pi_\pi(\lambda)| \leq \prod_{j=1}^{\infty} \exp \left[\beta_{p_j} \frac{|\lambda|^{p_j}}{|z_j|^{p_j}}\right] =$$

$$exp \left[\sum_{j=1}^{\infty} \frac{\beta_{p_j} |\lambda|^{p_j}}{|z_j|^{p_j}}\right].$$

As claimed. Q. E. D.

10.7   Perturbations of canonical products

Consider the products

$$\Pi_p(z) := \prod_{j=1}^{\infty} \left(1 - \frac{z}{z_j}\right) exp \left[\sum_{m=1}^{p-1} \frac{z^m}{m z_j^m}\right] \quad (p = 1, 2, ...) \qquad (7.1)$$

and

$$\tilde{\Pi}_p(z) := \prod_{j=1}^{\infty} \left(1 - \frac{z}{\tilde{z}_j}\right) exp\left[\sum_{m=1}^{p-1} \frac{z^m}{m\tilde{z}_j^m}\right], \tag{7.2}$$

were the numbers $z_j, \tilde{z}_j$ satisfy the conditions

$$\sum_{k=1}^{\infty} \frac{1}{|z_k|^p} < \infty, \ \sum_{k=1}^{\infty} \frac{1}{|\tilde{z}_k|^p} < \infty.$$

Let

$$\Pi_p(z) = \sum_{k=0}^{\infty} \frac{a_k \lambda^k}{(k!)^\gamma} \tag{7.3}$$

and

$$\tilde{\Pi}_p(z) = \sum_{k=0}^{\infty} \frac{b_k \lambda^k}{(k!)^\gamma} \quad (0 < \gamma \le 1/p) \tag{7.4}$$

be the corresponding Taylor expansions satisfying the conditions

$$\overline{\lim} \sqrt[k]{|a_k|} < 1 \text{ and } \overline{\lim} \sqrt[k]{|b_k|} < 1. \tag{7.5}$$

Introduce the notations

$$q = \left[\sum_{k=1}^{\infty} |a_k - b_k|^2\right]^{1/2},$$

$$\theta := \left[\sum_{k=1}^{\infty} |a_k|^2\right]^{1/2} \text{ and } \tilde{\theta} := \left[\sum_{k=1}^{\infty} |b_k|^2\right]^{1/2}.$$

In addition, put

$$\tau_p := \left[\left(\theta + \frac{1}{2\gamma}\right)^p + \zeta(\gamma p) - \frac{1}{2^{\gamma p}} - 1\right]^{1/p}$$

and

$$\tilde{\tau}_p := \left[\left(\tilde{\theta} + \frac{1}{2\gamma}\right)^p + \zeta(\gamma p) - \frac{1}{2^{\gamma p}} - 1\right]^{1/p}.$$

Theorem 10.7.1 *Let relations (7.3)-(7.5) hold. Then the canonical products* $\Pi_p(z)$ *and* $\tilde{\Pi}_p(z)$ *defined by (7.1) and (7.2) are subject to the inequality*

$$|\Pi_p(z) - \tilde{\Pi}_p(z)| \le q|z| \, exp\left[\beta_p\left(1 + \frac{|z|}{2}(q + \tau_p + \tilde{\tau}_p)\right)^p\right],$$

where β_p *is defined by (6.1).*

Proof: Recall that $det_p \, (I - A)$ denotes the regularized determinant of an operator $A \in SN_p$ (see Section 2.14). From Lemma 10.6.2, it follows that

$$\Pi_p(f, z) = det_p \, (I - Az) \text{ and } \Pi_p(h, z) = det_p \, (I - Bz),$$

where A is defined as in Lemma 10.6.2 and B is the compact operator defined in l^2 by the infinite matrix

$$B = \begin{pmatrix} -b_1 & -b_2 & -b_3 & \cdots \\ 1/2^\gamma & 0 & 0 & \cdots \\ 0 & 1/3^\gamma & 0 & \cdots \\ 0 & 0 & 1/4^\gamma & \cdots \\ \cdot & \cdot & \cdot & \cdots \end{pmatrix}.$$

Besides, $\lambda_k(A) = \frac{1}{z_k}$ and $\lambda_k(B) = \frac{1}{\hat{z}_k}$. By Corollary 2.14.4,

$$|det_p(I - zA) - det_p(I - zB)|$$

$$\leq |z| N_p(A - B) exp \left[\beta_p \left(1 + \frac{|z|}{2}(N_p(A-B) + N_p(A+B))\right)^p\right].$$

Hence,

$$|det_p(I - zA) - det_p(I - zB)|$$

$$\leq |z| N_p(A - B) exp \left[\beta_p \left(1 + \frac{|z|}{2}(N_p(A-B) + N_p(A) + N_p(B))\right)^p\right]. \quad (7.6)$$

Moreover, let $A_n(\gamma)$ be as in Section 10.6. As it was proved in Section 5.7,

$$\sum_{k=1}^{j} s_k(A_n(\gamma)) \leq \left[\sum_{k=1}^{n} |a_k|^2\right]^{1/2} + \sum_{k=1}^{j} \frac{1}{(k+1)^\gamma}$$

$$\leq \theta + \sum_{k=1}^{j} \frac{1}{(k+1)^\gamma} \quad (j = 1, ..., n).$$

Hence,

$$N_p^p(A_n(\gamma)) = \sum_{k=1}^{n} s_k^p(A_n(\gamma)) \leq \left(\theta + \frac{1}{2^\gamma}\right)^p + \sum_{k=2}^{n} \frac{1}{(k+1)^{p\gamma}}.$$

But $N_p(A_n(\gamma)) \to N_p(A)$ as $n \to \infty$. So

$$N_p^p(A) \leq \left(\theta + \frac{1}{2^\gamma}\right)^p + \sum_{k=2}^{\infty} \frac{1}{(k+1)^{p\gamma}}$$

$$= \left(\theta + \frac{1}{2^\gamma}\right)^p + \zeta(\gamma p) - \frac{1}{2^{\gamma p}} - 1 = \tau_p^p.$$

Similarly, $N_p(B) \leq \tilde{\tau}_p$. In addition,

$$B - A = \begin{pmatrix} -b_1 + a_1 & -b_2 + a_2 & -b_3 + a_3 & \cdots \\ 0 & 0 & 0 & \cdots \\ 0 & 0 & 0 & \cdots \\ 0 & 0 & 0 & \cdots \\ \cdot & \cdot & \cdot & \cdots \end{pmatrix}.$$

Hence,

$$(B - A)(B^* - A^*) = \begin{pmatrix} q^2 & 0 & 0 & \dots \\ 0 & 0 & 0 & \dots \\ 0 & 0 & 0 & \dots \\ 0 & 0 & 0 & \dots \\ \cdot & \cdot & \cdot & \dots \end{pmatrix}.$$

So $s_1(A - B) = q$, $s_k(A - B) = 0, k = 2, 3, \dots$. Hence,

$$N_p(A - B) = q.$$

Now (7.6) yields the required result. Q. E. D.

Corollary 10.7.2 *Under the hypothesis of the previous theorem, we have*

$$|\tilde{\Pi}_p(z)| \geq |\Pi_p(z)| - q|z| \, exp \, [\beta_p(1 + |z|\mu)^p],$$

where

$$\mu := \frac{1}{2}(q + \tau_p + \tilde{\tau}_p).$$

So, if some $z \in \mathbf{C}$ is not a zero of $\Pi_p(z)$ and

$$|\Pi_p(z)| > q|z| \, exp \, [\beta_p(1 + |z|\mu)^p],$$

then z is not a zero of $\tilde{\Pi}_p(z)$.

10.8 Comments to Chapter 10

As it was above mentioned, Theorem 10.1.1 is taken from the excellent book [Boas and Buck, 1964]. The results of Section 10.2 are probably new. Our definition of the Mittag-Leffler transform is different from the definition suggested in the book [Berenstein and Gay, 1995, p. 310]. Sections 10.3 - 10.6 are based on the paper [Gil', 2007c]. Theorem 10.7.1 is probably new.

Some important new inequalities for entire functions can be found in the papers [Foster and Krasikov, 2002] and [Gardner and Govil, 1995]. An interesting approach to the canonical products of infinite order entire functions has been suggested in the paper [Bergweiler, 1992].

Chapter 11

Polynomials with Matrix Coefficients

In this chapter, we investigate polynomials with matrix coefficients (polynomial matrix pencils). Bounds for the characteristic values are established. We also explore variations of the characteristic values of polynomial matrix pencils under perturbations. In addition, we establish the multiplicative representation for a class of rational matrix-valued functions, namely, for the inverse polynomial matrix pencils. Applications of our results to coupled systems of polynomial equations and vector difference equations are also discussed.

11.1   Partial sums of moduli of characteristic values

Let $\mathbf{C}^n$ be a Euclidean space with a scalar product $(.,.)$, the Euclidean norm

$$\|h\| = \sqrt{(h, h)} \ \ (h \in \mathbf{C}^n)$$

and the unit matrix $I = I_n$. For an $n \times n$-matrix A, $\|A\|$ means the operator norm:

$$\|A\| := \sup_{h \in \mathbf{C}^n} \frac{\|Ah\|}{\|h\|},$$

and $\lambda_k(A)$ $(k = 1, ..., n)$ are the eigenvalues with multiplicities taken into account. Our main object in this chapter is the polynomial with matrix coefficients (the polynomial matrix pencil)

$$\tilde{P}(\lambda) = \sum_{k=0}^{\mu} \lambda^{\mu-k} A_k \quad (\lambda \in \mathbf{C}; \ A_0 = I_n), \tag{1.1}$$

where A_k, $k = 1, 2, ..., \mu$ are constant $n \times n$-matrices. That is,

$$\tilde{P}(\lambda) = (p_{jk}(\lambda))_{j,k=1}^{n}$$

is a matrix dependent on a complex argument λ whose entries $p_{jk}(\lambda)$ are polynomials. A zero $z_k(\tilde{P})$ of $det\ \tilde{P}(z)$ is called *a characteristic value of* $\tilde{P}$. The collection of all the characteristic values of $\tilde{P}$ with their multiplicities is called *the spectrum of* $\tilde{P}$ *and is denoted by* $\Sigma(\tilde{P})$. Let the characteristic values of $\tilde{P}$ taken with their multiplicities be ordered in the non-increasing way:

$$|z_k(\tilde{P})| \geq |z_{k+1}(\tilde{P})| \quad (k = 1, ..., \mu n - 1).$$

Introduce the block matrix

$$\hat{A}(\tilde{P}) := \begin{pmatrix} -A_1 & -A_2 & ... & -A_{\mu-1} & -A_\mu \\ I_n & 0 & ... & 0 & 0 \\ 0 & I_n & ... & 0 & 0 \\ . & . & ... & . & . \\ 0 & 0 & ... & I_n & 0 \end{pmatrix}.$$

The the following result plays an essential role in this chapter.

Lemma 11.1.1 *The equality*

$$det\ \tilde{P}(\lambda) = det\ (\lambda I_{\mu n} - \hat{A}(\tilde{P}))$$

is true.

This lemma is a particular case of Lemma 12.3.1 proved below.
    Set

$$\Theta_{\tilde{P}} := \sum_{k=1}^{\mu} A_k A_k^*,$$

where the the asterisk means the conjugate. Denote

$$\omega_k(\tilde{P}) = \begin{cases} \sqrt{\lambda_k(\Theta_{\tilde{P}})} & \text{if } k = 1, ..., n, \\ 0 & \text{if } k = n+1, ..., \mu n \end{cases}.$$

Theorem 11.1.2 *The characteristic values of* $\tilde{P}$ *satisfy the inequalities*

$$\sum_{k=1}^{j} |z_k(\tilde{P})| < j + \sum_{k=1}^{j} \omega_k(\tilde{P}) \quad (j = 1, ..., \mu n).$$

Proof: For simplicity, put $\hat{A}(\tilde{P}) = \hat{A}$. Because of the previous lemma,

$$\lambda_k(\hat{A}) = z_k(\tilde{P}) \quad (k = 1, 2, ..., n\mu). \tag{1.2}$$

Take into account that by the Weyl inequalities

$$\sum_{k=1}^{j} |\lambda_k(\hat{A})| < \sum_{k=1}^{j} s_k(\hat{A}) \quad (j = 1, ..., n\mu), \tag{1.3}$$

where $s_k(\hat{A})$ are the singular numbers of $\hat{A}$ with their multiplicities ordered in the non-increasing way. But $\hat{A} = M + C_0$, where

$$M = \begin{pmatrix} -A_1 & -A_2 & \cdots & -A_{\mu-1} & -A_\mu \\ 0 & 0 & \cdots & 0 & 0 \\ \cdot & \cdot & \cdots & \cdot & \cdot \\ 0 & 0 & \cdots & 0 & 0 \end{pmatrix}$$

and

$$C_0 = \begin{pmatrix} 0 & 0 & \cdots & 0 & 0 \\ I_n & 0 & \cdots & 0 & 0 \\ 0 & I_n & \cdots & 0 & 0 \\ \cdot & \cdot & \cdots & \cdot & \cdot \\ 0 & 0 & \cdots & I_n & 0 \end{pmatrix}.$$

We have

$$MM^* = \begin{pmatrix} \Theta_{\tilde{P}} & 0 & \cdots & 0 & 0 \\ 0 & 0 & \cdots & 0 & 0 \\ \cdot & \cdot & \cdots & \cdot & \cdot \\ 0 & 0 & \cdots & 0 & 0 \end{pmatrix}$$

and

$$C_0 C_0^* = \begin{pmatrix} 0 & 0 & \cdots & 0 & 0 \\ 0 & I_n & \cdots & 0 & 0 \\ 0 & 0 & \cdots & 0 & 0 \\ \cdot & \cdot & \cdots & \cdot & \cdot \\ 0 & 0 & \cdots & 0 & I_n \end{pmatrix}.$$

Consequently, $s_k(M^*) = \omega_k(\tilde{P})$. In addition,

$$s_k(C_0^*) = 1 \quad (k = jn + l; j = 0, ..., \mu - 2; l = 1, ..., n)$$

and

$$s_k(C_0^*) = 0 \quad (k = (\mu - 1)n + l; l = 1, ..., n).$$

So

$$s_k(C_0^*) \leq 1 \quad (k = 1, ..., n\mu).$$

Thanks to Lemma 1.1.2

$$\sum_{k=1}^{j} s_k(\hat{A}^*) = \sum_{k=1}^{j} s_k(M^* + C_0^*) \leq \sum_{k=1}^{j} s_k(M^*) + \sum_{k=1}^{j} s_k(C_0^*).$$

Thus,

$$\sum_{k=1}^{j} s_k(\hat{A}) = \sum_{k=1}^{j} s_k(\hat{A}) \le \sum_{k=1}^{j} \omega_k(\tilde{P}) + j \quad (j = 1, 2, .., \mu n).$$

Now (1.2) and (1.3) yield the required result. Q. E. D.

From the previous theorem, it follows that

$$j|z_j(\tilde{P})| < \sum_{k=1}^{j} \omega_k(\tilde{P}) + j \quad (j = 1, ..., n\mu).$$

Therefore, $\tilde{P}$ has in the circle

$$|z| \le 1 + \frac{1}{j} \sum_{k=1}^{j} \omega_k(\tilde{P}) \quad (j < n\mu)$$

no more than $\mu n - j + 1$ characteristic values.

Let $\nu_{\tilde{P}}(r)$ denote the counting function of $\tilde{P}$. That is, $\nu_{\tilde{P}}(r)$ is the number of the characteristic values of $\tilde{P}$ in $|z| \le r$.

We thus get the following result.

Corollary 11.1.3 *The counting function $\nu_{\tilde{P}}(r)$ of the characteristic values of $\tilde{P}$ satisfies the inequality*

$$\nu_{\tilde{P}}(r) \le \mu n - j + 1 \quad (j = 1, 2, ..., \mu n)$$

for any

$$r \le 1 + \frac{1}{j} \sum_{k=1}^{j} \omega_k(\tilde{P}).$$

11.2 An identity for sums of characteristic values

Again consider the polynomial matrix pencil $\tilde{P}(\lambda)$ defined by (1.1). In this section, we derive identities for the sums

$$\tilde{s}_m(\tilde{P}) := \sum_{k=1}^{n\mu} z_k^m(\tilde{P}) \tag{2.1}$$

for an integer $m \le \mu$.

To formulate the result, introduce the $mn \times mn$-block matrix

$$\hat{B}_m = \begin{pmatrix} -A_1 & -A_2 & \cdots & -A_{m-1} & -A_m \\ I_n & 0 & \cdots & 0 & 0 \\ 0 & I_n & \cdots & 0 & 0 \\ \cdot & \cdot & \cdots & \cdot & \cdot \\ 0 & 0 & \cdots & I_n & 0 \end{pmatrix} \qquad (1 \le m \le \mu) \qquad (2.2)$$

and

$$\hat{B}_1 = -A_1.$$

Clearly, $\hat{B}_\mu = \hat{A}(\tilde{P})$.

Theorem 11.2.1 *Let $\tilde{P}$ be defined by (1.1). Then for any $m \le \mu$, one has*

$$\sum_{k=1}^{\mu n} z_k^m(\tilde{P}) = Trace \; \hat{B}_m^m. \qquad (2.3)$$

To prove this theorem, we need the following result.

Lemma 11.2.2 *For any natural $m \le \mu$, we have the equality* $Trace \; \hat{B}_\mu^m = Trace \; \hat{B}_m^m$.

Proof: Consider the $n\mu \times n\mu$ block matrix

$$T_{\mu,m} = \begin{pmatrix} b_{11} & b_{12} & \cdots & b_{1,\mu-m} & \cdots & b_{1,\mu-1} & b_{1,\mu} \\ b_{21} & b_{22} & \cdots & b_{2,\mu-m} & \cdots & b_{2,\mu-1} & b_{2,\mu} \\ \cdot & \cdot & \cdots & \cdot & \cdots & \cdot & \cdot \\ b_{m1} & b_{m2} & \cdots & b_{m,\mu-m} & \cdots & b_{m,\mu-1} & b_{m,\mu} \\ I_n & 0 & \cdots & 0 & \cdots & 0 & 0 \\ 0 & I_n & \cdots & 0 & \cdots & 0 & 0 \\ \cdot & \cdot & \cdots & \cdot & \cdots & \cdot & \cdot \\ 0 & 0 & \cdots & I_n & \cdots & 0 & 0 \end{pmatrix} \qquad (2.4)$$

with some $n \times n$-matrices b_{jk}. Direct calculations show that the matrix

$$D_{\mu,m+1} := T_{\mu,m}\hat{B}_\mu \quad (m < \mu)$$

has the form

$$D_{\mu,m+1} = \begin{pmatrix} d_{11} & d_{12} & \cdots & d_{1,\mu-m-1} & \cdots & d_{1,\mu-1} & d_{1,\mu} \\ d_{21} & d_{22} & \cdots & d_{2,\mu-m-1} & \cdots & d_{2,\mu-1} & d_{2,\mu} \\ \cdot & \cdot & \cdots & \cdot & \cdots & \cdot & \cdot \\ d_{m1} & d_{m2} & \cdots & d_{m,\mu-m-1} & \cdots & d_{m,\mu-1} & d_{m,\mu} \\ d_{m+1,1} & d_{m+1,2} & \cdots & d_{m+1,\mu-m-1} & \cdots & d_{m+1,\mu-1} & d_{m+1,\mu} \\ I_n & 0 & \cdots & 0 & \cdots & 0 & 0 \\ 0 & I_n & \cdots & 0 & \cdots & 0 & 0 \\ \cdot & \cdot & \cdots & \cdot & \cdots & \cdot & \cdot \\ 0 & 0 & \cdots & I_n & \cdots & 0 & 0 \end{pmatrix},$$

where

$$d_{jk} = -b_{j1}A_k + b_{j,k+1} \quad (1 \leq j \leq m, k < \mu),$$

$$d_{j\mu} = -b_{j1}A_\mu \quad (1 \leq j \leq m), d_{m+1,k} = -A_k \quad (1 \leq k \leq \mu). \tag{2.5}$$

Hence,

$$Trace\ D_{\mu,m+1} = \sum_{k=1}^{m+1} Trace\ d_{kk} = Trace\ (-A_{m+1} + \sum_{k=1}^{m} -b_{k1}A_k + b_{k,k+1}). \tag{2.6}$$

According to (2.2), $\hat{B}_\mu$ has the form $T_{\mu,1}$, $\hat{B}_\mu^2$ has the form $T_{\mu,2}$, etc. Take

$$T_{\mu,m} = \hat{B}_\mu^m.$$

Then

$$D_{\mu,m+1} = \hat{B}_\mu^{m+1}.$$

Let $c_{jk}^{(m)}$ be the entries of $\hat{A}_\mu^m(P_\gamma) = B_\mu^m$. Then according to (2.3) and (2.4),

$$c_{jk}^{(m+1)} = -c_{j1}^{(m)} A_k + c_{j,k+1}^{(m)} \quad (k < \mu);$$

$$c_{j\mu}^{(m+1)} = -c_{j1}^{(m)} A_\mu \quad (1 \leq j \leq m); \ c_{m+1,k}^{(m+1)} = -A_k.$$

Thus, by (2.6),

$$Trace\ B_\mu^{m+1} = Trace\ \big(-A_{m+1} + \sum_{k=1}^{m} -c_{k1}^{(m)} A_k + c_{k,k+1}^{(m)} \big).$$

Taking $m = 2, 3, ...$, we can assert that $Trace\ \hat{B}_\mu^m$ depends on $A_1, ..., A_m$, only, and

$$Trace\ \hat{B}_m^m = Trace\ \hat{B}_{m+1}^m, m < \mu.$$

This proves the required result. Q. E. D.

The assertion of Theorem 11.2.1 follows from Lemmas 11.1.1 and 11.2.2. In particular,

$$\tilde{s}_1(\tilde{P}) = Trace\ \hat{B}_1 = -Trace\ A_1$$

and

$$\tilde{s}_2(\tilde{P}) = Trace\ \hat{B}_2^2 = Trace\ (A_1^2 - 2A_2). \tag{2.7}$$

11.3 Imaginary parts of characteristic values of polynomial pencils

Again consider the polynomial pencil

$$\tilde{P}(z) = \sum_{k=0}^{\mu} A_k \lambda^{\mu-k}. \tag{3.1}$$

Put

$$\tau(\tilde{P}) := \sum_{k=1}^{\mu} N_2^2(A_k) + n(\mu - 1),$$

where $N_2(A)$ is the Hilbert-Schmidt norm of a matrix A: $N_2^2(A) = Trace\ (A^*A)$. In addition, for a $t \in [0, 2\pi)$ put

$$\chi(\tilde{P}, t) := \tau(\tilde{P}) + Re\ Trace\ [e^{2it}\ (A_1^2 - 2A_2)].$$

Theorem 11.3.1 *For any $t \in [0, 2\pi)$, the relations*

$$\tau(\tilde{P}) - \sum_{k=1}^{\mu n} |z_k(\tilde{P})|^2 = \chi(\tilde{P}, t) - 2 \sum_{k=1}^{\mu n} (Re\ e^{it} z_k(\tilde{P}))^2 \geq 0$$

are true.

To prove this theorem, we need the following result.

Lemma 11.3.2 *Let $B = (b_{jk})_{j,k=1}^{m}$ be a block matrix with submatrices b_{jk}. Then*

$$N_2^2(B) = \sum_{j,k=1}^{m} N_2^2(b_{jk}).$$

Proof: We have

$$N_2^2(B) = Trace\ B^*B = Trace\ \sum_{j,k=1}^{m} b_{jk}^* b_{kj} = \sum_{j,k=1}^{m} N_2^2(b_{kj}).$$

This proves the lemma. Q. E. D.

Proof of Theorem 11.3.1: For simplicity, again put $\hat{A}(\tilde{P}) = \hat{A}$. Thanks to Lemma 1.9.2, for any $n \times n$-matrix A, we have

$$N_2^2(A) - \sum_{k=1}^{n} |\lambda_k(A)|^2 = 2N_2^2(A_I) - 2 \sum_{k=1}^{n} |Im\ \lambda_k(A)|^2,$$

where $A_I = (A - A^*)/2i$. We use that lemma with $A = ie^{it}\hat{A}$. Then

$$N_2^2(\hat{A}) - \sum_{k=1}^{\mu n} |\lambda_k(\hat{A})|^2 = \frac{1}{2} N_2^2(e^{it}\hat{A} + e^{-it}\hat{A}^*)$$

$$-\frac{1}{2} \sum_{k=1}^{\mu n} |e^{it}\lambda_k(\hat{A}) + e^{-it}\overline{\lambda}_k(\hat{A})|^2. \tag{3.2}$$

In addition,

$$e^{it}\hat{A} + \hat{A}^* e^{-it} = \begin{pmatrix} -A_1 e^{it} - e^{-it}A_1^* & I_n e^{-it} - A_2 e^{it} & \cdots & -A_{\mu-1}e^{it} & -A_\mu e^{it} \\ I_n e^{it} - A_2^* e^{-it} & 0 & \cdots & 0 & 0 \\ -A_3^* e^{-it} & e^{it}I_n & \cdots & 0 & 0 \\ \cdot & \cdot & \cdots & \cdot & \cdot \\ -A_\mu^* e^{-it} & 0 & \cdots & e^{it}I_n & 0 \end{pmatrix}.$$

Simple calculations and the previous lemma show that $N_2^2(\hat{A}) = \tau(\tilde{P})$ and

$$N_2^2(e^{it}\hat{A} + e^{-it}\hat{A}^*)/2 = N_2^2(A_1 e^{it} + A_1^* e^{-it})/2$$

$$+N_2^2(I_n - A_2 e^{2it}) + \sum_{k=3}^{\mu} N_2^2(A_k) + (\mu - 2)n.$$

So

$$N_2^2(e^{it}\hat{A} + e^{-it}\hat{A}^*)/2 = N_2^2(A_1) + \text{Trace } (A_1^2 e^{2it} + (A_1^*)^2 e^{-2it})/2 + n + N_2^2(A_2)$$

$$-\text{Trace } (A_2 e^{2it} + A_2^* e^{-2it}) + \sum_{k=3}^{\mu} N_2^2(A_k) + n(\mu - 2) = \chi(\tilde{P}, t).$$

Hence (3.2) and Lemma 11.1.1 imply the required result. Q. E. D.

Note that

$$\chi(\tilde{P}, \pi/2) = -\text{Re Trace } (A_1^2 - 2A_2) + \tau(\tilde{P})$$

and

$$\chi(\tilde{P}, 0) = \text{Re Trace } (A_1^2 - 2A_2) + \tau(\tilde{P}).$$

Now Theorem 11.3.1 yields

Corollary 11.3.3 *The following relations are valid:*

$$\tau(\tilde{P}) - \sum_{k=1}^{\mu n} |z_k(\tilde{P})|^2 = \chi(\tilde{P}, \pi/2) - 2\sum_{k=1}^{\mu n} (\text{Im } z_k(\tilde{P}))^2$$

$$= \chi(\tilde{P}, 0) - 2\sum_{k=1}^{\mu n} (\text{Re } z_k(\tilde{P}))^2 \geq 0.$$

From this corollary it follows that

$$\sum_{k=1}^{\mu n} |z_k(\tilde{P})|^2 \le \tau(\tilde{P}), \tag{3.3}$$

$$2 \sum_{k=1}^{\mu n} (Im \ z_k(\tilde{P}))^2 \le \chi(\tilde{P}, \pi/2)$$

and

$$2 \sum_{k=1}^{\mu n} (Re \ z_k(\tilde{P}))^2 \le \chi(\tilde{P}, 0). \tag{3.4}$$

Note that inequalities (3.3) and (3.4) can be directly proved by Lemma 11.1.1 on the linearization and the Weyl inequalities, but Theorem 11.3.1 gives us an additional proof.

11.4 Perturbations of polynomial pencils

Consider the pencils

$$\tilde{P}(\lambda) = \sum_{k=0}^{\mu} A_k \lambda^{\mu-k} \ \text{ and } \ \tilde{Q}(\lambda) = \sum_{k=0}^{\mu} B_k \lambda^{\mu-k} \ \ (A_0 = B_0 = I_n) \tag{4.1}$$

with constant $n \times n$-matrices A_k, B_k $(k = 1, ..., \mu)$. Put

$$g_1(\tilde{P}) = [Trace \ (\sum_{k=1}^{\mu} A_k A_k^*) + n(\mu - 1) - \sum_{k=1}^{\mu n} |z_k(\tilde{P})|^2]^{1/2}$$

and

$$\xi(\tilde{P}, y) := \frac{1}{y} \Big[1 + \frac{1}{\mu n - 1} \Big(1 + \frac{g_1^2(\tilde{P})}{y^2} \Big) \Big]^{(\mu n-1)/2} \ \ (y > 0).$$

Since,

$$\sum_{k=1}^{\mu n} |z_k(\tilde{P})|^2 \ge |\sum_{k=1}^{\mu n} z_k^2(\tilde{P})| = |\tilde{s}_2(\tilde{P})|$$

by (2.7), we have

$$g_1(\tilde{P}) \le [Trace \ (\sum_{k=1}^{\mu} A_k A_k^*) - |Trace \ (A_1^2 - 2A_2)| + n(\mu - 1)]^{1/2}.$$

Finally, put

$$q(\tilde{P}, \tilde{Q}) = [\ \sum_{k=1}^{\mu} \|A_k - B_k\|^2 \ ]^{1/2}.$$

Theorem 11.4.1 *Let $\tilde{P}$ and $\tilde{Q}$ be defined by (4.1). Then for any character-istic value $z(\tilde{Q})$ of $\tilde{Q}(z)$, there is a characteristic value $z(\tilde{P})$ of $\tilde{P}(z)$, such that*

$$|z(\tilde{P}) - z(\tilde{Q})| \leq r(\tilde{P}, \tilde{Q}),$$

where $r(\tilde{P}, \tilde{Q})$ is the unique (positive) root of the equation

$$q(\tilde{P}, \tilde{Q})\xi(\tilde{P}, y) = 1. \tag{4.2}$$

Proof: Take the matrix $\hat{A}(\tilde{P})$ defined in Section 11.1 and the matrix

$$\hat{B}(\tilde{Q}) := \begin{pmatrix} -B_1 & -B_2 & \ldots & -B_{\mu-1} & -B_\mu \\ I_n & 0 & \ldots & 0 & 0 \\ 0 & I_n & \ldots & 0 & 0 \\ . & . & \ldots & . & . \\ 0 & 0 & \ldots & I_n & 0 \end{pmatrix}.$$

It is not hard to check that

$$q(\tilde{P}, \tilde{Q}) = \|\hat{A}(\tilde{P}) - \hat{B}(\tilde{Q})\|_{n\mu},$$

where $\|.\|_{n\mu}$ means the Euclidean norm in $\mathbf{C}^{n\mu}$.

Because of Theorem 1.7.3, for any eigenvalue $\lambda(\hat{B}(\tilde{Q}))$ of matrix $\hat{B}(\tilde{Q})$, there is an eigenvalue $\lambda_i(\hat{A}(\tilde{P})$ of matrix $\hat{A}(\tilde{P})$, such that

$$|\lambda_j(\hat{B}(\tilde{Q})) - \lambda_i(\hat{A}(\tilde{P}))| \leq x(Q, P),$$

where $x(Q, P)$ is the unique positive root of the equation

$$\frac{q(\tilde{P}, \tilde{Q})}{y}\Big[1 + \frac{1}{\mu n - 1}\Big(1 + \frac{g^2(\hat{A}(\tilde{P}))}{y^2}\Big)\Big]^{(\mu n-1)/2} = 1.$$

Simple calculations show that

$$g(\hat{A}(\tilde{P})) = g_1(\tilde{P}).$$

This and Lemma 11.1.1 prove the theorem. Q. E. D.

Denote

$$p_0 := q(\tilde{P}, \tilde{Q})\xi(\tilde{P}, 1)$$

and

$$\delta_0(\tilde{P}, \tilde{Q}) := \begin{cases} \sqrt[\mu n - 1]{p_0} & \text{if } p_0 \leq 1 \\ p_0 & \text{if } p_0 > 1 \end{cases}.$$

Because of Lemma 1.6.1,

$$r(\tilde{P}, \tilde{Q}) \leq \delta_0(\tilde{P}, \tilde{Q}).$$

Now Theorem 11.4.1 implies

Corollary 11.4.2 *Let $\tilde{P}$ and $\tilde{Q}$ be defined by (4.1). Then for any characteristic value $z(\tilde{Q})$ of $\tilde{Q}(z)$, there is a characteristic value $z(\tilde{P})$ of $\tilde{P}(z)$, such that*

$$|z(\tilde{P}) - z(\tilde{Q})| \le \delta_0(\tilde{P}, \tilde{Q}).$$

Let us consider perturbations of polynomials with triangular matrix coefficients. Let

$$B_j = (b_{sk}^{(j)})_{s,k=1}^n \quad (j = 1, ..., \mu)$$

be arbitrary $n \times n$-matrices. In addition, let v_j, w_j and d_j be the *upper nilpotent, lower nilpotent,* and *diagonal parts* of B_j. So

$$B_j = v_j + d_j + w_j.$$

Take $A_j = v_j + d_j$. That is, A_j is the upper triangular part of B_j, and according to the notations of the previous section $\tilde{P}(z)$ is the upper triangular part of $\tilde{Q}(z)$. Therefore

$$q(\tilde{P}, \tilde{Q}) = \tilde{q} := \Big[\, \sum_{k=1}^{\mu} \|w_k\|^2 \,\Big]^{1/2}.$$

In addition, since $\tilde{P}$ is triangular, according to Lemma 11.3.2, we have

$$g_1(\tilde{P}) = \tilde{g}_1 := \sqrt{\sum_{j=1}^{\mu} N_2^2(v_j)}.$$

Under consideration, we have

$$p_0 = \tilde{p} := \Big[1 + \frac{1}{\mu n - 1}(1 + \tilde{g}_1^2)\Big]^{(\mu n - 1)/2}$$

and $\delta_0(\tilde{P}, \tilde{Q}) := \tilde{\delta}$, where

$$\tilde{\delta} := \begin{cases} \sqrt[\mu n - 1]{\tilde{p}} & \text{if } \tilde{p} \le 1, \\ \tilde{p} & \text{if } \tilde{p} > 1 \end{cases}.$$

Since $\tilde{P}$ is a triangular matrix, the characteristic values of $\tilde{P}$ are the roots $R_{1k}, ..., R_{\mu k}$ of the diagonal polynomials.

Because of Corollary 11.4.2, we get

Corollary 11.4.3 *All the characteristic values of $\tilde{Q}$ lie in the union of the sets*

$$\{z \in \mathbf{C} : |z - R_{jk}| \le \tilde{\delta}\} \quad (j = 1, ..., \mu; \; k = 1, ..., n),$$

where R_{jk} $(j = 1, ..., \mu)$ are the roots of the diagonal polynomials

$$\lambda^{\mu} + b_{kk}^{(1)} \lambda^{\mu-1} + ... + b_{kk}^{(\mu)} \quad (k = 1, ..., n).$$

For instance, let us consider the polynomial pencil $\tilde{Q}(\lambda) = \lambda^2 I + B_1 \lambda + B_2$, where

$$B_j = (b_{sk}^{(j)})_{s,k=1}^n \quad (j = 1, 2).$$

Then

$$R_{1k} = -b_{kk}^{(1)}/2 + \sqrt{(b_{kk}^{(1)}/2)^2 - b_{kk}^{(2)}}$$

and

$$R_{2k} = -b_{kk}^{(1)}/2 - \sqrt{(b_{kk}^{(1)}/2)^2 - b_{kk}^{(2)}}.$$

Now, we can directly apply Corollary 11.4.3.

Finally, let us point a corollary of Theorem 1.7.5.

Corollary 11.4.4 *Let $\tilde{P}$ and $\tilde{Q}$ be defined by (4.1). Then*

$$|det\ (\tilde{P}(z)) - det\ (\tilde{Q}(z))|$$

$$\leq \frac{N_2(\tilde{P}(z) - \tilde{Q}(z))}{n^{n/2}} \left(1 + \frac{1}{2} N_2(\tilde{P}(z) - \tilde{Q}(z)) + \frac{1}{2} N_2(\tilde{P}(z) + \tilde{Q}(z))\right)^n.$$

11.5   Multiplicative representations of rational pencils

Let $B(\mathbf{C}^n)$ be the set of linear operators in $\mathbf{C}^n$ (the set of $n \times n$-matrices). Let E_k $(k = 1, \ldots, n)$ be *the maximal chain of the invariant projections of an* $A \in B(\mathbf{C}^n)$. That is, E_k are orthogonal projections,

$$AE_k = E_k AE_k \ (k = 1, \ldots, n)$$

and

$$0 = E_0 \mathbf{C}^n \subset E_1 \mathbf{C}^n \subset \ldots \subset E_n \mathbf{C}^n = \mathbf{C}^n.$$

So $dim\ \Delta E_k = 1$. Here

$$\Delta E_k = E_k - E_{k-1} \ (k = 1, ..., n).$$

We use the Schur triangular representation

$$A = D + V. \tag{5.1}$$

(see Section 1.5). Here V is the nilpotent part of A and

$$D = \sum_{k=1}^n \lambda_k(A)\Delta E_k$$

is the diagonal part. For $X_1, X_2, ..., X_m \in B(\mathbf{C}^n)$ denote

$$\overrightarrow{\prod_{1 \le k \le m}} X_k := X_1 X_2 ... X_m.$$

That is, the arrow over the symbol of the product means that the indexes of the co-factors increase from left to right.

Lemma 11.5.1 *For any $A \in B(\mathbf{C}^n)$, the equality*

$$\lambda R_\lambda(A) = - \overrightarrow{\prod_{1 \le k \le n}} \left(I + \frac{A \Delta E_k}{\lambda - \lambda_k(A)} \right) \quad (\lambda \notin \sigma(A), \lambda \ne 0)$$

is valid, where E_k, $k = 1, ..., n$ is the maximal chain of the invariant projections of A.

Proof: Denote $\tilde{E}_k = I - E_k$. Since

$$A = (\tilde{E}_k + E_k) A (\tilde{E}_k + E_k) \text{ for any } k = 1, ..., n$$

and $\tilde{E}_1 A E_1 = 0$, we get the relation

$$A = E_1 A \tilde{E}_1 + E_1 A E_1 + \tilde{E}_1 A \tilde{E}_1.$$

Take into account that $\Delta E_1 = E_1$ and

$$E_1 A E_1 = \lambda_1(A) \Delta E_1.$$

Then

$$A = \lambda_1(A) \Delta E_1 + E_1 A \tilde{E}_1 + \tilde{E}_1 A \tilde{E}_1$$
$$= \lambda_1(A) \Delta E_1 + A \tilde{E}_1. \tag{5.2}$$

Now, we check the equality

$$R_\lambda(A) = \eta(\lambda), \tag{5.3}$$

where

$$\eta(\lambda) := \frac{\Delta E_1}{\lambda_1(A) - \lambda} - \frac{\Delta E_1}{\lambda_1(A) - \lambda} A \tilde{E}_1 R_\lambda(A) \tilde{E}_1 + \tilde{E}_1 R_\lambda(A) \tilde{E}_1.$$

In fact, multiplying this equality from the left by $A - I\lambda$ and taking into account equality (5.2), we obtain the relation

$$(A - I\lambda)\eta(\lambda) = \Delta E_1 - \Delta E_1 A \tilde{E}_1 R_\lambda(A) \tilde{E}_1 + (A - I\lambda) \tilde{E}_1 R_\lambda(A) \tilde{E}_1.$$

But $\tilde{E}_1 A \tilde{E}_1 = \tilde{E}_1 A$ and thus

$$\tilde{E}_1 R_\lambda(A) \tilde{E}_1 = \tilde{E}_1 R_\lambda(A).$$

That is, we can write out

$$(A - I\lambda)\eta(\lambda) = \Delta E_1 + (-\Delta E_1 A + A - I\lambda)\tilde{E}_1 R_\lambda(A)$$

$$= \Delta E_1 + \tilde{E}_1(A - I\lambda)R_\lambda(A) = \Delta E_1 + \tilde{E}_1 = I.$$

Similarly, we multiply (5.3) by $A - I\lambda$ from the right and take into account (5.2). This gives I. Therefore, (5.3) is correct.

Because of (5.3)

$$I - AR_\lambda(A)$$
$$= \big(I - (\lambda_1(A) - \lambda)^{-1} A\Delta E_1\big)\big(I - A\tilde{E}_1 R_\lambda(A)\tilde{E}_1\big). \tag{5.4}$$

Now we apply the above arguments to operator $A\tilde{E}_1$. We obtain the following expression which is similar to (5.4):

$$I - A\tilde{E}_1 R_\lambda(A)\tilde{E}_1 =$$

$$(I - (\lambda_2(A) - \lambda)^{-1} A\Delta E_2)(I - A\tilde{E}_2 R_\lambda(A)\tilde{E}_2).$$

For any $k < n$, it similarly follows that

$$I - A\tilde{E}_k R_\lambda(A)\tilde{E}_k$$

$$= \big(I - \frac{A\Delta E_{k+1}}{\lambda_{k+1}(A) - \lambda}\big)(I - A\tilde{E}_{k+1} R_\lambda(A)\tilde{E}_{k+1}).$$

Substitute this in (5.4), as long as $k = 1, 2, ..., n - 1$. We have

$$I - AR_\lambda(A) = \overset{\longrightarrow}{\prod_{1 \le k \le n-1}} (I + \frac{A\Delta E_k}{\lambda - \lambda_k(A)})(I - A\tilde{E}_{n-1} R_\lambda(A)\tilde{E}_{n-1}). \tag{5.5}$$

It is clear that $\tilde{E}_{n-1} = \Delta E_n$. That is ,

$$I - A\tilde{E}_{n-1} R_\lambda(A)\tilde{E}_{n-1} = I + \frac{A\Delta E_n}{\lambda - \lambda_n(A)}.$$

Now the identity

$$I - AR_\lambda(A) = -\lambda R_\lambda(A)$$

and (5.5) imply the result. Q. E. D.

Let A be a normal matrix. Then

$$A = \sum_{k=1}^{n} \lambda_k(A)\Delta E_k.$$

Hence, $A\Delta E_k = \lambda_k(A)\Delta E_k$. Since $\Delta E_k \Delta E_j = 0$ for $j \ne k$, Lemma 11.5.1 gives us the equality

$$\lambda R_\lambda(A) = -\sum_{k=1}^{n} (I + (\lambda - \lambda_k(A))^{-1}\lambda_k(A)\Delta E_k).$$

But

$$I = \sum_{k=1}^{n} \Delta E_k.$$

The result is

$$\lambda R_\lambda(A) = -\sum_{k=1}^{n} [1 + (\lambda - \lambda_k(A))^{-1} \lambda_k(A)] \Delta E_k)$$

$$= -\sum_{k=1}^{n} \lambda \frac{\Delta E_k}{\lambda - \lambda_k(A)}.$$

Or

$$R_\lambda(A) = \sum_{k=1}^{n} \frac{\Delta E_k}{\lambda_k(A) - \lambda}.$$

We have obtained the well-known spectral representation for the resolvent of a normal matrix.

Thus the latter lemma generalizes the spectral representation for the resolvent of a normal matrix. In particular it implies

$$(I - A)^{-1} = \overrightarrow{\prod_{1 \le k \le n}} \left(I + \frac{A \Delta E_k}{1 - \lambda_k(A)}\right) \quad (1 \notin \sigma(A)). \tag{5.6}$$

Again let

$$\tilde{P}(z) = \sum_{k=0}^{\mu} z^{\mu-k} A_k \quad (z \in \mathbf{C}; \ A_0 = I_n), \tag{5.7}$$

where A_k, $k = 1, 2, ..., \mu$ are constant $n \times n$-matrices. Put

$$\Psi(z) := \tilde{P}(z) - z^\mu I = \sum_{k=1}^{n} A_k z^{\mu-k}.$$

Then the matrix inverse to $\tilde{P}(z)$ for a fixed z equals

$$\tilde{P}^{-1}(z) = \frac{1}{z^\mu} \left(I + \frac{1}{z^\mu} \Psi(z)\right)^{-1}.$$

By (5.6), we get the main result of this section.

Theorem 11.5.2 *For any $z \notin \Sigma(\tilde{P}) \cup \{0\}$, the equality*

$$\tilde{P}^{-1}(z) = \frac{1}{z^\mu} \overrightarrow{\prod_{1 \le k \le n}} \left(I - \frac{\Psi(z) \Delta E_k(z)}{\lambda_k(\tilde{P}(z))}\right) \quad (\Delta E_k(z) = E_k(\tilde{P}, z) - E_{k-1}(\tilde{P}, z)),$$

is true, where $E_0(\tilde{P}, z) = 0$; $E_k(\tilde{P}, z)$, $k = 1, ..., n$ is the maximal chain of the invariant projections of $\tilde{P}(z)$ and $\lambda_k(\tilde{P}(z))$ are the eigenvalues of matrix $\tilde{P}(z)$.

Let us establish another multiplicative representation for the considered class of rational matrix functions.

Lemma 11.5.3 *Let $V \in B(\mathbf{C}^n)$ be a nilpotent operator and E_k, $k = 1, ..., n$, be the maximal chain of its invariant projections. Then*

$$(I - V)^{-1} = \overrightarrow{\prod_{2 \leq k \leq n}} (I + V\Delta E_k). \qquad (5.8)$$

Proof: In fact, all the eigenvalues of V are equal to zero, and $V\Delta E_1 = 0$. Now Lemma 11.5.1 gives us relation (5.8). Q. E. D.

Relation (5.8) allows us to prove the second multiplicative representation of the resolvent of A.

Lemma 11.5.4 *Let D and V be the diagonal and nilpotent parts of $A \in B(\mathbf{C}^n)$, respectively. Then*

$$R_\lambda(A) = R_\lambda(D) \overrightarrow{\prod_{2 \leq k \leq n}} \left[I - \frac{V\Delta E_k}{\lambda_k(A) - \lambda}\right] \quad (\lambda \notin \sigma(A)), \qquad (5.9)$$

where E_k, $k = 1, ..., n$, is the maximal chain of invariant projections of A.

Proof: Because of (5.1),

$$R_\lambda(A) = (A - \lambda I)^{-1} = (D + V - \lambda I)^{-1} = R_\lambda(D)(I + VR_\lambda(D))^{-1}.$$

But $VR_\lambda(D)$ is a nilpotent operator. Take into account that

$$R_\lambda(D)\Delta E_k = (\lambda_k(A) - \lambda)^{-1}\Delta E_k.$$

Now (5.8) ensures the relation (5.9). Q. E. D.

The previous lemma gives us the equality

$$A^{-1} = D^{-1} \overrightarrow{\prod_{2 \leq k \leq n}} \left[I - \frac{V\Delta E_k}{\lambda_k(A)}\right]$$

for any invertible operator A.

Now replace A by $\tilde{P}(z)$ and denote by $\tilde{D}(z)$ and $\tilde{V}(z)$ be the diagonal and nilpotent parts of $\tilde{P}(z)$, respectively. Recall that $E_k(\tilde{P}, z)$, $k = 1, ..., n$, is the maximal chain of invariant projections of $\tilde{P}(z)$, and

$$\Delta E_k(z) = E_k(\tilde{P}, z) - E_{k-1}(\tilde{P}, z).$$

Then the previous equality with $A = \tilde{P}(z)$ implies.

Theorem 11.5.5 *Let $\tilde{P}$ be defined by (5.7). Then*

$$\tilde{P}^{-1}(z) = \tilde{D}^{-1}(z) \overrightarrow{\prod_{2 \leq k \leq n}} \left[I - \frac{\tilde{V}(z)\Delta E_k(z)}{\lambda_k(\tilde{P}(z))}\right]$$

for any $z \notin \Sigma(\tilde{P})$.

11.6 The Cauchy type theorem for polynomial pencils

Consider in $\mathbf{C}^{\mu n}$ the operator defined by the block matrix

$$T = \begin{pmatrix} A_{11} & A_{12} & \cdots & A_{1\mu} \\ A_{21} & A_{22} & \cdots & A_{2\mu} \\ \cdot & \cdot & \cdots & \cdot \\ A_{\mu 1} & A_{\mu 2} & \cdots & A_{\mu\mu} \end{pmatrix}, \tag{6.1}$$

where A_{jk} are $n \times n$-matrices. Put

$$m_{jk} = \|A_{jk}\| \quad (j, k = 1, 2, ..., \mu),$$

where the norm is understood as the operator norm with respect to the Euclidean norm. Consider the matrix

$$M = (m_{jk})_{j,k=1}^{\mu}.$$

Lemma 11.6.1 *The spectral radius of T is less than or is equal to the spectral radius of M.*

Proof: Let $A_{jk}^{(\nu)}$ and $m_{jk}^{(\nu)}$ $(\nu = 2, 3, ...)$ be the entries of T^ν and M^ν, respectively. We have

$$\|A_{jk}^{(2)}\| = \|\sum_{l=1}^{\mu} A_{jl} A_{lk}\|$$

$$\leq \sum_{l=1}^{\mu} \|A_{jl}\| \|A_{lk}\| = \sum_{l=1}^{\mu} m_{jl} m_{lk} = m_{jk}^{(2)}.$$

Similarly, we get the inequality

$$\|A_{jk}^{(\nu)}\| \leq m_{jk}^{(\nu)} \quad (\nu = 2, 3, ...; \ j, k = 1, ..., n).$$

But for any $h = \{h_k \in \mathbf{C}^n\}_{k=1}^{\mu}$, we have

$$\|Th\|_{n\mu}^2 \leq \sum_{j=1}^{\mu} \left(\sum_{k=1}^{\mu} \|A_{jk} h_k\|\right)^2 \leq \sum_{j=1}^{\mu} \left(\sum_{k=1}^{\mu} m_{jk} \|h_k\|\right)^2 = \|M\tilde{h}\|_{\mu}^2,$$

where

$$\tilde{h} = \{\|h_k\|\}_{k=1}^{\mu}.$$

and $\|.\|_{\mu}$ is the Euclidean norm in $\mathbf{C}^{\mu}$. Since

$$\|h\|_{n\mu}^2 = \sum_{k=1}^{\mu} \|h_k\|^2 = \|\tilde{h}\|_{\mu}^2,$$

we obtain

$$\|T^\nu\|_{n\mu} \le \|M^\nu\|_\mu \quad (\nu = 2, 3, ...).$$

Now the Gel'fand formula for the spectral radius yields the required result.
Q. E. D.

Furthermore, Lemma 11.1.1 and the previous lemma imply the main result of this section.

Theorem 11.6.2 *Let*

$$m_k = \|A_k\| \quad (k = 1, 2, ..., \mu).$$

Then the spectrum of the pencil $\tilde{P}(\lambda) = \lambda^n + A_1\lambda^{n-1} + ... + A_n$ *lies in the set*

$$\{z \in \mathbf{C} : |z| \le \tilde{r}\},$$

where $\tilde{r}$ *is the unique positive root of the equation*

$$z^\mu = \sum_{k=1}^\mu m_k z^{\mu-k}.$$

11.7   The Gerschgorin type sets for polynomial pencils

Let T be the block matrix defined as in the previous section and

$$\eta_j := \sum_{k=1,k\neq j}^\mu \|A_{jk}\|.$$

We need the following generalized Hadamard criterion for the invertibility of block matrices, cf. [Gantmacher, 1967].

Lemma 11.7.1 *Let* T *be defined by (6.1) and* A_{jj} $(j = 1, ..., \mu)$ *be invertible matrices. In addition, let*

$$\|A_{jj}^{-1}\|^{-1} > \eta_j \quad (j = 1, 2, ...). \tag{7.1}$$

Then T *is invertible. Moreover, if*

$$\max_j \|(A_{jj} - \lambda)^{-1}\|\eta_j < 1,$$

then λ *is a regular point of* T.

Proof: Put

$$\tilde{D} = diag\,[A_{11}, ..., A_{\mu\mu}]$$

and $W = \tilde{A} - \tilde{D}$. That is, W is the off-diagonal part of T, and

$$T = \tilde{D} + W = \tilde{D}(I + \tilde{D}^{-1}W). \tag{7.2}$$

Clearly,

$$\sum_{k=1,k\neq j}^{\infty} \|A_{jj}^{-1}A_{jk}\| \leq \eta_j \|A_{jj}^{-1}\|.$$

From (7.1), it follows

$$1 - \eta_j \|A_{jj}^{-1}\| \geq \|A_{jj}^{-1}\|\epsilon \quad (j = 1, 2, ...)$$

for a sufficiently small $\epsilon > 0$. Therefore,

$$\max_{j} \sum_{k=1,k\neq j}^{\infty} \|A_{jj}^{-1}A_{jk}\| < 1. \tag{7.3}$$

Then thanks to the well-known bound for the spectral radius (see Section 1.1) and Lemma 11.6.1, the spectral radius $r_s(\tilde{D}^{-1}W)$ of $\tilde{D}^{-1}W$ is less than one. Therefore, $I + \tilde{D}^{-1}W$ is invertible. Now (7.2) implies that T is also invertible. This proves the lemma. Q. E. D.

From the previous lemma, our next result immediately follows.

Corollary 11.7.2 *The spectrum of T lies in the union of the sets*

$$\|(A_{jj} - \lambda)^{-1}\|\eta_j \geq 1 \quad (j = 1, ..., \mu).$$

The latter corollary and Lemma 11.1.1 yield the main result of this section.

Theorem 11.7.3 *Let $\tilde{P}$ be defined by (5.7). Then its spectrum lies in the union of the sets*

$$|\lambda| \leq 1 \text{ and } \|(A_1 + \lambda)^{-1}\| \sum_{k=2}^{\mu} \|A_k\| \geq 1.$$

11.8   Estimates for rational matrix functions

Let $A = (a_{jk})$ be an $n \times n$-matrix. The quantity

$$g(A) = \left(N^2(A) - \sum_{k=1}^{n} |\lambda_k(A)|^2\right)^{1/2} \tag{8.1}$$

introduced in Section 1.5 plays a key role in this section. Recall that the relations

$$g^2(A) \leq \frac{1}{2} N^2(A^* - A), \tag{8.2}$$

$$g^2(A) \leq N^2(A) - |Trace\, A^2|, \tag{8.3}$$

and

$$g(e^{i\tau} A + zI) = g(A) \tag{8.4}$$

are true for all $\tau \in \mathbf{R}$ and $z \in \mathbf{C}$ (see Section 1.5).

Thanks to Corollary 1.8.5, the resolvent $R_\lambda(A) = (A - \lambda I)^{-1}$ of a linear operator A in $\mathbf{C}^n$ satisfies the inequality

$$\|R_\lambda(A)\| \leq \sum_{k=0}^{n-1} \frac{g^k(A)}{\sqrt{k!}\rho^{k+1}(A, \lambda)} \quad \text{for any regular point } \lambda \text{ of } A, \tag{8.5}$$

where $\rho(A, \lambda) = \min_{k=1,\ldots,n} |\lambda - \lambda_k(A)|$.

Let A be an invertible $n \times n$-matrix. Then by (8.5)

$$\|A^{-1}\| \leq \sum_{k=0}^{n-1} \frac{g^k(A)}{\sqrt{k!}\rho_0^{k+1}(A)}, \tag{8.6}$$

where $\rho_0(A) = \rho(A, 0)$ is the smallest modulus of the eigenvalues of A:

$$\rho_0(A) = \inf_{k=1,\ldots,n} |\lambda_k(A)|.$$

Again consider the polynomial

$$\tilde{P}(z) = \sum_{k=0}^{\mu} A_k z^{\mu-k} \quad (A_0 = I_n) \tag{8.7}$$

with matrix coefficients. According to (8.4),

$$g(\tilde{P}(z)) = g\Big(\sum_{k=1}^{\mu} \lambda^{\mu-k} A_k \Big),$$

since $A_0 = I_n$. Because of (8.6), we can write out

$$\|\tilde{P}^{-1}(z)\| \leq \sum_{k=0}^{n-1} \frac{g^k(\tilde{P}(z))}{\sqrt{k!}\rho_0^{k+1}(\tilde{P}(z))}, \tag{8.8}$$

where

$$\rho_0(\tilde{P}(z)) = \inf_{k=1,\ldots,n} |\lambda_k(\tilde{P}(z))|$$

for a fixed z.

Directly from the definition of $g(.)$ and (8.4), it follows that

$$g(\tilde{P}(z)) \le \sum_{k=1}^{\mu} |\lambda|^{\mu-k} N_2(A_k).$$

Moreover, thanks to (8.2), with the notation

$$A_{Ik} = (A_k - A_k^*)/2i,$$

we have

$$g(\tilde{P}(z)) \le \sqrt{2} \sum_{k=1}^{\mu} |\lambda|^{\mu-k} N_2(A_{Ik})$$

provided λ is real.

Since

$$det\ \tilde{P}(z) = \prod_{k=1}^{n} \lambda_k(\tilde{P}(z)),$$

a number z is a characteristic value of $\tilde{P}(z)$ if and only if $\rho_0(\tilde{P}(z)) = 0$. We thus have proved the following result.

Theorem 11.8.1 *Let $\tilde{P}$ be the polynomial matrix pencil defined by (8.7). Then for any $z \notin \Sigma(\tilde{P})$, inequality (8.8) holds.*

Now let us use Theorem 1.5.1, which gives us the inequality

$$\|(I\lambda - A)^{-1}\| \le \frac{1}{\rho(A, \lambda)} \left[1 + \frac{1}{n-1}\left(1 + \frac{g^2(A)}{\rho^2(A, \lambda)}\right)\right]^{(n-1)/2}$$

for any regular λ of an $n \times n$-matrix A. Let A be invertible. Then

$$\|A^{-1}\| \le \frac{1}{\rho_0(A)}\left[1 + \frac{1}{n-1}\left(1 + \frac{g^2(A)}{\rho_0^2(A)}\right)\right]^{(n-1)/2}. \tag{8.9}$$

Hence, we arrive at our next result.

Theorem 11.8.2 *Let $\tilde{P}$ be the polynomial matrix pencil defined by (8.7). Then for any $z \notin \Sigma(\tilde{P})$ the inequality*

$$\|\tilde{P}^{-1}(z)\| \le \frac{1}{\rho_0(\tilde{P}(z))}\left[1 + \frac{1}{n-1}\left(1 + \frac{g^2(\tilde{P}(z))}{\rho_0^2(\tilde{P}(z))}\right)\right]^{(n-1)/2}$$

holds.

Furthermore, as it is proved in Section 1.4,

$$\|(I\lambda - A)^{-1} det\ (\lambda I - A)\|$$

$$\le \left[\frac{N_2^2(A) - 2Re\ (\bar{\lambda}\ Trace\ (A))\ + n|\lambda|^2}{n-1}\right]^{(n-1)/2} \quad (\lambda \notin \sigma(A)).$$

Hence, it follows

$$\|A^{-1} \det A\| \le \frac{N_2^{n-1}(A)}{(n-1)^{(n-1)/2}} \quad (0 \notin \sigma(A)).$$

We thus get the inequality

$$\|\tilde{P}^{-1}(z)\| \le \frac{1}{|\det \tilde{P}(z)|} \frac{N_2^{n-1}(\tilde{P}(z))}{(n-1)^{(n-1)/2}} \quad (z \notin \Sigma(\tilde{P})). \tag{8.10}$$

Consider the pencils

$$\tilde{P}(\lambda) = \sum_{k=0}^{\mu} A_k \lambda^{\mu-k} \text{ and } \tilde{Q}(\lambda) = \sum_{k=0}^{\mu} B_k \lambda^{\mu-k} \quad (A_0 = B_0 = I_n) \tag{8.11}$$

with constant $n \times n$-matrices A_k, B_k $(k = 1, ..., \mu)$.

Put

$$q(\lambda) := \|\tilde{P}(\lambda) - \tilde{Q}(\lambda)\|,$$

where $\|.\|$ denotes the Euclidean norm, again.

Let us prove the following simple lemma.

Lemma 11.8.3 *Each characteristic value μ of $\tilde{Q}(.)$ either coincides with a characteristic value of $\tilde{P}(.)$, or satisfies the inequality*

$$q(\mu)\|\tilde{P}^{-1}(\mu)\| \ge 1.$$

Proof: Suppose that for some characteristic value μ of $B(.)$, the inequality

$$q(\mu)\|\tilde{P}^{-1}(\mu)\| < 1 \tag{8.12}$$

holds. We can write

$$\tilde{P}^{-1}(\mu) - \tilde{Q}^{-1}(\mu) = \tilde{Q}^{-1}(\mu)(\tilde{P}(\mu) - \tilde{P}(\mu))\tilde{P}^{-1}(\mu).$$

Therefore,

$$\|\tilde{Q}^{-1}(\mu)\| \le \|\tilde{P}^{-1}(\mu)\| + q(\mu)\|\tilde{Q}^{-1}(\mu)\|\|\tilde{P}^{-1}(\mu)\|.$$

Condition (8.12) yields the relation

$$\|\tilde{Q}^{-1}(\mu)\| \le \|\tilde{P}^{-1}(\mu)\|(1 - q(\mu))^{-1}.$$

Thus, (8.12) implies that μ is a regular point of $B(.)$. That contradiction proves the result. Q. E. D.

Now Theorem 11.8.1 implies our next result.

Corollary 11.8.4 *Each characteristic value μ of $\tilde{Q}(.)$ either coincides with a characteristic value of $\tilde{P}(.)$, or satisfies the inequality*

$$q(\mu) \sum_{k=0}^{n-1} \frac{g^k(\tilde{P}(\mu))}{\rho_0^{k+1}(\tilde{P}(\mu))\sqrt{k!}} \ge 1.$$

Similarly, one can apply Theorem 11.8.2 and inequality (8.10).

11.9 Coupled systems of polynomial equations

Let us consider the coupled system

$$f(x,y) = g(x,y) = 0, \tag{9.1}$$

where

$$f(x,y) = \sum_{j=0}^{m_1}\sum_{k=0}^{n_1} a_{jk}x^{m_1-j}y^{n_1-k} \quad (a_{m_1 n_1} \neq 0)$$

and

$$g(x,y) = \sum_{j=0}^{m_2}\sum_{k=0}^{n_2} b_{jk}x^{m_2-j}y^{n_2-k} \quad (b_{m_2 n_2} \neq 0).$$

The coefficients a_{jk}, b_{jk} are complex, in general.

The classical Bézout and Bernstein theorems give us bounds for the total number of solutions of a polynomial system, cf. [Gel'fand et al., 1994], [Sturmfels, 2002]. But for many applications, it is very important to know the number of solutions in a given domain. In the present section, we establish an estimate for the sums of absolute values of the roots of (9.1) in some domains. By that estimate, we get a bound for the number of the roots of (9.1).

A pair of complex numbers $(\tilde{y}, \tilde{x})$ is a solution of (9.1) if $f(\tilde{x}, \tilde{y}) = g(\tilde{x}, \tilde{y}) = 0$. Besides $\tilde{x}$ will be called an X-root coordinate (corresponding to $\tilde{y}$) and $\tilde{y}$ a Y-root coordinate (corresponding to $\tilde{x}$). All the considered roots are counted with their multiplicities.

Put

$$a_j(z) = \sum_{k=0}^{n_1} a_{jk}z^k \ (j=0,...,m_1), \ b_j(z) = \sum_{k=0}^{n_2} b_{jk}z^k \ (j=0,...,m_2) \tag{9.2}$$

and $m = m_1 + m_2$. The following matrix pencil (the $m \times m$ Sylvester matrix) plays an essential role in the theory of polynomial coupled systems:

$$S(z) = \begin{pmatrix} a_0 & a_1 & a_2 & \cdots & a_{m_1-1} & a_{m_1} & 0 & 0 & \cdots & 0 \\ 0 & a_0 & a_1 & \cdots & a_{m_1-2} & a_{m_1-1} & a_{m_1} & 0 & \cdots & 0 \\ \cdot & \cdot & \cdot & \cdots & \cdot & \cdot & \cdot & \cdot & \cdots & \cdot \\ 0 & 0 & 0 & \cdots & a_0 & a_1 & a_2 & a_3 & \cdots & a_{m_1} \\ b_0 & b_1 & b_2 & \cdots & b_{m_2} & 0 & 0 & 0 & \cdots & 0 \\ 0 & b_0 & b_1 & \cdots & b_{m_2-1} & b_{m_2} & 0 & 0 & \cdots & 0 \\ \cdot & \cdot & \cdot & \cdot & \cdot & \cdot & \cdot & \cdot & \cdots & \cdot \\ 0 & 0 & 0 & \cdots & b_0 & b_1 & b_2 & b_3 & \cdots & b_{m_2} \end{pmatrix}$$

with $a_k = a_k(z)$ and $b_k = b_k(z)$. Put $R(z) = det\ S(z)$ and assume that

$$R(0) \neq 0. \tag{9.3}$$

Furthermore, denote

$$\theta(R) := \left[\frac{1}{2\pi} \int_0^{2\pi} \left|\frac{R(e^{it})}{R(0)}\right|^2 dt - 1\right]^{1/2}.$$

Clearly,

$$\theta(R) \le \sup_{|z|=1} \left|\frac{R(z)}{R(0)}\right|. \tag{9.4}$$

Theorem 11.9.1 *Let condition (9.3) hold. Then the Y-root coordinates y_k of (9.1), ordered in the decreasing way: $|y_k| \ge |y_{k+1}|$, satisfy the a priori estimates*

$$\sum_{k=1}^{j} |y_k| < \theta(R) + j \quad (j = 1, 2, ..., n_R), \text{ where } n_R := \deg R(z).$$

Proof: Let $f_n(z) := 1 + c_1 z + ... + c_n z^n$. Then by Theorem 4.3.1 its zeros $z_k(f_n)$ ordered in the increasing way satisfy the inequalities

$$\sum_{k=1}^{j} \frac{1}{|z_k(f_n)|} < j + \left[\frac{1}{2\pi} \int_0^{2\pi} |f_n(e^{it})|^2 dt - 1\right]^{1/2}. \tag{9.5}$$

Put

$$F(x, z) = \sum_{j=0}^{m_1} \sum_{k=0}^{n_1} a_{jk} x^{m_1 - j} z^k, \, G(x, z) = \sum_{j=0}^{m_2} \sum_{k=0}^{n_2} b_{jk} x^{m_2 - j} z^k.$$

Rewrite (9.1) as

$$y^{n_1} F(x, 1/y) = y^{n_2} G(x, 1/y) = 0.$$

Any nonzero Y-root $\tilde{y}$ of (9.1) is connected with a nonzero Y-roots $\tilde{z}$ of the system

$$F(x, z) = G(x, z) = 0 \tag{9.6}$$

as $\tilde{y} = 1/\tilde{z}$. But

$$F(x, z) = a_0(z) x^{m_1} + ... + a_{m_1}(z), \, G(x, z) = b_0(z) x^{m_2} + ... + b_{m_2}(z).$$

Therefore, the resultant corresponding to (9.6) is $R(z)$ [Milovanović et al., 1994]. That is, any possible Y-root coordinate $\tilde{z}$ of (9.6) is a root of $R(z)$. Now (9.5) with

$$f_n(z) = \frac{R(z)}{R(0)}$$

and the relation $\tilde{y} = 1/\tilde{z}$, where $\tilde{y}$ is a Y-root coordinate to (9.1) imply the required inequality. Q. E. D.

By the previous theorem, the Y-root coordinate y_j of system (9.1) satisfies the inequality

$$|y_j| < 1 + \frac{\theta(R)}{j}.$$

Thus, (9.1) has in the set

$$\{z \in \mathbf{C} : |z| \leq 1 + \frac{\theta(R)}{j}\}$$

no more than $n_R - j + 1$ Y-root coordinates. Similarly, the X-root coordinates can be investigated.

11.10   Vector difference equations

The present section is devoted to the vector difference equation

$$\sum_{k=0}^{m} A_{m-k} v(t+k) = f(t) \quad (A_0 = I;\ t = m, m+1, ...) \qquad (10.1)$$

with constant matrix coefficients A_k, a given vector sequence $f = \{f(t) \in \mathbf{C}^n\}_{t=0}^{\infty}$ and the zero initial condition

$$v(k) = 0, \quad k = 0, 1, ..., m-1. \qquad (10.2)$$

Introduce the characteristic pencil

$$\tilde{P}(z) := z^m + A_1 z^{m-1} + ... + A_m.$$

The Z-transform $F(z) = (Zf)(z)$ of sequence $f = \{f(k)\}_{k=0}^{\infty}$ under the condition

$$\|f(j)\| \leq const\, j^{\nu} \quad (\nu = const, j = 1, 2, ...)$$

is defined by

$$F(z) = \sum_{k=0}^{\infty} \frac{f(k)}{z^k}.$$

The inverse Z-transform is given by

$$f(j) = \frac{1}{2\pi i} \int_{C_0} \lambda^{j-1} F(\lambda)\, d\lambda.$$

Here C_0 is a closed Jordan contour containing all the singularities of $F(z)$. Applying the Z-transform to problem (10.1), (10.2), we get

$$V(z) = \tilde{P}^{-1}(z) F(z),$$

where $V(z)$ is the Z-transform of the solution $v(t)$. By the inverse Z-transform, a solutions of problem (10.1), (10.2) can be represented by

$$v(t) = \frac{1}{2\pi i} \int_C \lambda^{t-1} \tilde{P}^{-1}(\lambda) F(\lambda) \, d\lambda.$$

Here C is a closed Jordan contour containing all the zeros of $det\ \tilde{P}(z)$ and singularities of $F(z)$. Now Theorems 11.8.1 and 11.8.2 allow us to estimate $\|\tilde{P}^{-1}(z)\|$ and therefore, to estimate the norm of the solution of problem (10.1), (10.2).

The homogeneous equation

$$\sum_{k=0}^{m} A_{m-k} x(t+k) = 0 \quad (t = m, m+1, ...) \tag{10.3}$$

is said to be *asymptotically stable* if any its solution tends to zero as $t \to \infty$. It is not hard to check that equation (10.3) is asymptotically stable, provided

$$\tilde{r}_s(\tilde{P}) := \max_k |z_k(\tilde{P})| < 1. \tag{10.4}$$

Here $z_k(\tilde{P})$ $(k = 1, ..., nm)$ are the characteristic values of $\tilde{P}$. By Theorem 11.6.2 and Lemma 1.6.1, and Corollary 1.6.2, we can assert that *equation (10.3) is asymptotically stable provided one of the following inequalities*

$$\sum_{k=1}^{m} \|A_k\| < 1, \tag{10.5}$$

or

$$2 \max_k \sqrt[k]{\|A_k\|} < 1 \tag{10.6}$$

holds.

Now let us return to problem (10.1), (10.2). Let

$$\|F(e^{it})\|_{L^2}^2 = \frac{1}{2\pi} \int_0^{2\pi} \|F(e^{it})\|^2 dt < \infty.$$

If condition (10.4) holds, then taking into account that

$$\|V(z)\| \le \|\tilde{P}^{-1}(z)\| \|F(z)\|$$

we have

$$\|V(e^{it})\|_{L^2} \le \sup_{|z|=1} \|\tilde{P}^{-1}(z)\| \|F(e^{it})\|_{L^2}.$$

Furthermore, let $f \in l^2(\mathbf{C}^n)$. That is,

$$\|f\|_{l^2}^2 = \sum_{k=0}^{\infty} \|f(k)\|^2 < \infty.$$

Then, by the Parseval equality, we have

$$\|f\|_{l^2} = \|F(e^{it})\|_{L^2}.$$

Therefore,

$$\|V(e^{it})\|_{L^2} \le \sup_{|z|=1} \|\tilde{P}^{-1}(z)\|\|f\|_{l^2}.$$

Again, applying the Parseval equality, we obtain

$$\|v\|_{l^2} \le \sup_{|z|=1} \|\tilde{P}^{-1}(z)\|\|f\|_{l^2}.$$

So, we can assert that *a solution of problem (10.1), (10.2) is in $l^2(\mathbf{C}^n)$ provided one of conditions (10.5) or (10.6) hold.*

11.11 Comments to Chapter 11

Lemma 11.1.1 is well known, cf. [Rodman, 1989]. Theorem 11.1.2 is taken from [Gil', 2003c]. Theorem 11.2.1 appears in the paper [Gil', 2008a]. Theorem 11.3.1 has been proved in the paper [Gil', 2008c] in the more general situation. The material of Sections 11.4 and 11.5 is adopted from the paper [Gil', 2005d]. Theorems 11.6.2 and 11.7.3 are probably new. The material of Section 11.8 is adopted from [Gil', 2003a]. Theorems 11.9.1 is probably new. In Section 11.9, we use the approach based on the resultant formulations, which has a long story; the literature devoted to this approach is very rich, for example, [Gel'fand et al., 1994], [Sturfmels, 2002]. We combine it with above derived estimates for the zeros of polynomials.

Polynomials with matrix coefficients arise in various applications, in particular, in the theory of vector difference equations, for instance, see [Kolmanovskii and Myshkis, 1998]. The theory of polynomial matrix pencils is well developed, for example, the well-known book [Gohberg, Lancaster, and Rodman, 1982] and references therein. The spectrum of polynomial matrix pencils was investigated in many works. In particular, in the paper [Psarrakos and Tsatsomeros, 2004], the authors present an extension of the Perron-Frobenius theory to spectra and numerical ranges of Perron polynomials. Their approach relies on a companion matrix linearization. They also examine the role of the matrix polynomial in the multistep version of difference equations and provide a multistep version of the fundamental theorem of demography. The paper [Psarrakos and Tsatsomeros, 2002] deals with the stability radius of a matrix polynomial $P(\lambda)$ relative to an open region Ω of the complex plane, and its relation to the numerical range of $P(\lambda)$ are investigated. Using an expression of the stability radius in term of λ on the boundary of Ω and $\|P(\lambda)^{-1}\|$, a lower bound is obtained. This bound for the stability radius involves the distances of Ω to the connected components of the numerical range. The special case of hyperbolic matrix polynomials is also considered in that paper. Many papers are devoted to perturbations of the spectrum of polynomial matrix pencils. In particular, the paper [Ahues and Limaye, 1998] deals with linear matrix pencils. Besides, an error bound for eigenvalues is established. In the paper [Gracia and Velasco, 1995], stability of invariant subspaces of regular matrix pencils is considered. In the paper [Higham and Tisseur, 2003], upper and lower bounds are derived for the absolute values of the eigenvalues of matrix polynomials. The bounds are based on norms of coefficient matrices. They generalize some well-known bounds for scalar polynomials and single matrices. In the paper [Li, R.-C., 2003] and references given therein, perturbations of eigenvalues of diagonalizable matrix pencils with real spectra are investigated.

Chapter 12

Entire Matrix-Valued Functions

In this chapter, we consider entire matrix-valued functions of a complex argument (entire matrix pencils). Bounds for the characteristic values are derived in terms of the Taylor coefficients. A matrix-valued function of a complex argument is called a meromorphic matrix function if its determinant is a meromorphic function. In the present chapter, we establish the multiplicative representation for a class of meromorphic matrix functions, namely, for the inverse entire matrix pencils.

12.1 Preliminaries

Our main object in this chapter is the matrix

$$F(\lambda) = (f_{jk}(\lambda))_{j,k=1}^n$$

dependent on the complex argument λ. If all the entries $f_{jk}(\lambda)$ are analytic in a domain W of $\mathbf{C}$, then $F(\lambda)$ is said to be analytic in W. A zero $z_k(F)$ of $det\ F(z)$ is called *a characteristic value of F*. The collection of all the characteristic values of F with their multiplicities is called *the spectrum of F and is denoted by $\Sigma(F)$*.

The following Cauchy formula holds: if F is analytic in W and L is a simple rectifiable closed curve in W, the interior of which is contained in W, then

$$F(z_0) = \frac{1}{2\pi i} \int_L \frac{F(z)dz}{z - z_0} \quad (z_0 \in W).$$

If

$$L = \{z \in \mathbf{C} : |z - z_0| = r\},$$

then, we have

$$F(z_0) = \frac{1}{2\pi} \int_0^{2\pi} F(z_0 + re^{it})dt.$$

Consequently,

$$\|F(z_0)\| \leq \frac{1}{2\pi} \int_0^{2\pi} \|F(z_0 + re^{it})\|dt. \tag{1.1}$$

Recall that $\|.\|$ means the Euclidean norm in $\mathbf{C}^n$ although in this section and sometimes below one can use an arbitrary norm in $\mathbf{C}^n$.

From (1.1), it easily follows that the function $z \to \|F(z)\|$ can have no maximum interior W except $F(z)$ is constant.

Recall that a scalar-valued function u defined on a domain W of $\mathbf{C}$ is upper semicontinuous if for all $z_0 \in W$

$$\lim_{z \to z_0} \sup u(z) \leq u(z_0).$$

Besides, a function u is said to be subharmonic in W, if it is upper semicontinuous and satisfies the mean inequality

$$u(z_0) \leq \frac{1}{2\pi} \int_{-\pi}^{\pi} u(z_0 + re^{it})dt$$

for any closed disc $\Omega(z_0, r)$ contained in W. As above $z_0 \in \mathbf{C}$ is the center and r is the radius of $\Omega(z_0, r)$. A function u is harmonic if both u and $-u$ are subharmonic.

From (1.1), it follows that $\|F(z)\|$ is subharmonic in W. Moreover, as it is well known, if F is a matrix-valued function analytic in W, then the function $\ln \|F(z)\|$ is subharmonic in W, cf. Theorem 3.3.1 from the book [O. Nevanlinna, 2003, p. 38].

Recall also that $r_s(A)$ means the spectral radius of an operator A. So

$$r_s(F(z)) = \max_k |\lambda_k(F(z))|,$$

where $\lambda_k(F(z))$ $(k = 1, ..., n)$ are the eigenvalues of matrix $F(z)$ for a fixed z.

Let F be a matrix-valued function analytic in W. Then the functions $\ln r_s(F(z))$ and $r_s(F(z))$ are subharmonic in W, cf. the well-known Theorem 14.4.1 from the book [Istratesku, 1981, p. 509].

Furthermore, if all the entries $f_{jk}(\lambda)$ are entire functions, then $F(\lambda)$ is said to be an entire matrix-valued function of a complex argument *(an entire matrix pencil)*.

For an entire matrix pencil F, let $\rho\ (det\ F(.))$ be the order of its determinant. Then

$$\rho\ (F) := \rho\ (det\ F(.))$$

will be called *the order of F*. That is, the order of $F(z)$ coincides with the order of its determinant. Clearly, an entire pencil F can be represented by the Taylor series

$$F(\lambda) = \sum_{k=0}^{\infty} \lambda^k C_k,$$

where C_k, $k = 1, 2, ...$ are constant $n \times n$-matrices and the series converges for all finite $\lambda \in \mathbf{C}$.

12.2   Partial sums of moduli of characteristic values

Let A_k, $k = 1, 2, ...$ be $n \times n$-matrices. Our main object in this section is the entire matrix pencil

$$F(\lambda) = \sum_{k=0}^{\infty} \frac{\lambda^k}{(k!)^\gamma} A_k \ \ (A_0 = I_n, \lambda \in \mathbf{C}) \tag{2.1}$$

with a finite $\gamma > 0$. The characteristic values of F with their multiplicities are enumerated in the nondecreasing way: $|z_k(F)| \le |z_{k+1}(F)|$ $(k = 1, 2, ...)$. If F has a finite number l of the characteristic values $z_k(F)$, we put

$$\frac{1}{z_k(F)} = 0 \text{ for } k = l+1, l+2, ...$$

We restrict ourselves by the root-factorial representation (2.1) for the brevity, only, but our arguments below are valid in the case of the ψ-representation, in particular, in the case of the Mittag-Leffler representation (see Chapter 5).
  Suppose that

$$\text{the series } \Theta_F := [\sum_{k=1}^{\infty} A_k A_k^*]^{1/2} \text{ converges.} \tag{2.2}$$

Recall that the asterisk means the adjointness. So Θ_F is an $n \times n$-matrix and under (2.1) and (2.2), by the Hólder inequality, it follows that

$$\|F(\lambda)\| \le m_0 \sum_{k=0}^{\infty} \frac{|\lambda|^k}{(k!)^\gamma} \le m_0 \ [\sum_{k=0}^{\infty} 2^{-p'k}]^{1/p'} \ [\sum_{k=0}^{\infty} \frac{|2\lambda|^{k/\gamma}}{k!}]^\gamma$$

$$\le m_1 e^{\gamma|2\lambda|^{1/\gamma}} \ (\gamma + 1/p' = 1),$$

where

$$m_0 = \sup_k \|A_k\|, \quad m_1 = m_0 \Big[\sum_{k=0}^{\infty} 2^{-kp'}\Big]^{1/p'}.$$

So function F has order no more than $1/\gamma$.

Furthermore, put

$$\omega_k(F) = \lambda_k(\Theta_F) \text{ for } k = 1, ..., n \text{ and } \omega_k(F) = 0 \text{ for } k \geq n+1.$$

Theorem 12.2.1 *Let condition (2.2) hold. Then the characteristic values of the pencil F defined by (2.1) satisfy the inequalities*

$$\sum_{k=1}^{j} \frac{1}{|z_k(F)|} < \sum_{k=1}^{j}[\omega_k(F) + \frac{n^\gamma}{(k+n)^\gamma}] \quad (j = 1, 2, ...).$$

The proof of this theorem is presented in the next section.

Note that the case

$$\alpha(F) := \overline{\lim}_{k\to\infty} \sqrt[k]{\|A_k\|} < \infty,$$

formally more general than (2.2), can be easily reduced to (2.2) by the substitution

$$\lambda = \frac{z}{2\alpha(F)}.$$

From Theorem 12.2.1, it follows that

$$\frac{j}{|z_j(F)|} < \sum_{k=1}^{j}[\omega_k(F) + \frac{n^\gamma}{(k+n)^\gamma}].$$

Therefore, F has in

$$|z| \leq \frac{j}{\sum_{k=1}^{j}[\omega_k(F) + \frac{n^\gamma}{(k+n)^\gamma}]}$$

no more that than $j - 1$ zeros. Let $\nu_F(r)$ be the counting function of the characteristic values of F in $|z| \leq r$. We thus get.

Corollary 12.2.2 *Let the pencil $F(z)$ defined by (2.1) satisfy condition (2.2). Then the inequality $\nu_F(r) \leq j - 1$ is valid for any*

$$r \leq \frac{j}{\sum_{k=1}^{j}[\omega_k(F) + \frac{n^\gamma}{(k+n)^\gamma}]}.$$

Put

$$\chi_k = \omega_k(F) + \frac{n^\gamma}{(k+n)^\gamma} \quad (k = 1, 2, ...).$$

The following result is due to Lemma 1.2.1 and Theorem 12.2.1.

Corollary 12.2.3 *Let $\phi(t)$ $(0 \leq t < \infty)$ be a continuous convex scalar-valued function, such that $\phi(0) = 0$. Let F be defined by (2.1) under condition (2.2). Then the inequalities*

$$\sum_{k=1}^{j} \phi\left(\frac{1}{|z_k(F)|}\right) < \sum_{k=1}^{j} \phi(\chi_k) \quad (j = 1, 2, ...)$$

are valid. In particular, for any $r \geq 1$,

$$\sum_{k=1}^{j} \frac{1}{|z_k(F)|^r} < \sum_{k=1}^{j} \chi_k^r$$

and thus

$$\left[\sum_{k=1}^{j} \frac{1}{|z_k(F)|^r}\right]^{1/r} < \left[\sum_{k=1}^{j} \omega_k^r(F)\right]^{1/r} + n^{\gamma}\left[\sum_{k=1}^{j} \frac{1}{(k+n)^{r\gamma}}\right]^{1/r} \quad (j = 1, 2, ...).$$

$$(2.3)$$

Furthermore, assume that

$$r\gamma > 1, \ r \geq 1. \tag{2.4}$$

Then

$$\zeta_n(\gamma r) := \sum_{k=1}^{\infty} \frac{1}{(k+n)^{r\gamma}} < \infty.$$

Relation (2.3) with

$$N_r(\Theta_F) = \left[\sum_{k=1}^{n} \lambda_k^r(\Theta_F)\right]^{1/r}$$

yields our next result.

Corollary 12.2.4 *Let the pencil $F(z)$ defined by (2.1) satisfy conditions (2.2) and (2.4). Then the inequality*

$$\left(\sum_{k=1}^{\infty} \frac{1}{|z_k(F)|^r}\right)^{1/r} < N_r(\Theta_F) + n^{\gamma}\zeta_n^{1/r}(\gamma r)$$

is valid.

In particular, if $\gamma > 1$, then taking into account that $\zeta_1(\gamma) := \zeta(\gamma) - 1$, where ζ is the Riemann zeta function, we get the inequality

$$\sum_{k=1}^{\infty} \frac{1}{|z_k(F)|} < N_1(\Theta_F) + n^{\gamma}(\zeta(\gamma) - 1).$$

Now consider a positive scalar-valued function $\Phi(t_1, t_2, ..., t_j)$ with an integer j, defined on the domain

$$0 \leq t_j \leq t_{j-1} \leq t_2 \leq t_1 < \infty$$

and satisfying

$$\frac{\partial \Phi}{\partial t_1} > \frac{\partial \Phi}{\partial t_2} > ... > \frac{\partial \Phi}{\partial t_j} > 0 \text{ for } t_1 > t_2 > ... > t_j. \tag{2.5}$$

Corollary 12.2.5 *Under the hypothesis of Theorem 12.2.1, let condition (2.5) hold. Then*

$$\Phi\left(\frac{1}{|z_1(F)|}, \frac{1}{|z_2(F)|}, ..., \frac{1}{|z_j(F)|}\right) < \Phi(\chi_1, \chi_2, ..., \chi_j).$$

Indeed, this result is due to Theorem 12.2.1 and Lemma 1.2.2.

In particular, let $\{d_k\}_{k=1}^{\infty}$ be a decreasing sequence of non-negative numbers. Take

$$\Phi(t_1, t_2, ..., t_j) = \sum_{k=1}^{j} d_k t_k.$$

Then the previous corollary yields the inequalities

$$\sum_{k=1}^{j} d_k \frac{1}{|z_k(F)|} < \sum_{k=1}^{j} \chi_k d_k = \sum_{k=1}^{j} d_k \left[\omega_k(F) + \frac{n^\gamma}{(k+n)^\gamma}\right] \ (j = 1, 2, ...).$$

12.3   Proof of Theorem 12.2.1

Consider the polynomial matrix pencil

$$\tilde{P}_\gamma(\lambda) = \sum_{k=0}^{m} \frac{\lambda^{m-k}}{(k!)^\gamma} A_k \ (A_0 = I_n) \tag{3.1}$$

with the characteristic values ordered in the non-increasing way:

$$|z_k(\tilde{P}_\gamma)| \geq |z_{k+1}(\tilde{P}_\gamma)| \ (k = 1, ..., mn - 1).$$

For a $\gamma \geq 0$, let us introduce the block matrix

$$T(\gamma) = \begin{pmatrix} -A_1 & -A_2 & ... & -A_{m-1} & -A_m \\ \frac{1}{2^\gamma} I_n & 0 & ... & 0 & 0 \\ 0 & \frac{1}{3^\gamma} I_n & ... & 0 & 0 \\ . & & ... & . & . \\ 0 & 0 & ... & \frac{1}{m^\gamma} I_n & 0 \end{pmatrix}. \tag{3.2}$$

We need the following lemma.

Lemma 12.3.1 *The relation* $\det \tilde{P}_\gamma(\lambda) = \det(\lambda I_{mn} - T(\gamma))$ *is true.*

Proof: Let z_0 be a characteristic value of $\tilde{P}(\gamma)$. Then

$$\sum_{k=0}^{m} \frac{z_0^{m-k}}{(k!)^\gamma} A_k h = 0,$$

where $h \in \mathbf{C}^n$ is the eigenvector of $F(z_0)$. Put

$$x_k = \frac{z_0^{m-k}}{(k!)^\gamma} h \quad (k = 1, ..., m).$$

Then

$$z_0 x_k = \frac{x_{k-1}}{k^\gamma} \quad (k = 2, ..., m)$$

and

$$\sum_{k=0}^{m} \frac{z_0^{m-k}}{(k!)^\gamma} A_k h = \sum_{k=1}^{m} A_k x_k + z_0 x_1 = 0.$$

So the vector $x = (x_1, ..., x_m) \in \mathbf{C}^{nm}$ satisfies the equation $T(\gamma)x = z_0 x$. If the spectrum of $T(\gamma)$ is simple, the lemma is proved. If $det\, T(\gamma)$ has non-simple roots, then the required result can be proved by a small perturbation. Q. E. D.

Lemma 12.3.2 *The characteristic values of $\tilde{P}_\gamma$ satisfy the inequalities*

$$\sum_{k=1}^{j} |z_k(\tilde{P}_\gamma)| < \sum_{k=1}^{j} \left[\omega_k(F) + \frac{n^\gamma}{(k+n)^\gamma} \right] \quad (j = 1, ..., mn).$$

Proof: Because of the previous lemma,

$$\lambda_k(T(\gamma)) = z_k(\tilde{P}_\gamma) \quad (k = 1, 2, ..., nm). \tag{3.3}$$

Take into account that

$$\sum_{k=1}^{j} |\lambda_k(T(\gamma))| < \sum_{k=1}^{j} s_k(T(\gamma)) \quad (j = 1, ..., nm), \tag{3.4}$$

where $s_k(T(\gamma))$ are the singular numbers of $T(\gamma)$ ordered in the non-increasing way (see Section 1.1). But $T(\gamma) = M + C(\gamma)$, where

$$M = \begin{pmatrix} -A_1 & -A_2 & \cdots & -A_{m-1} & -A_m \\ 0 & 0 & \cdots & 0 & 0 \\ . & . & \cdots & . & . \\ 0 & 0 & \cdots & 0 & 0 \end{pmatrix}$$

and

$$C(\gamma) = \begin{pmatrix} 0 & 0 & \cdots & 0 & 0 \\ \frac{1}{2^\gamma} I_n & 0 & \cdots & 0 & 0 \\ 0 & \frac{1}{3^\gamma} I_n & \cdots & 0 & 0 \\ . & . & \cdots & . & . \\ 0 & 0 & \cdots & \frac{1}{m^\gamma} I_n & 0 \end{pmatrix}.$$

We have

$$
C(\gamma)C^*(\gamma) = \begin{pmatrix}
0 & 0 & \cdots & 0 & 0 \\
0 & I_n/2^{2\gamma} & \cdots & 0 & 0 \\
0 & 0 & \cdots & 0 & 0 \\
\cdot & \cdot & \cdots & \cdot & \cdot \\
0 & 0 & \cdots & 0 & I_n/m^{2\gamma}
\end{pmatrix}.
$$

Take into account that $s_k(M^*) = \omega_k(F)$. In addition,

$$
s_k(C^*(\gamma)) = \frac{1}{(j+2)^\gamma} \quad (k = jn + l; j = 0, ..., m-2; l = 1, ..., n)
$$

and

$$
s_k(C^*(\gamma)) = 0 \quad (k = (m-1)n + l; l = 1, ..., n).
$$

Since $j = (k - l)/n$ and

$$
(k-l)/n + 2 \geq k/n + 1 = (k+n)/n,
$$

we can write out,

$$
s_k(C^*(\gamma)) < \frac{n^\gamma}{(k+n)^\gamma} \quad (k = 1, ..., nm).
$$

Thanks to Lemma 1.1.2,

$$
\sum_{k=1}^{j} s_k(A_m^*(\gamma)) = \sum_{k=1}^{j} s_k(M^* + C^*(\gamma)) < \sum_{k=1}^{j} s_k(M^*) + \sum_{k=1}^{j} s_k(C^*(\gamma)).
$$

So

$$
\sum_{k=1}^{j} s_k(T(\gamma)) = \sum_{k=1}^{j} s_k(T^*(\gamma)) < \sum_{k=1}^{j} \left[\omega_k(F) + \frac{n^\gamma}{(k+n)^\gamma}\right]
$$

$(j = 1, 2, ..., mn)$. Now (3.3) and (3.4) yield the required result. Q. E. D.

Proof of Theorem 12.2.1: Consider the polynomial pencil

$$
F_m(\lambda) = \sum_{k=0}^{m} \frac{\lambda^k}{(k!)^\gamma} A_k. \tag{3.5}
$$

Clearly, $\lambda^m F_m(1/\lambda) = \tilde{P}_\gamma(\lambda)$. So

$$
z_k(\tilde{P}_\gamma) = \frac{1}{z_k(\tilde{F}_m)}.
$$

Now the latter lemma yields the inequalities

$$
\sum_{k=1}^{j} \frac{1}{|z_k(\tilde{F}_m)|} < \omega_k(F) + n^\gamma \sum_{k=1}^{j} \frac{1}{(k+n)^\gamma} \quad (j = 1, ..., nm). \tag{3.6}
$$

But the characteristic values of entire matrix functions continuously depend on its coefficients. So for any $j = 1, 2, ...,$

$$\sum_{k=1}^{j} \frac{1}{|z_k(F_m)|} \to \sum_{k=1}^{j} \frac{1}{|z_k(F)|}$$

as $m \to \infty$. Now (3.6) implies the required result. Q. E. D.

12.4 Imaginary parts of characteristic values of entire pencils

Consider the entire matrix pencil

$$F(\lambda) = \sum_{k=0}^{\infty} \frac{A_k \lambda^k}{(k!)^\gamma} \quad (\gamma \in (1/2, 1]; \ A_0 = I_n) \tag{4.1}$$

with $n \times n$-matrices A_k. Again suppose that condition (2.2) holds. So in the considered case $\rho(F) \le 1/\gamma < 2$. It is not hard to show that any entire matrix pencil F of order less than two can be written in the form (4.1) under (2.2), provided $F(0) = I_n$.

Denote

$$\tau_\gamma(F) := \sum_{k=1}^{\infty} N_2^2(A_k) + (\zeta(2\gamma) - 1)n$$

and

$$\chi_\gamma(F, t) := \tau_\gamma(F) + Re\, Trace\, \left[e^{2it} \left(A_1^2 - 2^{1-\gamma} A_2 \right) \right] \quad (t \in [0, 2\pi)).$$

Theorem 12.4.1 *Let F be defined by (4.1) and condition (2.2) hold. Then for any $t \in [0, 2\pi)$ the relations*

$$\tau_\gamma(F) - \sum_{k=1}^{\infty} \frac{1}{|z_k(F)|^2} = \chi_\gamma(F, t) - 2 \sum_{k=1}^{\infty} \left(Re\, \frac{e^{it}}{z_k(F)} \right)^2 \ge 0$$

are valid.

This theorem is proved below in the present section.
    Note that

$$\chi_\gamma\left(F, \frac{\pi}{2}\right) = \tau_\gamma(F) - Re\, Trace\, (A_1^2 - 2^{1-\gamma} A_2)$$

and

$$\chi_\gamma(F, 0) = \tau_\gamma(F) + Re\, Trace\, (A_1^2 - 2^{1-\gamma} A_2).$$

Now the latter theorem yields

Corollary 12.4.2 *If F is defined by (4.1) and condition (2.2) holds, then*

$$\tau_\gamma(F) - \sum_{k=1}^{\infty} \frac{1}{|z_k(F)|^2} = \chi_\gamma\left(F, \frac{\pi}{2}\right) - 2\sum_{k=1}^{\infty}\left(Im\ \frac{1}{z_k(F)}\right)^2 =$$

$$\chi_\gamma(F, 0) - 2\sum_{k=1}^{\infty}\left(Re\ \frac{1}{z_k(F)}\right)^2 \geq 0.$$

Consequently,

$$\sum_{k=1}^{\infty} \frac{1}{|z_k(F)|^2} \leq \tau_\gamma(F), \quad 2\sum_{k=1}^{\infty}\left(Im\ \frac{1}{z_k(F)}\right)^2 \leq \chi_\gamma\left(F, \frac{\pi}{2}\right)$$

and

$$2\sum_{k=1}^{\infty}\left(Re\ \frac{1}{z_k(F)}\right)^2 \leq \chi_\gamma(F, 0).$$

To prove Theorem 12.4.1, consider the polynomial matrix pencil $\tilde{P}_\gamma(\lambda)$ and block matrix $T(\gamma)$ defined as in the previous section. Denote

$$\tau_\gamma(\tilde{P}_\gamma) := \sum_{k=1}^{m} N_2^2(A_k) + n\sum_{k=2}^{m} k^{-2\gamma}$$

and for a $t \in [0, 2\pi)$ put

$$\chi_\gamma(\tilde{P}_\gamma, t) := Re\ Trace\ e^{2it}\left(A_1^2 - 2^{1-\gamma}A_2\right) + \tau_\gamma(\tilde{P}_\gamma).$$

Lemma 12.4.3 *For any $t \in [0, 2\pi)$, we have the relations*

$$\tau_\gamma(\tilde{P}_\gamma) - \sum_{k=1}^{mn} |z_k(\tilde{P}_\gamma)|^2 = \chi_\gamma(\tilde{P}_\gamma, t) - 2\sum_{k=1}^{mn}(Re\ e^{it}z_k(\tilde{P}_\gamma))^2 \geq 0.$$

Proof: Let $T(\gamma)$ be defined as in the previous section. For simplicity, put $T(\gamma) = T$. Let us apply Lemma 1.9.1 to the matrix $ie^{it}T$. Then

$$N_2^2(T) - \sum_{k=1}^{mn} |\lambda_k(T)|^2$$

$$= \frac{1}{2}N_2^2(e^{it}T + e^{-it}T^*) - \frac{1}{2}\sum_{k=1}^{mn} |e^{it}\lambda_k(T) + e^{-it}\overline{\lambda}_k(T)|^2.$$

Simple calculations and Lemma 11.3.2 show that $N_2^2(T) = \tau_\gamma(\tilde{P}_\gamma)$ and

$$\frac{1}{2}N_2^2(e^{it}T + e^{-it}T^*) = \frac{1}{2}N_2^2(A_1e^{it} + A_1^*e^{-it})$$

$$+ N_2^2 (\frac{1}{2\gamma} I_n - A_2 e^{2it}) + \sum_{k=3}^{m} (N_2^2(A_k) + nk^{-2\gamma}).$$

So

$$\frac{1}{2} N_2^2 (e^{it}T + e^{-it}T^*)$$

$$= N_2^2(A_1) + \frac{1}{2} Trace \ (A_1^2 e^{2it} + (A_1^*)^2 e^{-2it}) + n2^{-2\gamma} -$$

$$Trace \ 2^{-\gamma} \left[e^{2it} A_2 + (e^{2it}A_2)^* \right] + N^2(A_2)$$

$$+ \sum_{k=3}^{m} (N_2^2(A_k) + nk^{-2\gamma}) = \chi_\gamma(\tilde{P}, t).$$

Hence, (3.3) implies the required result. Q. E. D.

Proof of Theorem 12.4.1:   Consider the polynomial matrix pencil $F_m(\lambda)$ defined by (3.5). Thanks to the latter lemma and (3.6), we obtain

$$\tau_\gamma(\tilde{P}_\gamma) - \sum_{k=1}^{mn} \frac{1}{|z_k(F_m)|^2} = \chi_\gamma(\tilde{P}_\gamma, t) - 2 \sum_{k=1}^{mn} \left(Re \ \frac{e^{it}}{z_k(F_m)} \right)^2 \geq 0.$$

Since the characteristic values continuously depend on the coefficients, letting $m \to \infty$ in the latter relation, we get the required result. Q. E. D.

12.5   Variations of characteristic values of entire pencils

Let A_k, B_k $(k = 1, 2, ...)$ be $n \times n$-matrices. Consider the matrix pencils

$$F(\lambda) = \sum_{k=0}^{\infty} \frac{\lambda^k}{(k!)^\gamma} A_k \ (A_0 = I_n) \qquad (5.1a)$$

and

$$\tilde{F}(\lambda) = \sum_{k=0}^{\infty} \frac{\lambda^k}{(k!)^\gamma} B_k \ (B_0 = I_n, \lambda \in \mathbb{C}) \qquad (5.1b)$$

with a positive γ. Assume that

the series $\Theta_F = \left[\sum_{k=1}^{\infty} A_k A_k^* \right]^{1/2}$ and $\Theta_{\tilde{F}} := \left[\sum_{k=1}^{\infty} B_k B_k^* \right]^{1/2}$ converge. (5.2)

Our main problem in this section is: if A_k and B_k are close, how close are the characteristic values of $\tilde{F}$ to those of F ? The quantity

$$rv_F(\tilde{F}) = \max_j \min_k \left| \frac{1}{z_k(F)} - \frac{1}{z_j(\tilde{F})} \right|$$

will be called *the relative variation of characteristic values* of pencil $\tilde{F}$ with respect to pencil F.

Everywhere in this section p is a natural number satisfying the inequality $p > 1/2\gamma$. Put

$$w_p(F) := 2N_{2p}(\Theta_F) + 2[n(\zeta(2\gamma p) - 1)]^{1/2p},$$

where $\zeta(.)$ is the Riemann zeta function. Denote also

$$\xi_p(F,y) := \sum_{k=0}^{p-1} \frac{w_p^k(F)}{y^{k+1}} exp \left[\frac{1}{2} + \frac{w_p^{2p}(F)}{2y^{2p}} \right] \quad (y > 0) \tag{5.3}$$

and

$$q = \left[\sum_{k=1}^{\infty} \|A_k - B_k\|^2 \right]^{1/2},$$

where $\|.\|$ is the Euclidean norm.

Theorem 12.5.1 *Let F and $\tilde{F}$ be defined by (5.1) and condition (5.2) hold. Then*

$$rv_F(\tilde{F}) \le r_p(F,q),$$

where $r_p(F,q)$ is the unique (positive) root of the equation

$$q\xi_p(F,y) = 1. \tag{5.4}$$

The proof of this theorem is presented in the next section.

If we substitute the equality $y = xw_p(F)$ into (5.4) and use Lemma 1.6.4, we get $r_p(F,q) \le \delta_p(F,q)$, where

$$\delta_p(F,q) := \begin{cases} e\, p\, q & \text{if } w_p(F) \le e\, p\, q \\ w_p(F)\, [ln\, (w_p(F)/(q\, p)]^{-1/2p} & \text{if } w_p(F) > ep\, q \end{cases}.$$

Therefore, $rv_F(\tilde{F}) \le \delta_p(F,q)$.
    Put

$$W_j = \left\{ z \in \mathbf{C} : q\xi_p\left(F, \left| \frac{1}{z} - \frac{1}{z_j(F)} \right| \right) \ge 1 \right\} \quad (j = 1, ..., l),$$

where $l \le \infty$ is the number of the zeros of F. Since $\xi_p(F,y)$ is a monotone decreasing function with respect to $y > 0$, Theorem 12.5.1 yields the following result.

Corollary 12.5.2 *Under the hypothesis of Theorem 12.5.1, all the characteristic values of $\tilde{F}$ lie in the set*

$$\cup_{j=1}^{\infty} W_j$$

provided $l = \infty$. If $l < \infty$, then all the characteristic values of $\tilde{F}$ are in the set

$$\cup_{j=0}^{l} W_j,$$

where

$$W_0 = \{z \in \mathbb{C} : q\xi_p(F, 1/|z|) \geq 1\}.$$

Let us consider approximations of an entire function $\tilde{F}$ by the polynomial pencil

$$h_m(\lambda) = \sum_{k=0}^{m} \frac{B_k \lambda^k}{(k!)^{\gamma}} \quad (B_0 = I, \lambda \in \mathbb{C}).$$

Put

$$q_m(\tilde{F}) := \Big[\sum_{k=m+1}^{\infty} \|B_k\|^2 \Big]^{1/2}$$

and

$$w_p(h_m) = 2\, N_p^{1/2} \Big(\sum_{k=1}^{m} B_k B_k^* \Big) + 2\, [n(\zeta(2\gamma p) - 1)]^{1/2p}.$$

Define $\xi_p(h_m, y)$ according to (5.3). Taking h_m instead of F in Theorem 12.4.1, we get

Corollary 12.5.3 *Let $\tilde{F}$ be defined by (5.1b) and satisfy (5.2). Let $r_m(\tilde{F})$ be the unique (positive) root of the equation*

$$q_m(\tilde{F})\xi_p(h_m, y) = 1.$$

Then either for any characteristic value $z(\tilde{F})$ of $\tilde{F}$ there is a characteristic value $z(h_m)$ of polynomial pencil h_m, such that

$$\Big| \frac{1}{z(\tilde{F})} - \frac{1}{z(h_m)} \Big| \leq r_m(\tilde{F})$$

or

$$|z(\tilde{F})| \geq \frac{1}{r_m(\tilde{F})}.$$

Suppose that under (5.1), there is a constant $d_0 \in (0, 1)$, such that

$$\overline{\lim}_{k \to \infty} \sqrt[k]{\|A_k\|} < 1/d_0 \text{ and } \overline{\lim}_{k \to \infty} \sqrt[k]{\|B_k\|} < 1/d_0,$$

and consider the functions

$$F_d(\lambda) = \sum_{k=0}^{\infty} \frac{A_k(d_0\lambda)^k}{(k!)^{\gamma}} \text{ and } \tilde{F}_d(\lambda) = \sum_{k=0}^{\infty} \frac{B_k(d_0\lambda)^k}{(k!)^{\gamma}}.$$

That is, $F_d(\lambda) \equiv F(d_0\lambda)$ and $\tilde{F}_d(\lambda) \equiv \tilde{F}(d_0\lambda)$. So functions $\tilde{f}(\lambda)$ and $\tilde{h}(\lambda)$ satisfy conditions (5.2). Moreover,

$$w_p(F_d) = 2\Big[\sum_{k=1}^{\infty} d_0^{2k}|A_k|^2\Big]^{1/2} + 2[n(\zeta(2\gamma p) - 1)]^{1/2p}.$$

Since

$$\sum_{k=1}^{\infty} d_0^{2k}\|A_k - B_k\|^2 < \infty$$

we can directly apply Theorem 12.5.1 and its corollaries taking into account that $d_0 z_k(F_d) = z_k(F)$, $d_0 z_k(\tilde{F}_d) = z_k(\tilde{F})$.

Example 12.5.4 *Let us consider the pencil*

$$\tilde{F}(z) = I_n + c_1 z + z^2 e^{z\mu} c_2 \ (0 < \mu = const < 1)$$

with $n \times n$-matrices c_1, c_2. As it is well-known, such matrix quasipolynomials play an essential role in the theory of differential-difference equations, cf. [Kolmanovskii and Myshkis, 1999]. Rewrite this function in the form (5.1b) with $\gamma = 1$, and

$$B_1 = c_1, \ B_k = c_2\mu^{k-2}k(k - 1) \ (k = 2, 3, ...).$$

Put

$$h_2(\lambda) = I_n + c_1 z + c_2 z^2.$$

We have

$$q_2(\tilde{F}) = \|c_2\|\Big[\sum_{k=3}^{\infty} \mu^{2(k-2)}k^2(k - 1)^2\Big]^{1/2}.$$

Clearly, this series converges. Furthermore, put

$$w_1(h_2) = 2Trace\,(c_1 c_1^* + 4c_2 c_2^*) + 2[n(\zeta(2) - 1)]^{1/2}.$$

Now we can apply Corollary 12.5.3.

Furthermore, from Theorem 1.7.5 at once we get our next result.

Corollary 12.5.5 *Let $F(z)$ and $\tilde{F}(z)$ be two $n \times n$-matrix pencils. Then*

$$|det\,F(z) - det\,\tilde{F}(z)|$$

$$\leq \frac{1}{n^{n/2}}N_2(F(z) - \tilde{F}(z))\Big[1 + \frac{1}{2}N_2(F(z) - \tilde{F}(z)) + \frac{1}{2}N_2(F(z) + \tilde{F}(z))\Big]^n.$$

12.6 Proof of Theorem 12.5.1

For a finite integer m, consider the matrix polynomials

$$P(\lambda) = \sum_{k=0}^{m} \frac{\lambda^{m-k}}{(k!)^\gamma} A_k \text{ and } Q(\lambda) = \sum_{k=0}^{m} \frac{\lambda^{m-k}}{(k!)^\gamma} B_k \ (A_0 = B_0 = I). \quad (6.1)$$

So $P(\lambda) = \tilde{P}_\gamma(\lambda)$. In addition, $\{z_k(P)\}_{k=1}^n$ and $\{z_k(Q)\}_{k=1}^m$ are the sets of all the characteristic values of P and Q, respectively, taken with their multiplicities. We need the block matrix $T(\gamma)$ defined by (3.2) and the block matrix

$$\tilde{T}(\gamma) := \begin{pmatrix} -B_1 & -B_2 & \ldots & -B_{m-1} & -B_m \\ \frac{1}{2^\gamma} I_n & 0 & \ldots & 0 & 0 \\ 0 & \frac{1}{3^\gamma} I_n & \ldots & 0 & 0 \\ . & . & \ldots & . & \\ 0 & 0 & \ldots & \frac{1}{m^\gamma} I_n & 0 \end{pmatrix}.$$

Thanks to Lemma 12.3.1 the relations

$$det\ P(\lambda) = det\ (\lambda I_{mn} - T(\gamma)) \text{ and } det\ Q(\lambda) = det\ (\lambda I_{mn} - \tilde{T}(\gamma)) \quad (6.2)$$

are true. Put

$$q(P,Q) := \left[\sum_{k=1}^{m} \|A_k - B_k\|^2 \right]^{1/2}$$

and

$$w_p(P) := 2\ N_{2p}(\Theta_P) + 2[n(\zeta(2\gamma p) - 1)]^{1/2p},$$

where

$$\Theta_P = \left[\sum_{k=1}^{m} A_k A_k^* \right]^{1/2};$$

$\xi_p(P, y)$ is defined according to (5.3).

Lemma 12.6.1 *For any characteristic value $z(Q)$ of $Q(z)$, there is a characteristic value $z(P)$ of $P(z)$, such that*

$$|z(P) - z(Q)| \le r_p(P, Q),$$

where $r_p(P, Q,)$ is the unique (positive) root of the equation

$$q(F, Q)\xi_p(P, y) = 1.$$

Proof: For the brevity put $T(\gamma) = T$ and $\tilde{T}(\gamma) = \tilde{T}$. Clearly,

$$\|T - \tilde{T}\|_{mn} = q(P, Q).$$

Here $\|.\|_{mn}$ is the Euclidean norm in $\mathbf{C}^{mn}$. Because of Theorem 2.12.4, for any $\lambda_j(\tilde{T})$, there is a $\lambda_i(T)$, such that

$$|\lambda_j(T) - \lambda_i(\tilde{T})|_{mn} \le y_p(T, \tilde{T}),$$

where $y_p(T, \tilde{T})$ is the unique positive root of the equation

$$\|T - \tilde{T}\|_{mn} \sum_{k=0}^{p-1} \frac{(2N_{2p}(T))^k}{y^{k+1}} exp\left[(1 + \frac{(2N_{2p}(T))^{2p}}{y^{2p}})/2\right] = 1.$$

But $T = M + C(\gamma)$, where M and $C(\gamma)$ are defined in Section 12.3. So

$$MM^* = \begin{pmatrix} \Theta_P^2 & 0 & ... & 0 & 0 \\ 0 & 0 & ... & 0 & 0 \\ . & . & ... & . & . \\ 0 & 0 & ... & 0 & 0 \end{pmatrix} \text{ and } CC^* = \begin{pmatrix} 0 & 0 & ... & 0 & 0 \\ 0 & I_n/2^{2\gamma} & ... & 0 & 0 \\ 0 & 0 & ... & 0 & 0 \\ . & . & ... & . & . \\ 0 & 0 & ... & 0 & I_n/m^{2\gamma} \end{pmatrix}.$$

Hence,

$$N_{2p}(M^*) = N_{2p}(\Theta_P).$$

In addition,

$$Trace\,(CC^*)^p = n \sum_{k=2}^{m} \frac{1}{k^{2p\gamma}}.$$

Thus,

$$N_{2p}(C^*) = [n \sum_{k=2}^{m} \frac{1}{k^{2p\gamma}}]^{1/2p}.$$

Hence,

$$N_{2p}(T) \le N_{2p}(\Theta_P) + [n \sum_{k=2}^{m} \frac{1}{k^{2p\gamma}}]^{1/2p}.$$

This and (6.2) prove the lemma. Q. E. D.

Proof of Theorem 12.5.1:   Consider the polynomial pencils

$$F_m(\lambda) = \sum_{k=0}^{m} \frac{A_k \lambda^k}{(k!)^\gamma} \text{ and } \tilde{F}_m(\lambda) = \sum_{k=0}^{m} \frac{B_k \lambda^k}{(k!)^\gamma}.$$

Clearly, $\lambda^m F_m(1/\lambda) = P(\lambda)$ and $\tilde{F}_m(1/\lambda)\lambda^m = Q(\lambda)$. So

$$z_k(P) = 1/z_k(F_m); \quad z_k(Q) = 1/z_k(\tilde{F}_m). \tag{6.3}$$

Take into account that the roots continuously depend on coefficients, we have the required result, letting in the previous lemma $m \to \infty$. Q. E. D.

12.7   An identity for powers of characteristic values

Consider the entire matrix pencil

$$F(\lambda) = \sum_{k=0}^{\infty} C_k \lambda^k \ (C_0 = I_n) \tag{7.1}$$

with complex in general $n \times n$-matrices C_k, $k = 1, 2, ...$ In this section, we derive identities for the sums

$$\tilde{s}_m(F) = \sum_{k=1}^{\infty} \frac{1}{z_k^m(F)} \ (m > \rho(F)).$$

To formulate the result, for an integer m, introduce the $mn \times mn$-block matrix

$$\hat{B}_m = \begin{pmatrix} -C_1 & -C_2 & \cdots & -C_{m-1} & -C_m \\ I_n & 0 & \cdots & 0 & 0 \\ 0 & I_n & \cdots & 0 & 0 \\ \cdot & \cdot & \cdots & \cdot & \cdot \\ 0 & 0 & \cdots & I_n & 0 \end{pmatrix} \ (m > 1)$$

and $\hat{B}_1 = -C_1$.

Theorem 12.7.1 *Let $\rho(F) < \infty$. Then for any integer $m > \rho(F)$, we have*

$$\tilde{s}_m(F) = Trace\ \hat{B}_m^m.$$

Proof: Consider the polynomial pencil

$$F_\mu(\lambda) = \sum_{k=0}^{\mu} \lambda^k C_k.$$

Clearly, $\lambda^\mu F_\mu(1/\lambda) = \tilde{P}(\lambda)$, where

$$\tilde{P}(\lambda) = \sum_{k=0}^{\mu} \lambda^{\mu-k} C_k.$$

So

$$z_k(P) = \frac{1}{z_k(F_\mu)}.$$

Now Theorem 11.2.1 yields the relation

$$\sum_{k=1}^{\mu} \frac{1}{z_k^m(F_\mu)} = Trace\ \hat{B}_m^m$$

for any $m \leq \mu$. Taking into account that under consideration, thanks to the Hadamard theorem, $\tilde{s}_m(F) < \infty$ and that the characteristic values of entire matrix functions continuously depend on its coefficients, we get the required result. Q. E. D.

12.8 Multiplicative representations of meromorphic matrix functions

Again denote by $B(\mathbf{C}^n)$ the set of all $n \times n$-matrices. Let

$$A = D + V \ (A \in B(\mathbf{C}^n))$$

be the Schur triangular representation of A (see Section 1.5). Here V is the nilpotent part of A and

$$D = \sum_{k=1}^n \lambda_k(A) \Delta E_k$$

is its diagonal part, $\Delta E_k = E_k - E_{k-1}$ $(k = 1, ..., n)$, $E_0 = 0$ and E_k $(k = 1, ..., n)$ is the maximal chain of the invariant projections of A (see Section 11.5). Again set

$$\overrightarrow{\prod_{1 \le k \le m}} X_k = X_1 X_2 ... X_m$$

for $X_1, X_2, ..., X_m \in B(\mathbf{C}^n)$. As it was proved in Section 11.5,

$$(I - A)^{-1} = \overrightarrow{\prod_{1 \le k \le n}} \left(I + \frac{A\Delta E_k}{1 - \lambda_k(A)}\right), \tag{8.1}$$

for any $A \in B(\mathbf{C}^n)$ with $1 \notin \sigma(A)$.

Now let $F(z)$ be an entire matrix pencil. Put

$$\Psi(z) := F(z) - I.$$

Let $E_k(F, z)$, $k = 1, ..., n$, be the maximal chain of invariant projections of matrix $F(z)$ for a fixed z, and $\Delta E_k(z) = E_k(F, z) - E_{k-1}(F, z)$. By (8.1) with $F(z)$ instead of $I - A$, we get our next result.

Theorem 12.8.1 *Let $F(z)$ be an entire matrix-valued function. Then for any $z \notin \Sigma(F)$, the equality*

$$F^{-1}(z) = \overrightarrow{\prod_{1 \le k \le n}} \left(I - \frac{\Psi(z)\Delta E_k(z)}{\lambda_k(F(z))}\right)$$

is true, where $E_k(z)$, $k = 1, ..., n$ is the maximal chain of the invariant projections of $F(z)$ and $\lambda_k(F(z))$ are the eigenvalues of matrix $F(z)$.

Furthermore, in Section 11.5, the equality

$$A^{-1} = D^{-1} \overrightarrow{\prod_{2 \le k \le n}} \left[I - \frac{V\Delta E_k}{\lambda_k(A)}\right]$$

has been proved for any non-singular matrix A. Now replace A by $F(z)$ and denote by $\tilde{D}_F(z)$ and $\tilde{V}_F(z)$ be the diagonal and nilpotent parts of $F(z)$, respectively. Then the previous equality at once yields the following result.

Theorem 12.8.2 *Let $F(z)$ be an entire matrix pencil. Then*

$$F^{-1}(z) = \tilde{D}_F^{-1}(z) \overrightarrow{\prod_{2 \leq k \leq n}} \Big[I - \frac{\tilde{V}_F(z) \Delta E_k(z)}{\lambda_k(F(z))} \Big]$$

for any $z \notin \Sigma(F)$.

12.9 Estimates for meromorphic matrix functions

As it was shown in Section 11.8, for any invertible $n \times n$-matrix A we have

$$\|A^{-1}\| \leq \sum_{k=0}^{n-1} \frac{g^k(A)}{\sqrt{k!}\rho_0^{k+1}(A)}, \tag{9.1}$$

where $\|.\|$ is the Euclidean norm, $\rho_0(A) = \rho(A, 0)$ is the smallest modulus of the eigenvalues of A:

$$\rho_0(A) = \inf_{k=1,\dots,n} |\lambda_k(A)|,$$

and

$$g(A) = \Big(N_2^2(A) - \sum_{k=1}^{n} |\lambda_k(A)|^2 \Big)^{1/2}.$$

Besides,

$$g^2(A) \leq \frac{1}{2} N_2^2(A^* - A), \tag{9.2}$$

$$g^2(A) \leq N_2^2(A) - |Trace\ A^2| \text{ and } g(e^{i\tau}A + zI) = g(A) \quad (\tau \in \mathbf{R}, z \in \mathbf{C}). \tag{9.3}$$

In addition, in Section 11.8 the inequality

$$\|A^{-1}\| \leq \frac{1}{\rho_0(A)} \Big[1 + \frac{1}{n-1}\Big(1 + \frac{g^2(A)}{\rho_0^2(A)}\Big) \Big]^{(n-1)/2} \tag{9.4}$$

has been proved. Put

$$\rho_0(F(z)) = \inf_{k=1,\dots,n} |\lambda_k(F(z))|.$$

Because of (9.1) and (9.4), we arrive at the following result.

Theorem 12.9.1 *Let F be an entire matrix pencil. Then for any $z \notin \Sigma(F)$, the inequalities*

$$\|F^{-1}(z)\| \le \sum_{k=0}^{n-1} \frac{g^k(F(z))}{\sqrt{k!}\rho_0^{k+1}(F(z))}$$

and

$$\|F^{-1}(z)\| \le \frac{1}{\rho_0(F(z))}\Big[1 + \frac{1}{n-1}\Big(1 + \frac{g^2(F(z))}{\rho_0^2(F(z))}\Big)\Big]^{(n-1)/2}$$

hold.

Furthermore, as it is proved in Section 11.8,

$$\|A^{-1} \det A\| \le \frac{N_2^{n-1}(A)}{(n-1)^{(n-1)/2}}$$

for any invertible $n \times n$-matrix A. Hence, we easily get the inequality

$$\|F^{-1}(z)\| \le \frac{1}{|\det F(z)|} \frac{N_2^{n-1}(F(z))}{(n-1)^{(n-1)/2}} \quad (z \notin \Sigma(F)).$$

Again consider the entire pencil in the root-factorial form

$$F(\lambda) = \sum_{k=0}^{\infty} \frac{\lambda^k}{(k!)^\gamma} A_k \quad (A_0 = I_n, \lambda \in \mathbf{C}) \tag{9.5}$$

with a finite $\gamma > 0$ and $n \times n$-matrices A_k, $k = 1, 2, \dots$. According to (9.3),

$$g(F(\lambda)) = g\Big(\sum_{k=1}^{\infty} \frac{\lambda^k}{(k!)^\gamma} A_k\Big),$$

since $A_0 = I_n$. Directly from (9.3), it follows that

$$g(F(\lambda)) \le \sum_{k=1}^{\infty} \frac{|\lambda|^k}{(k!)^\gamma} N_2(A_k) \quad (\lambda \in \mathbf{C}).$$

Moreover, thanks to (9.2),

$$g(F(x)) \le \sqrt{2} \sum_{k=1}^{\infty} \frac{|x|^k}{(k!)^\gamma} N_2(A_{Ik})$$

for all real x. Here $A_{Ik} = (A_k - A_k^*)/2i$.

Now let us consider perturbations of entire pencils. Let $\tilde{F}(z)$ be another entire pencil. Put

$$q(\lambda) := \|F(\lambda) - \tilde{F}(\lambda)\|.$$

Lemma 12.9.2 *Each characteristic value μ of $\tilde{F}(.)$ either satisfies the inequality*

$$q(\mu)\|F^{-1}(\mu)\| \geq 1,$$

or is equal to a characteristic value of $F(.)$.

The proof of this lemma is absolutely similar to the proof of Lemma 11.8.3. Q. E. D.

This lemma and Theorem 12.9.1 imply

Corollary 12.9.3 *Any characteristic value μ of $\tilde{F}(.)$ either satisfies the inequality*

$$q(\mu) \sum_{k=0}^{n-1} \frac{g^k(F(\mu))}{\rho_0^{k+1}(F(\mu))\sqrt{k!}} \geq 1,$$

or is equal to a characteristic value of $F(.)$.

Now, let $F_+(\lambda) = (f_{jk}(\lambda))_{j,k}^n$ be the upper-triangular part of an entire matrix-valued function $\tilde{F}(\lambda) = (\tilde{f}_{jk}(\lambda))_{j,k=1}^n$. That is,

$$f_{jk}(\lambda) = \tilde{f}_{jk}(\lambda) \text{ if } j \leq k \text{ and } f_{jk}(t) = 0 \text{ if } j > k.$$

Since diagonal entries of a triangular matrix are its eigenvalues, we get

$$\rho_0(F_+(\lambda)) = \min_{k=1,\dots,n} |\tilde{f}_{kk}(\lambda)|.$$

According to the definition of $g(A)$, we have the equality

$$g(F_+(\lambda)) = \Big[\sum_{1 \leq j < k \leq n} |\tilde{f}_{jk}(\lambda)|^2 \Big]^{1/2}.$$

Furthermore, denote

$$q_+(\lambda) = \|F_+(\lambda) - \tilde{F}(\lambda)\|.$$

Then Corollary 12.9.3 with $F(\mu) = F_+(\mu)$ yields the next result.

Corollary 12.9.4 *Any characteristic value μ of $\tilde{F}(.)$ either satisfies the inequality*

$$q_+(\mu) \sum_{k=0}^{n-1} \frac{g^k(F_+(\mu))}{\min_j |\tilde{f}_{jj}(\mu)|^{k+1}\sqrt{k!}} \geq 1,$$

or is equal to a zero of some $\tilde{f}_{jj}(.)$, $j = 1, \dots, n$.

Let us investigate the characteristic values of the matrix quasipolynomial

$$K(\lambda) = \sum_{\nu=1}^{m} A_\nu e^{-\lambda h_\nu} - \lambda I \quad (m < \infty, \; \lambda \in \mathbf{C}), \tag{9.6}$$

where A_ν $(\nu = 1, ..., m)$ are real constant $n \times n$-matrices, and

$$0 \leq h_1 < ... < h_m < \infty$$

are constants.

We will say that $K(.)$ *is stable,* if all its characteristic values lie in the open left half-plane C_-. By the previous corollary, we can establish stability conditions for $K(.)$.

Indeed, let $T_\nu = (T_{jk}^{(\nu)})_{j,k=1}^n$ be the upper-triangular part of the matrix $A_\nu = (a_{jk}^{(\nu)})_{j,k=1}^n$. That is,

$$T_{jk}^{(\nu)} = a_{jk}^{(\nu)} \text{ if } j \leq k \text{ and } T_{jk}^{(\nu)} = 0 \text{ if } j > k \ (\nu = 1, ..., m).$$

Put

$$T(\lambda) = \sum_{\nu=1}^m T_\nu e^{-\lambda h_\nu} - \lambda I.$$

Thus, $T(\lambda)$ is the upper-triangular part of $K(\lambda)$, and the diagonal entries of $T(\lambda)$ are

$$b_{kk}(\lambda) = \sum_{\nu=1}^m a_{kk}^{(\nu)} e^{-h_\nu \lambda} - \lambda. \tag{9.7}$$

Besides, for all complex λ, according to (9.3),

$$g(T(\lambda)) \leq \sum_{\nu=1}^m N_2(T_\nu) e^{-h_\nu \, \mathrm{Re} \, \lambda}.$$

Here

$$N_2(T_\nu) = \Big[\sum_{1 \leq j < k \leq n} |a_{jk}^{(\nu)}|^2 \Big]^{1/2}.$$

Consequently, for $\mathrm{Re} \, \lambda \geq 0$, we have

$$g(T(\lambda)) \leq g_T := \sum_{\nu=1}^m N_2(T_\nu).$$

In addition, for all complex λ,

$$\|K(\lambda) - T(\lambda)\| \leq \sum_{\nu=1}^m \|W_\nu\| e^{-\mathrm{Re} \, \lambda h_\nu},$$

where $W_\nu = (W_{jk}^{(\nu)})_{j,k=1}^n$ are the lower-triangular parts of matrices

$$A_\nu \ (\nu = 1, ..., m).$$

That is,

$$W_{jk}^{(\nu)} = a_{jk}^{(\nu)} \text{ if } j > k \text{ and } W_{jk}^{(\nu)} = 0 \text{ if } j \leq k.$$

Thus, for $Re\ \lambda \geq 0$, we have

$$\|K(\lambda) - T(\lambda)\| \leq q_T := \sum_{\nu=1}^{m} \|W_\nu\|.$$

Let each function $b_{kk}(\lambda)$ be stable, that is all its zeros lie in the open left half-plane C_-. Then $T(.)$ is stable and therefore

$$\rho_T := \inf_{j=1,\ldots,n; y \in \mathbf{R}} b_{jj}(iy) > 0.$$

Assume that

$$q_T \sum_{k=0}^{n-1} \frac{g_T^k}{\rho_T^{k+1}\sqrt{k!}} < 1. \tag{9.8}$$

Then

$$\|K(z) - T(z)\| \sum_{k=0}^{n-1} \frac{g^k(T(z))}{\min_j |b_{jj}(z)|^{k+1}\sqrt{k!}}$$

$$\leq q_T \sum_{k=0}^{n-1} \frac{g_T^k}{\rho_T^{k+1}\sqrt{k!}} < 1,$$

provided $Re\ z \geq 0$. Thanks to the previous corollary, $K(.)$ does not have characteristic values in the open right half-plane. We thus have established the following theorem.

Theorem 12.9.5 *Let the scalar quasipolynomials $b_{jj}(.)$, $j = 1, \ldots, n$, defined by (9.7) be stable. Let condition (9.8) be fulfilled. Then the matrix quasipolynomial defined by (9.6) is stable.*

12.10   Zero-free domains

Theorem 12.10.1 *The spectrum of the entire pencil*

$$F(\lambda) = \sum_{k=0}^{\infty} C_k \lambda^k, C_0 = I \tag{10.1}$$

lies in the set $\{z \in \mathbf{C} : |z| \geq \tilde{r}\}$, where $\tilde{r}$ is the unique positive root of the equation

$$\sum_{k=1}^{\infty} \|C_k\| z^k = 1.$$

Proof: Introduce the pencil

$$F_m(\lambda) = \sum_{k=0}^{m} C_k \lambda^k. \tag{10.2}$$

Clearly,

$$\lambda^m F_m(1/\lambda) = \sum_{k=0}^{m} C_k \lambda^{m-k}. \tag{10.3}$$

Now the required result directly follows from Theorem 11.6.2. Q. E. D.

The latter result can be considered as a generalization of the Cauchy theorem (see Chapter 4).

Let $z_j(F)$, $j = 1, 2, ..., l$ $(l \leq \infty)$ be the characteristic values of F.

In particular, Theorem 12.10.1 and Corollary 5.12.2 imply the inequalities

$$\inf_j |z_j(F)| \geq \frac{1}{1 + \sup_j \|C_j\|}$$

and

$$\inf_j |z_j(F)| \geq \frac{1}{\max\{1, \sum_{j=1}^{\infty} \|C_j\|\}}.$$

Theorem 12.10.2 *Let $F(z)$ be the entire pencil defined by (10.1). Then its spectrum lies in the union of the sets*

$$|\lambda| > 1 \ and \ \|(C_1 + \frac{1}{\lambda})^{-1}\| \sum_{k=2}^{\infty} \|C_k\| \geq 1.$$

Proof: Again use the function defined by (10.2) and relation (10.3).

Now the required result easily follows from Theorem 11.7.3 . Q. E. D.

12.11   Matrix-valued functions of a matrix argument

Let A be an $n \times n$-matrix and $\sigma(A)$ be its spectrum. Let $M \supset \sigma(A)$ be an open set whose boundary L consists of a finite number of rectifiable Jordan curves, oriented in the positive sense customary in the theory of complex variables. Suppose that $M \cup L$ is contained in the domain of analycity of a scalar-valued function f. Then $f(A)$ is defined by

$$f(A) = -\frac{1}{2\pi i} \int_L f(\lambda) R_\lambda(A) d\lambda.$$

Recall that $R_\lambda(A) = (A - \lambda I)^{-1}$ ($\lambda \notin \sigma(A)$) is the resolvent, I is the unit operator. Recall also that the numbers

$$\gamma_{n,k} = \sqrt{\frac{C_{n-1}^k}{(n-1)^k}} \quad (k = 1, ..., n-1, n \geq 2) \text{ and } \gamma_{n,0} = 1$$

are introduced in Section 1.10. Here

$$C_{n-1}^k = \frac{(n-1)!}{(n-k-1)!k!}$$

are binomial coefficients. Evidently, for all $n > 2$,

$$\gamma_{n,k}^2 = \frac{(n-1)(n-2)\ldots(n-k)}{(n-1)^k k!} \leq \frac{1}{k!} \quad (k = 1, 2, ..., n-1). \tag{11.1}$$

Theorem 12.11.1 *Let A be an $n \times n$-matrix and let f be a function regular on a neighborhood of the closed convex hull $co(A)$ of the eigenvalues of A. Then*

$$\|f(A)\| \leq \sum_{k=0}^{n-1} \sup_{\lambda \in co(A)} |f^{(k)}(\lambda)| g^k(A) \frac{\gamma_{n,k}}{k!}.$$

Here $\|.\|$ is the Euclidean norm in $\mathbf{C}^n$, again, and $g(A)$ is defined as in Section 12.9.

The proof of this theorem is divided into two lemmas, which are presented in this section below.

Theorem 12.11.1 is sharp: if A is a normal matrix, and

$$\sup_{\lambda \in co(A)} |f(\lambda)| = \sup_{\lambda \in \sigma(A)} |f(\lambda)|,$$

then, we have the equality $\|f(A)\| = \sup_{\lambda \in \sigma(A)} |f(\lambda)|$.

Theorem 12.11.1 and inequalities (11.1) yield our next result.

Corollary 12.11.2 *Under the hypothesis of Theorem 12.11.1 the inequality*

$$\|f(A)\| \leq \sum_{k=0}^{n-1} \sup_{\lambda \in co(A)} |f^{(k)}(\lambda)| \frac{g^k(A)}{(k!)^{3/2}}$$

is valid.

Example 12.11.3 *For a linear operator A in $\mathbf{C}^n$, Theorem 12.11.1 and Corollary 12.11.2 give us the estimates*

$$\|exp(At)\| \leq e^{\alpha(A)t} \sum_{k=0}^{n-1} g^k(A) t^k \frac{\gamma_{n,k}}{k!} \leq e^{\alpha(A)t} \sum_{k=0}^{n-1} \frac{g^k(A) t^k}{(k!)^{3/2}} \quad (t \geq 0),$$

where $\alpha(A) = \max_{k=1,...,n} Re \, \lambda_k(A)$.

To prove Theorem 12.11.1, we need the following technical result.

Lemma 12.11.4 *Suppose* $\{d_k\}$ *is an orthogonal normal basis in* $\mathbf{C}^n$, $(.,.)$ *the scalar product in* $\mathbf{C}^n$, $A_1, \ldots, A_j$ *are* $n \times n$-*matrices and* $\phi(k_1, \ldots, k_{j+1})$ *a scalar-valued function of arguments*

$$k_1, \ldots, k_{j+1} = 1, 2, \ldots, n; \; j < n.$$

In addition, define projections $Q(k)$ *by* $Q(k)h = (h, d_k)d_k$ $(h \in \mathbf{C}^n, \; k = 1, \ldots, n)$, *and set*

$$T = \sum_{1 \leq k_1, \ldots, k_{j+1} \leq n} \phi(k_1, \ldots, k_{j+1})Q(k_1)A_1Q(k_2) \ldots A_jQ(k_{j+1}).$$

Then $\|T\| \leq a(\phi)\||A_1||A_2|\ldots|A_j|\|$, *where*

$$a(\phi) = \max_{1 \leq k_1, \ldots, k_{j+1} \leq n} |\phi(k_1, \ldots, k_{j+1})|$$

and $|A_k|$ $(k = 1, \ldots, j)$ *are the matrices, whose entries in* $\{d_k\}$ *are the absolute values of the entries of* A_k *in* $\{d_k\}$.

Proof: For any entry $T_{sm} = (Td_s, d_m)$ $(s, m = 1, \ldots, n)$ of operator T, we have

$$T_{sm} = \sum_{1 \leq k_2, \ldots, k_j \leq n} \phi(s, k_2, \ldots, k_j, m)a_{sk_2}^{(1)} \ldots a_{k_j,m}^{(j)},$$

where $a_{jk}^{(i)} = (A_i d_k, d_j)$ are the entries of A_i. Hence,

$$|T_{sm}| \leq a(\phi) \sum_{1 \leq k_2, \ldots, k_j \leq n} |a_{sk_2}^{(1)} \ldots a_{k_j m}^{(j)}|.$$

This relation and the equality

$$\|Tx\|^2 = \sum_{j=1}^{n} |(Tx)_j|^2 \; (x \in \mathbf{C}^n),$$

where $(.)_j$ means the j-th coordinate, imply the required result. Q. E. D.

Let $\{e_k\}$ is the orthonormal Schur basis (the basis of the triangular representation) of A. That is,

$$A = D + V \; (\sigma(D) = \sigma(A)),$$

where D is a diagonal matrix and V is an upper-diagonal nilpotent matrix.

Furthermore, let $|V|$ be the operator whose entries in the Schur basis are the absolute values of the entries of V in that basis. That is,

$$|V| = \sum_{k=2}^{n} \sum_{j=1}^{k-1} |a_{jk}|(., e_k)e_j,$$

where $a_{jk} = (Ae_k, e_j)$.

Lemma 12.11.5 *Under the hypothesis of Theorem 12.11.1, the estimate*

$$\|f(A)\| \leq \sum_{k=0}^{n-1} \sup_{\lambda \in co(A)} |f^{(k)}(\lambda)| \frac{\| \, |V|^k \, \|}{k!}$$

is true, where V is the nilpotent part of A.

Proof: It is not hard to see that the representation (3.4) implies the equality

$$R_\lambda(A) = (A - I\lambda)^{-1} = (D + V - \lambda I)^{-1}$$
$$= (I + R_\lambda(D)V)^{-1} R_\lambda(D)$$

for all regular λ. But $R_\lambda(D)V$ is a nilpotent operator, since V and $R_\lambda(D)$ have common invariant subspaces. Hence,

$$(R_\lambda(D)V)^n = 0.$$

Therefore,

$$R_\lambda(A) = \sum_{k=0}^{n-1} (R_\lambda(D)V)^k (-1)^k R_\lambda(D). \tag{11.2}$$

Because of the representation for functions of matrices

$$f(A) = -\frac{1}{2\pi i} \int_L f(\lambda) R_\lambda(A) d\lambda = \sum_{k=0}^{n-1} C_k, \tag{11.3}$$

where

$$C_k = (-1)^{k+1} \frac{1}{2\pi i} \int_L f(\lambda) (R_\lambda(D)V)^k R_\lambda(D) d\lambda.$$

Here L is a closed contour surrounding $\sigma(A)$. Since D is a diagonal matrix with respect to the Schur basis $\{e_k\}$ and its diagonal entries are the eigenvalues of A, then

$$R_\lambda(D) = \sum_{j=1}^{n} \frac{Q_j}{\lambda_j(A) - \lambda},$$

where $Q_k = (., e_k)e_k$. We have

$$C_k = \sum_{j_1=1}^{n} Q_{j_1} V \sum_{j_2=1}^{n} Q_{j_2} V \dots V \sum_{j_k=1}^{n} Q_{j_k} I_{j_1 j_2 \dots j_{k+1}}.$$

Here

$$I_{j_1 \dots j_{k+1}} = \frac{(-1)^{k+1}}{2\pi i} \int_L \frac{f(\lambda) d\lambda}{(\lambda_{j_1} - \lambda) \dots (\lambda_{j_{k+1}} - \lambda)}.$$

Lemma 12.11.4 gives us the estimate

$$\|C_k\| \leq \max_{1 \leq j_1 \leq \dots \leq j_{k+1} \leq n} |I_{j_1 \dots j_{k+1}}| \| \, |V|^k \, \|.$$

Because of Lemma 8.6.1,

$$|I_{j_1\dots j_{k+1}}| \le \sup_{\lambda \in co(A)} \frac{|f^{(k)}(\lambda)|}{k!}.$$

This inequality and (11.3) prove the lemma. Q. E. D.

Proof of Theorem 12.11.1: Lemma 1.10.4 yields the inequality

$$\| \, |V|^k \, \| \le \gamma_{n,k} N_2^k(|V|) \ (k = 1, \dots, n-1).$$

But $N_2(|V|) = N_2(V)$. Moreover, thanks to Lemma 1.5.2, $N_2(V) = g(A)$. Thus,

$$\| \, |V|^k \, \| \le \gamma_{n,k} g^k(A) \ (k = 1, \dots, n-1).$$

Now the previous lemma yields the required result. Q. E. D.

12.12   Green's functions of differential equations

For an integer $\nu \ge 1$, we consider the scalar function $G_\nu(\lambda, t)$ ($\lambda \in \mathbf{C}, t \ge 0$), satisfying the equation

$$\frac{\partial^\nu G_\nu(\lambda, t)}{\partial t^\nu} = \eta \lambda^\nu G_\nu(\lambda, t) \ (t > 0) \tag{12.1}$$

with a real constant η and the initial conditions

$$\frac{\partial^j G_\nu(\lambda, 0)}{\partial t^j} = 0 \ (j = 0, \dots, \nu - 2), \frac{\partial^{\nu-1} G_\nu(\lambda, 0)}{\partial t^{\nu-1}} = 1. \tag{12.2}$$

For example, if $\nu = 1$, $\eta = 1$, then $G_1(\lambda, t) = e^{\lambda t}$. If $\nu = 2$, $\eta = -1$, then

$$G_2(\lambda, t) = \frac{sin\ (\lambda t)}{\lambda},$$

and so on.

A function satisfying (12.1), (12.2) is the Green functions to equation (12.1). For any continuous function h, a solution $x(t)$ of the equation

$$\frac{d^\nu x(t)}{dt^\nu} - \eta \lambda^\nu x(t) = h(t)$$

with initial conditions

$$x^{(j)}(0) = 0 \ (j = 0, \dots, \nu - 1),$$

can be represented as

$$x(t) = \int_0^t G_\nu(\lambda, t - s)h(s)ds. \qquad (12.3)$$

Taking into account (12.1) and (12.2), we can write out,

$$\frac{d^\nu G_\nu(\lambda, t)}{dt^\nu} - \frac{d^\nu G_\nu(\mu, t)}{dt^\nu} = \eta(\lambda^\nu G_\nu(\lambda, t) - \mu^\nu G_\nu(\mu, t)) \quad (\lambda, \mu \in \mathbf{C}; \ t \geq 0).$$

So

$$\frac{d^\nu (G_\nu(\lambda, t) - G_\nu(\mu, t))}{dt^\nu} - \eta\lambda^\nu (G_\nu(\lambda, t) - G_\nu(\mu, t)) = \eta[\lambda^\nu - \mu^\nu]G_\nu(\mu, t).$$

According to (12.3), this gives

$$G_\nu(\lambda, t) - G_\nu(\mu, t) = \eta \int_0^t G_\nu(\lambda, t - s)[\lambda^\nu - \mu^\nu]G_\nu(\mu, s)ds.$$

Hence, we get the following result.

Lemma 12.12.1 *A and $\tilde{A}$ be $n \times n$ matrices. Let the Green function $G_\nu(\lambda, t)$ to equation (12.1) be regular in λ on $\sigma(A)$ for each $t \geq 0$. Then*

$$\frac{d^\nu G_\nu(A, t)}{dt^\nu} = \eta A^\nu G_\nu(A, t), t > 0 \qquad (12.4)$$

and

$$\frac{\partial^j G_\nu(A, 0)}{\partial t^j} = 0 \ (j = 0, ..., \nu - 2), \frac{\partial^{n-1} G_\nu(A, 0)}{\partial t^{n-1}} = I, \qquad (12.5)$$

where I is the unit matrix. If, in addition, $G_\nu(\lambda, t)$ is regular in λ on $\sigma(\tilde{A})$, then

$$G_\nu(A, t) - G_\nu(\tilde{A}, t) = \eta \int_0^t G_\nu(A, t - s)[A^\nu - \tilde{A}^\nu]G_\nu(\tilde{A}, s)ds. \qquad (12.6)$$

In particular,

$$e^{At} - e^{\tilde{A}t} = \int_0^t e^{A(t-s)}[A - \tilde{A}]e^{\tilde{A}s}ds$$

and

$$A^{-1}sin\ (At) - \tilde{A}^{-1}sin\ (\tilde{A}t) = \int_0^t A^{-1}sin\ (A(t - s))[A^2 - \tilde{A}^2]\tilde{A}^{-1}sin\ (\tilde{A}s)ds.$$

From (12.6), it follows that

$$\|G_\nu(A, t) - G_\nu(\tilde{A}, t)\| \leq |\eta| \int_0^t \|G_\nu(A, t - s)\| \|A^\nu - \tilde{A}^\nu\| \|G_\nu(\tilde{A}, s)\|ds.$$

This inequality and Corollary 12.11.2 imply

Corollary 12.12.2 *Let the Green function $G_\nu(\lambda, t)$ to equation (12.1) be regular on co $(A) \cup co(\tilde{A})$. Then with the notation*

$$\gamma_k(A, t) := \frac{1}{(k!)^{3/2}} \sup_{z \in co\,(A)} \left| \frac{\partial^k G_\nu(z, t)}{\partial z^k} \right|,$$

the inequality

$$\|G_\nu(A, t) - G_\nu(\tilde{A}, t)\| \le \|A^\nu - \tilde{A}^\nu\| \sum_{m,j=0}^{n-1} c_{mj} g^m(A) g^j(\tilde{A})$$

is valid, where

$$c_{mj} := |\eta| \int_0^t \gamma_m(A, t - s) \gamma_j(\tilde{A}, s) ds.$$

In particular, by Example 12.11.3 and the previous corollary, we arrive at the inequality

$$\|e^{At} - e^{\tilde{A}t}\|$$

$$\le \|A - \tilde{A}\| \int_0^t e^{\alpha(A)(t-s) + \alpha(\tilde{A})s} \sum_{j,m=0}^{n-1} \frac{(t-s)^m g^m(A)}{(m!)^{3/2}} \frac{s^j g^j(\tilde{A})}{(j!)^{3/2}} ds.$$

This integral is simply calculated. If A and $\tilde{A}$ are normal, then $g(A) = g(\tilde{A}) = 0$. So

$$\|e^{At} - e^{\tilde{A}t}\| \le \|A - \tilde{A}\| \int_0^t e^{\alpha(A)(t-s) + \alpha(\tilde{A})s} ds$$

and thus

$$\|e^{At} - e^{\tilde{A}t}\| \le \frac{\|A - \tilde{A}\|}{\alpha(A) - \alpha(\tilde{A})} \left(e^{\alpha(A)t} - e^{\alpha(\tilde{A})t} \right) \quad (t \ge 0).$$

12.13   Comments to Chapter 12

Sections 12.2 - 12.4 are based on the papers [Gil', 2003b, 2003c and 2004]. Theorem 12.5.1 appears in [Gil', 2005d]. Theorem 12.7.1 is taken from [Gil', 2008a], the more general case is considered in [Gil', 2008c]. Theorems 12.8.1 an 12.11.1 are adopted from [Gil', 2003a]. The material of Sections 12.9, 12.10, and 12.12 is probably new. Theorem 12.11.1 is taken from [Gil', 2003a]. Lemma 12.12.1 is probably new.

Numerous mathematical and physical problems lead to analytic matrix-valued functions of complex and matrix arguments, cf. [Diekmann, Gils, and Verduyn Lunel, 1995], [Kaashoek and Verduyn Lunel, 1992], [Kolmanovskii and Myshkis, 1999]. The theory of matrix functions is well developed, cf. [Bhatia, 1997] and [Higham, 2008]. Of course, we could not survey the whole subject here and refer reader to the interesting recent papers [Hryniv and Lancaster, 1999], [Lancaster, Markus, and Zhou, 2003], [Langer and Najman, 1992], [Lasarow, 2006], and [Mikhailova, Pavlov, and Prokhorov, 2007].

Bibliography

[1] Abian, A. (1981), A family of analytic functions whose members have equal number of zeros. *Acta Math. Vietnam, 6*, No. 1, 18 - 20.

[2] Agarwal, R. P., Grace, S. R. and O'Regan, D. (2000), *Oscillation Theory for Difference and Functional Differential Equations*. Kluwer, Dordrecht.

[3] Agarwal, R. P., O'Regan, D. and Wong, P. J. Y. (1999), *Positive Solutions of Differential, Difference and Integral Equations*, Kluwer, Dordrecht.

[4] Ahiezer, N. I. and Glazman, I. M. (1981), *Theory of Linear Operators in a Hilbert Space*. Pitman Advanced Publishing Program, Boston.

[5] Ahn, H. and Cho, H. R. (2002), Zero sets of holomorphic functions in the Nevanlinna type class on convex domains in C^2. *Jap. J. Math., New Ser., 28*, No. 2, 245 - 260.

[6] Ahues, M. and Limaye, B. (1998), On error bound for eigenvalues of a matrix pencil. *Linear Algebra Appl., 268*, 71 - 89.

[7] Alzer, H. (2001), Bounds for the zeros of polynomials, *Complex Variables, 45*, 143 - 150.

[8] Bakan, A. and Ruscheweyh S. (2002), On the existence of generalized Polya frequency functions corresponding to entire functions with zeros in angular sectors, *Complex Variables, 47*, No. 7, 565 - 576.

[9] Barnard, R. W., K. Pearce and Wheeler, W. (2001), Zeros of Cesaro sum approximations. *Complex Variables, 45*, 327 - 348.

[10] Baumgartel, H. (1985), *Analytic Perturbation Theory for Matrices and Operators*. Operator Theory, Advances and Appl., 52, Birkhauser Verlag, Basel.

[11] Berenstein C. and Gay R. (1995), *Complex Analysis and Special Topics in Harmonic Analysis*, Springer-Verlag, New York.

[12] Bergweiler, W. (1992), Canonical products of infinite order. *J. Reine Angew. Math., 430*, 85 - 107.

[13] Bhatia, R. (1997), *Matrix Analysis*, Springer-Verlag, Berlin.

[14] Boas, R. P. and Buck, R. C. (1964), *Polynomial Expansions of Analytic Functions*, Academic Press, Springer-Verlag, Berlin.

[15] Borwein, P. and Erdelyi, T. (1995), *Polynomials and Polynomial Inequalities*, Springer-Verlag, New York.

[16] Brauer, A. (1947), Limits of the characteristic roots of a matrix. *Duke Math Journal, 14*, 21 - 26.

[17] Brodskii, M. S. (1971), *Triangular and Jordan Representations of Linear Operators*, Transl. Math. Mongr., vol. 32, Amer. Math. Soc., Providence, RI.

[18] Buckholtz, J. D. (1966), A characterization of the exponential series. *Amer. Math. Monthly, 73*, 121 - 123.

[19] Buckholtz, J. D. and J. K. Shaw (1972), Zeros of partial sums and remainders of power series. *Trans. Amer. Math. Soc., 166*, 269 - 284.

[20] Cardon, D. A. and de Gaston, S. A. (2005), Differential operators and entire functions with simple real zeros. *J. Math. Anal. Appl., 301*, No. 2, 386 - 393.

[21] Chanane, B. (2003), Recovering entire functions with polynomial and exponential growth and approximation of their zeros. *Int. J. Appl. Math., 13*, No. 3, 261 - 277.

[22] Clunie, J. G. and Edrei, A. (1991), Zeros of successive derivatives of analytic functions having a single essential singularity. *J. Anal. Math., 56*, 141 - 185.

[23] Craven, T., Csordas, G. and Smith, W. (1987), Zeros of derivatives of entire functions. *Proc. Amer. Math. Soc., 101*, 323 - 326.

[24] Csordas, G. and Smith, W. (2000), Level sets and the distribution of zeros of entire functions. *Mich. Math. J., 47*, No. 3, 601 - 612.

[25] Dewan, K. K. and Govil, N. K. (1990), On the location of the zeros of analytic functions. *Int. J. Math. Math. Sci., 13*, No. 1, 67 - 72.

[26] Diekmann, O., Gils, S. A. van and Verduyn Lunel, S. M. (1995), *Delay Equations: Functional-, Complex- and Nonlinear Analysis*, Applied Mathematical Sciences, V. 110, Springer-Verlag, Berlin.

[27] Diestel, D., Jarchow and Tonge, A. (1995), *Absolutely Summing Operators*, Cambridge University Press, Cambridge.

[28] Döetsch, G. (1961), *Anleitung zum Praktishen Gebrauch der Laplacetransformation*, Oldenburg, Munchen.

[29] Dunford, N. and Schwartz, J. T. (1966), *Linear Operators, Part I. General Theory*, Interscience Publishers, New York.

[30] Dyakonov, K. M. (2000), Polynomials and entire functions: Zeros and geometry of the unit ball. *Math. Res. Lett.*, *7*, No. 4, 393 - 404.

[31] Edrei, A., E. B. Saff, and Varga R. S. (1983), *Zeros of Sections of Power Series*, Lecture Notes in Math., 1002, Springer-Verlag, Berlin.

[32] Edwards, S. and Hellerstein, S. (2002), Non-real zeros of derivatives of real entire functions and the Pólya-Wiman conjectures. *Complex Variables, Theory Appl.*, *47*, No. 1, 25 - 57.

[33] Feingold D. G. and Varga, R. S. (1962), Block diagonally dominant matrices and generalization of the Gershgorin circle theorem. *Pacific J. of Mathematics*, *12*, 1241 - 1250.

[34] Fiedler, M. (1960), Some estimates of spectra of matrices. In: *Symposium of the Numerical Treatment of Ordinary Differential Equations, Integral and Integro Differential Equations*, Birkháuser Verlag, Rome, 33 - 36.

[35] Foster, W. and Krasikov, I. (2002), Inequalities for real-root polynomials and entire functions. *Adv. Appl. Math.*, *29*, No. 1, 102 - 114.

[36] Fuchs, W. H. J. (1967), *Topics in the Theory of Functions of One Complex Variable*, D. Van Nostrand Company, Princeton, NJ.

[37] Ganelius, T. (1953), Sequences of analytic functions and their zeros. *Arkiv for Matematik*, *3*, 1 - 50.

[38] Ganelius, T. (1963), The zeros of partial sums of power series. *Duke Math. J.*, *30*, 533 - 540.

[39] Gantmacher, F. R. (1967), *Theory of Matrices*, Nauka, Moscow. In Russian.

[40] Gardner, R. B. and Govil, N. K. (1995), Some inequalities for entire functions of exponential type. *Proc. Am. Math. Soc.*, *123*, No. 9, 2757 - 2761.

[41] Gel'fand, I., Kapranov M. and A. Zelevinsky (1994), *Discriminants, Resultants and Multidimensional Determinants*, Birkháuser, Boston.

[42] Gel'fond A.O. (1967), *Calculations Of Finite Differences*, Nauka, Moscow. In Russian.

[43] Genthner, R. M. (1985), On the zeros of the derivatives of some entire functions of finite order. *Proc. Edinb. Math. Soc., II. Ser.*, *28*, 381 - 407.

[44] Gil', M. I. (2000a), Approximations of zeros of entire functions by zeros of polynomials. *J. of Approximation Theory*, *106*, 66 - 76.

[45] Gil', M. I. (2000b), Inequalities for imaginary parts of zeros of entire functions. *Results in Mathematics, 37,* 331 - 334.

[46] Gil', M. I. (2000c), Perturbations of zeros of a class of entire functions. *Complex Variables, 42,* 97 - 106.

[47] Gil', M. I. (2001), Inequalities for zeros of entire functions, *Journal of Inequalities, 6,* 463 - 471.

[48] Gil', M. I. (2002a), Boundedness of solutions of nonlinear differential delay equations with positive Green functions and the Aizerman - Myshkis problem. *Nonlinear Analysis, TMA, 49,* 1065 - 1078.

[49] Gil', M. I. (2002b), Invertibility and spectrum localization of nonselfadjoint operators. *Adv. Appl. Mathematics, 28,* 40 - 58.

[50] Gil', M. I. (2003a), *Operator Functions and Localization of Spectra,* Lecture Notes In Mathematics, vol. 1830, Springer-Verlag, Berlin.

[51] Gil', M. I. (2003b), On bounds for spectra of operator pencils in a Hilbert space. *Acta Mathematica Sinica, 19,* No. 1, 1 - 11.

[52] Gil', M. I. (2003c), Bounds for the spectrum of analytic quasinormal operator pencils in a Hilbert space. *Contemporary Mathematics, 5,* No. 1, 101 - 118.

[53] Gil', M. I. (2004), Bounds for characteristic values of entire matrix pencils. *Linear Algebra and Applications, 390,* 311 - 320.

[54] Gil', M. I. (2005a), On inequalities for zeros of entire functions. *Proceedings of AMS, 133,* 97 - 101.

[55] Gil', M. I. (2005b), On sums of roots of infinite order entire functions. *Complex Variables, 50,* No. 1, 27 - 33.

[56] Gil', M. I. (2005c), *Explicit Stability Conditions for Continuous Systems,* Lecture Notes in Control and Information Sciences, Vol. 314, Springer-Verlag.

[57] Gil', M. I. (2005d), On variations of characteristic values of entire matrix pencils. *J. of Approximation Theory, 131,* 115 - 128.

[58] Gil', M. I. (2006), Equalities for roots of entire functions of order less than two. *Arch. Math., 86,* No. 2, 133 - 137.

[59] Gil', M. I. (2007a), Bounds for zeros of entire functions. *Acta Appl. Math., 99,* 117 - 159.

[60] Gil', M. I. (2007b), Bounds for counting functions of finite order entire functions. *Complex Variables, 52,* No. 4, 273 - 286.

[61] Gil', M. I. (2007c), Positive solutions of equations with nonlinear causal mappings. *Positivity, 11*, N3, 523 - 535.

[62] Gil', M. I. (2007d), *Difference Equations in Normed Spaces*, Elsevier, Amsterdam.

[63] Gil', M. I. (2008a), Identities for sums of characteristic matrix pencils. *Linear Algebra Appl., 428*, 814 - 823.

[64] Gil', M. I. (2008b), Inequalities of the Carleman type for Neumann - Schatten operators. *Asian-European J. of Math., 1*, No. 2, 203 - 212.

[65] Gil', M. I. (2008c), Sums of characteristic values of compact operator pencils. *J. of Math. Anal. Appl., 338*, 1469 - 1476.

[66] Gohberg, I., Golberg, S. and Krupnik, N. (2000), *Traces and Determinants of Linear Operators*, Birkháuser-Verlag, Basel.

[67] Gohberg, I. C. and Krein, M. G. (1969), *Introduction to the Theory of Linear Nonselfadjoint Operators*, Trans. of Math. Monographs, vol. 18, Amer. Math. Soc., Rhode Island.

[68] Gohberg, I. C., Lancaster, P. and Rodman, L. (1982), *Matrix Polynomials*, Academic Press, New York.

[69] Gracia, J.M. and Velasco, F.E. (1995), Stability of invariant subspaces of regular matrix pencils. *Linear Algebra Appl., 221*, 219 - 226.

[70] Greene, R. E. and Krantz, S.G. (2006), *Function Theory of One Complex Variable*. 3rd ed. Graduate Studies in Mathematics 40. Amer. Math. Soc., Providence, Rhode Island,

[71] Hale, J. (1977), *Theory of Functional Differential Equations*, Springer-Verlag, New York.

[72] Hale, J. K. and S. M. Lunel (1996), *Introduction to Functional Differential Equations*, Springer-Verlag, New York.

[73] Hanson, B. (1985), The zero distribution of holomorphic functions on the unit disc. *Proc. Lond. Math. Soc., 51*, 339 - 368.

[74] Hardy, G. H, Litllwood, J. E. and Ploya, G. (1934), *Inequalities*, Cambridge University Press, Cambridge.

[75] Harel, A., Namn, S. and Sturm, J. (1999), Migration of zeros for successive derivatives of entire functions. *Proc. Am. Math. Soc., 127*, No. 2, 563 - 567.

[76] Hedstrom G. W. and Korevaar J. (1963), The zeros of partial sums of certain small entire functions. *Duke Math. J., 30*, 519 - 532.

[77] Higham, N. J. (2008), *Functions of Matrices: Theory and Computations*, Society for Industrial and Applied Mathematics, Philadelphia, PA.

[78] Higham, N. J. and Tisseur, F. (2003), Bounds for the eigenvalues of matrix polynomials. *Linear Algebra Appl., 358*, No. 1 - 3, 5 - 22.

[79] Horn, R. A. and Johnson Ch. R. (1991). *Topics in Matrix Analysis*, Cambridge University Press, Cambridge.

[80] Hryniv, R. and Lancaster, P. (1999), On the perturbation of analytic matrix functions. *Integral Equations Oper. Theory, 34*, No. 3, 325 - 338.

[81] Ioakimidis, N. I. (1988), A unified Riemann - Hilbert approach to the analytical determination of zeros of sectionally analytic functions. *J. Math. Anal. Appl., 129*, No. 1, 134 - 141.

[82] Ioakimidis, N. I. and Anastasselou, E. G. (1985), A modification of the Delves - Lyness method for locating the zeros of analytic functions. *J. Comput. Phys., 59*, 490 - 492.

[83] Istratescu, V. (1981), *Introduction to Linear Operator Theory*, Marcel Dekker, New York.

[84] Jeffrey, A. (2005), *Complex Analysis and Applications*. 2nd ed., Chapman & Hall/CRC, Boca Raton, FL.

[85] Kaashoek, M. A. and Verduyn Lunel, S. M. (1992), Characteristic matrices and spectral properties of evolutionary systems. *Trans. of the AMS, 334*, No. 2, 479 - 517.

[86] Kato, T. (1966), *Perturbation Theory for Linear Operators*, Springer-Verlag, New York.

[87] Kolmanovskii, V. and Myshkis, A. (1999), *Introduction to the Theory and Applications of Functional Differential Equations*, Kluwer Academic Publishers, Dordrecht.

[88] Krasnosel'skii, M. A., Lifshits, J. and Sobolev A. (1989), *Positive Linear Systems. The Method of Positive Operators*, Heldermann Verlag, Berlin.

[89] Kravanja, P. and Van Barel, M. (2000), *Computing the Zeros of Analytic Functions*, Lecture Notes in Mathematics, 1727, Springer-Verlag, Berlin.

[90] Krein S. G. (1972), *Functional Analysis*, Nauka, Moscow. In Russian.

[91] Kurdyka, K. and Paunescu, L. (2004), Arc-analytic roots of analytic functions are Lipschitz. *Proc. Am. Math. Soc. 132*, No. 6, 1693 - 1702.

[92] Lancaster, P., Markus, A. S. and Zhou, F. (2003), Perturbation theory for analytic matrix functions: The semisimple case. *SIAM J. Matrix Anal. Appl., 25*, No. 3, 606 - 626.

[93] Langer, H. and Najman, B. (1992), Remarks on the perturbation of analytic matrix functions. *Integral Equations Oper. Theory, 15*, No. 5, 796 - 806.

[94] Lasarow, A. (2006), Dual Szegő pairs of sequences of rational matrix-valued functions. *Int. J. Math. Math. Sci., 2006*, No. 5, 1 - 37.

[95] Leont'ev, A. F. (1976), *Exponential Series*, Nauka, Moscow. In Russian.

[96] Levin, B. Ya. (1980), *Distribution of Zeros of Entire Functions*, Amer. Math. Soc., Providence, RI.

[97] Levin, B. Ya. (1996), *Lectures on Entire Functions*, Trans. of Math. Monographs, vol. 150. Amer. Math. Soc., Providence, RI.

[98] Li, R.-C. (2003), On perturbations of matrix pencils with real spectra. *Mathematics of Computations, 72*, 715 - 728.

[99] Marcus, M. and Minc H. (1964), *A Survey of Matrix Theory and Matrix Inequalities*, Allyn and Bacon, Boston.

[100] Marden, M. (1985), *Geometry of Polynomials*, Amer. Math. Soc., Providence, RI.

[101] Markushevich, A. I. (1968), *The Theory of Analytic Functions*, vol. 1, Nauka, Moscow. In Russian.

[102] Mikhailova, A., Pavlov, B. and Prokhorov, L. (2007), Intermediate Hamiltonian via Glazman's splitting and analytic perturbation for meromorphic matrix-functions. *Math. Nachr., 280*, No. 12, 1376 - 1416.

[103] Milovanović, G. V., Mitrinović, D. S. and Rassias, Th. M. (1994), *Topics in Polynomials: Extremal Problems, Inequalities, Zeros*, World Scientific, Singapore.

[104] Mishina, A. P. and Proskuryakov, I. V. (1965), *Higher Algebra*, Nauka, Moscow. In Russian.

[105] Mitrinović, D. S. (1970), *Analytic Inequalities*, Springer-Verlag, New York.

[106] Nevanlinna, O. (2003), *Meromorhic Functions and Linear Algebra*, American Math. Soc. Providence, RI.

[107] Ostrovski, A. M. (1973), *Solution of Equations in Euclidean and Banach Spaces*, Academic Press, New York.

[108] Ostrovskii, I. V. (2000), On the zeros of tails of power series. *Operator Theory: Advances and Applications, 113*, 279 - 285.

Bibliography

[109] Pietsch, A. (1987), *Eigenvalues and s-Numbers*, Cambridge University Press, Cambridge.

[110] Pontryagin, L. S. (1955), On zeros of some elementary transcendental functions. *Trans. of AMS, Series 2, 1,* 96 - 110.

[111] Priestley, H. A. (2003), *Introduction to Complex Analysis.* 2nd ed., Oxford University Press, Oxford.

[112] Psarrakos, P. and Tsatsomeros, M. J. (2002), On the stability radius of matrix polynomials. *Linear Multilinear Algebra, 350,* No. 2, 151 - 165.

[113] Psarrakos, P. and Tsatsomeros, M. J. (2004), A primer of Perron - Frobenius theory for matrix polynomials. *Linear Algebra Appl., 393,* 333 - 351.

[114] Ram Murty. (2001), *Problems in Analytic Number Theory*, Springer, Berlin.

[115] Rodman, L. (1989), *An Introduction to Operator Polynomials,* Birkhäuser-Verlag, Basel.

[116] Ronkin, L. I. (1992), *Functions of Completely Regular Growth*, Kluwer, Dordrecht.

[117] Rosenbloom, P. C. (1969), Perturbation of zeros of analytic functions. I. *Journal of Approximation Theory, 2,* 111 - 126.

[118] Rubel, L. A. (1996), *Entire and Meromorphic Functions*, Series: Universitext VIII, Springer-Verlag, Berlin.

[119] Saks S. and Zygmund A. (1965), *Analytic Functions*, Polish Sci. Publishers, Warszawa.

[120] Schmieder, G. and Szynal J. (2002), On the distribution of the derivative zeros of a complex polynomial. *Complex Variables, 47,* No. 3, 239 - 241.

[121] Sheil-Small, T. (1989), On the zeros of the derivatives of real entire functions and Wiman's conjecture. *Ann. Math. (2) 129,* No. 1, 179 - 193.

[122] Shkalikov, A. A. (1984), Zero distribution for pairs of holomorphic functions with applications to eigenvalue distribution. *Trans. Amer. Math. Soc., 281,* 49 - 63.

[123] Simon, B. (1979), *Trace Ideals and their Applications*, Cambridge University Press, Cambridge.

[124] Stewart, G. W. and Sun J. (1990), *Matrix Perturbation Theory*, Academic Press, New York.

[125] Sturmfels, B. (2002), *Solving Systems of Polynomial Equations*, Regional Conference Series in Mathematics, No. 97, Amer. Math. Soc, RI.

[126] Titchmarsh, E. C. (1939), *The Theory of Functions*, 2nd ed., Calderon Press, Oxford.

[127] Varga, R. S. (2004), *Gerschgorin and his Cycles*, Springer Series in Computational Mathematics 36, Springer-Verlag, Berlin.

[128] Wang, Q.-G., Lee, T. H. and Tan, K. K. (1999), *Finite Spectrum Assignment for Time-delay Systems*, Springer, London.

[129] Wong, M. W. (2008), *Complex Analysis*, Series on Analysis, Applications and Computation, World Scientific, Singapore.

[130] Yildirim, C. Y. (1994), On the tails of exponential series. *Can. Math. Bull., 37*, 278 - 286.

[131] Zheng, J.-H. and Yang, C.-C. (1995), Further results on fixpoints and zeros of entire functions. *Trans. Amer. Math. Soc., 347*, No. 1, 37 - 50.

.

List of main symbols

$\|A\|$ operator norm of an operator A

$(.,.)$ scalar product

A^{-1} inverse to A

A^* conjugate to A

$A_I = (A - A^*)/2i$

$A_R = (A + A^*)/2$

β_p is defined on page 56

$\mathbf{C}^n$ complex Euclidean space

$det\,(A)$ determinant of A

$det_p\,(I - A)$ regularized determinant of A

$g(A)$ is defined on page 8

$\Gamma(.)$ Euler gamma function

H separable Hilbert space

$I = I_H$ identity operator (in a space H)

$\lambda_k(A)$ eigenvalue of A

$\nu_f(r)$ counting function of roots of f

$N_p(A)$ Schatten - von Neumann norm of A

$N(A) = N_2(A)$ Hilbert-Schmidt (Frobenius) norm of A

$\mathbf{R}^n$ real Euclidean space

$R_\lambda(A)$ resolvent of A

$r_s(A)$ spectral radius of A

$\rho(f)$ order of f

$\rho(A, \lambda)$ distance between a point lambda and the spectrum of A

$rv_f(h)$ relative variation of zeros of h with respect to zeros of f

SN_1 Trace class

SN_2 Hilbert-Schmidt ideal

SN_p Schatten - von Neumann ideal

$s_j(A)$ s-number (singular number) of A

$sv_A(B)$ spectral variation of B with respect to A

$\sigma(A)$ spectrum of A

$Tr\,A = Trace\,A$ trace of A

$\theta_k^{(p)} = \dfrac{1}{\sqrt{[k/p]!}}$

$[x]$ integer part of x

$var_P(Q)$ variation of zeros of Q with respect to zeros of P

$z_k(f)$ zero of f

$\zeta(.)$ Riemann zeta function

Index